Edward Robert Robson

School Architecture

Being Practical Remarks on the Planning, Designing, Building and Furnishing of

School Houses

Edward Robert Robson

School Architecture
Being Practical Remarks on the Planning, Designing, Building and Furnishing of School Houses

ISBN/EAN: 9783337112127

Printed in Europe, USA, Canada, Australia, Japan

Cover: Foto ©berggeist007 / pixelio.de

More available books at **www.hansebooks.com**

SCHOOL ARCHITECTURE.

BEING

Practical Remarks

ON

THE PLANNING, DESIGNING, BUILDING, AND
FURNISHING OF SCHOOL-HOUSES.

By EDWARD ROBERT ROBSON,
FELLOW OF THE ROYAL INSTITUTE OF BRITISH ARCHITECTS.

WITH MORE THAN 300 ILLUSTRATIONS.

LONDON:
JOHN MURRAY, ALBEMARLE STREET.
1874.

TO

THE RIGHT HON. LORD LAWRENCE,

G.C.B., G.C.S.I., F.R.G.S., &c. &c. &c.,

THE FORMER VICEROY OF INDIA,

WHO, AFTER HOLDING THE HIGHEST POSITION NEXT TO THE SOVEREIGN,

HESITATED NOT TO SPEND THREE YEARS

IN ORGANIZING THE ELEMENTARY EDUCATION OF THE POOREST OF

HIS FELLOW-SUBJECTS IN LONDON,

This Volume

IS DEDICATED.

PREFACE.

AMID the labour involved in the erection of a hundred new school-houses, this volume would certainly not have been undertaken except from a sense of its imperative necessity, as urged from many different sides. My principal motive for undertaking it arose from the singular want of English works on school-planning, and the circumstance that the subject had not previously been regarded by architects as possessing much importance. My professional connection with the School Board for London having made me the instrument of new developments in the planning of Elementary Schools, and afforded opportunities of a kind never before enjoyed by any one, seemed to carry with it a kind of duty. The appearance, in the public prints, however, of descriptions and illustrations of new school-buildings utterly unsuited to their purpose, became the determining reason.

These pages cannot fail to be full of imperfection, as they were written in the spare moments of the evening, by one who is in no sense a literary man. They are yet the result of much study; of a comparison of the unpublished opinions of many eminent men whose lives are devoted to school work; of personal examination of numerous buildings in the United Kingdom; of special journeys undertaken to different continental countries; and, lastly, of results as they have been applied to English wants and uses.

It is believed that the numerous plans and illustrations given, which have been taken in some cases from drawings specially prepared for this work, and in others from foreign works inaccessible to the general reader, will of themselves prove of great value to those engaged in building or managing schools, while the letter-press lays down the first principles which should guide the school-planner. If they together succeed in raising such discussion as will tend to carry the science of school-planning still further, I shall be gratified by the result. If they have the effect of inducing the Government to undertake a more decided lead in encouraging the erection of good and prohibiting that of unsuitable buildings, and of directing the attention of architects generally to the importance of securing good school-houses, thoroughly fitted for the purpose of teaching, and expressing that purpose in their architectural character, my object will be more than attained.

The aim of the work, therefore, is to be strictly practical and useful. With this view technicalities have been avoided as far as possible. Dealing chiefly with the subject of buildings suited to Elementary education, it directs attention, also, to Secondary, Technical, and other Schools.

To render the enquiry complete, some description of the principles of school-planning adopted in other countries of high educational repute became indispensable. A general survey of the principal foreign systems is therefore given. Germany naturally receives special attention, and is the subject of many illustrations.

My thanks are due in many quarters for valuable assistance rendered towards securing greater accuracy and completeness.

Especial acknowledgment should, however, be paid to the Sheffield School Board for acceding—with a rare public spirit—to my request to allow their clerk to accompany me on a tour of school inspection through Belgium, Germany, Austria, Switzerland, and France, undertaken in the spring of 1873. To the pen of Mr. Moss himself I am indebted for the concluding chapter on "School Furniture and Apparatus."

The views of London schools have been all drawn on the spot by Mr. H. W. Brewer, and, in almost all cases, have also been transferred to the wood by him. Several of the designs selected for illustration are from the pencil of my partner, Mr. J. J. Stevenson, who, although having no connection with the School Board has rendered much valuable assistance in their work. All the woodcuts have been engraved by Mr. James Cooper with that accuracy and clearness for which he is distinguished.

Those who may be able, from long acquaintance with school work, to detect errors either in fact or judgment, or to supply further information tending to elucidate the subject, will confer a favour by sending any notes to the author.

CONTENTS.

CHAPTER I.

INTRODUCTORY . . . 1

CHAPTER II.
EXISTING SCHOOLS.

Necessity of State Control over School Buildings—Want of Secondary Schools—Lancasterian System—Stow—Wesleyan Schools—The Education Department—Mixed System—Switzerland—Holland . . . 8

CHAPTER III.
AMERICA.

Coloured—Alphabet—Primary—Normal—Grammar—High Schools—Specimens—Girls' School at Boston—General 27

CHAPTER IV.
SCOTLAND AND IRELAND.

Religious Instruction—Compulsory Attendance—Educational Aspect—Buildings in the two Countries 47

CHAPTER V.
FRANCE.

General Varieties of Schools—Want of Elementary Education—Statistics—Ecoles Mixtes—The Crèche—Salles d'Asile—The Collége Chaptal at Paris 54

CHAPTER VI.

GERMANY.

Principles of School System—Classification of Schools—Prussia—Elementary Schools—Description of ordinary Parish School-house—Grammar Schools—King William Gymnasium in Berlin—Cottbus—Liegnitz—Marburg—Hildesheim—Commercial Schools—Sophien-Realschule in Berlin—Halberstadt—Cologne—Höhere Bürgerschule at Wiesbaden—Schools of Saxony—District, Parish, and Burgher Schools at Dresden—Königlische's Gymnasium at Chemnitz—Kreuzschule at Dresden—Polytechnic and Chemistry Schools at Aix-la-Chapelle—Education of Women—Defects in German System—General excellence . . 69

CHAPTER VII.

AUSTRIA.

Vienna Primary Schools—Theory of Austrian Education—The Handel's Academy at Vienna—The Imperial Gymnasium—Stadische Schools—Polytechnic Schools 147

CHAPTER VIII.

THEORY OF ENGLISH ELEMENTARY SCHOOLS.

Necessity of clear preliminary Arrangements—Controlling effect of the Code—Pædagogy—Size of Class-rooms—Size of School-room—Division of Departments—Points to be settled before building—Compactness of Plan—Economy—The Site 159

CHAPTER IX.

SCHOOL SEATS AND THEIR LIGHTING.

School Desks control Dimensions of School-room and Class-rooms—Difficulty in massing a Class conveniently and compactly—The Dual Desk with Lifting Flap—Extra width required to Rooms—Eye Diseases occurring during School Life—Proportion of School Desks—Summary 168

CHAPTER X.

INFANT SCHOOLS.

Age of Commencement—School for 120—School for 170—School for 300—Galleries, their Proportion, Arrangement, and Lighting—French Infant Gallery—Desks for Infants—The Playground or Uncovered Schoolroom—Five kinds of Apparatus for Infants' Playground . . . 180

CHAPTER XI.

ELEMENTARY GRADED SCHOOLS.

PAGE

Division of Schools into Infant and Graded Departments—Examinations—Number of Class-rooms in relation to School-room—Their Size—The School-room—Position of Class-rooms in reference thereto—Study of Internal Details 195

CHAPTER XII.

PRACTICAL DETAILS.

Temporary Schools—Walls—Entrances—Staircases—Lavatories—Hat, Cloak, and Bonnet-rooms—Their Arrangement in Mezzanines—Latrines—Playgrounds—Teachers' Rooms—The Master's House—The Caretaker—Windows—Sliding Partitions—Floors—The Dado . . 203

CHAPTER XIII.

MIDDLE SCHOOLS.

The Old Foundation Grammar Schools—Object of Middle Schools—Desirability of State Control—Sources of Inspiration for new Plans—German Higher Schools—Absence of recognised Code for English Secondary Schools—Girls' Schools—Boarding Schools—Milton Mount College . 232

CHAPTER XIV.

PHYSICAL EDUCATION.

Its Rise and Progress in Germany—Desirability in England—The Turnhalle and Turnplatz attached to Berlin Elementary Schools—Difference in Fittings respectively—Apparatus for the Playground—Description of its Construction—Extension of Turnhalle to Elementary Schools—Turnhalle at Hof—Apparatus for the Gymnastic Hall 244

CHAPTER XV.

WARMING AND VENTILATION.

Their inseparable Nature—Importance to Schools—Application in relation to School work—Cooling power of Glass—Methods of Warming to be avoided—General Principles to be adopted—Demand and Supply—First Cost and Annual Maintenance—The presence of a Caretaker as affecting System to be adopted—The open Fire—Different kinds—The Gill Stove—German Methods—The Hot-water Low-pressure System—Cases where it may be economically adopted—Practical Usefulness . 263

CHAPTER XVI.

THE BOARD SCHOOLS OF LONDON.

Exceptional case of London—Board formed without option by the Act—No Statistics—Proceedings taken to obtain these—Schools erected before their completion—Old Castle Street School—Harwood Road School—Its Architecture—Agitation for Prussian System—Johnson Street School—Its planning—Results—New North Street School—Its Plan and Abandonment—Winstanley Road School—The Slums in which some are placed—Eagle Court—Angler's Gardens—Style of Architecture suited to London Schools—Wornington Road, Aldenham Street, Orange Street, and West Street Schools—Variation in Style—Camden Street and Mansfield Place Schools—One-story School at Haverstock Hill . 291

CHAPTER XVII.

INDUSTRIAL SCHOOLS.

The Prison—The Reformatory—The Industrial School—Special object of the latter—Powers of the Industrial Schools Act which can be employed by School Boards—Programme of the kind of Building to be erected—Playground and Lavatory Arrangements—Barnes' Home and Industrial School 351

CHAPTER XVIII.

SCHOOL FURNITURE AND APPARATUS.

Advantages of good Furniture—School Desks, old and new—Dimensions suggested by Education Department—Convertible Desks—Specimens in other Countries—German Opinions—German Dimensions—The Dual Arrangement—Dutch and American Desks—English Designs—Swedish Desks—Graduation of Desks in Class—Drill for Dual Desks—Drawing—Teachers' Desks—School Cupboards—Fenders and Fire-irons—The Cooking Stove—Minor Details—Easels—Lesson-stands—Blackboards—Maps—Diagrams—Models—Abaci—The French "Compendium"—Kindergarten Apparatus—Infants' Hammock 359

APPENDIX A.—DESCRIPTION OF THE KÖNIG WILHELM GYMNASIUM AT BERLIN 403

APPENDIX B.—THE RULES OF THE EDUCATION DEPARTMENT . . 417

APPENDIX C.—REGULATIONS OF THE SCHOOL BOARD FOR LONDON, FOR THE MANAGEMENT OF ITS SCHOOLS 425

INDEX 441

LIST OF WOODCUTS.

NO.			PAGE
	West Street School, London Fields	*Frontispiece.*	
1.	Lancasterian School	*Plan*	11
2.	School on Stow's system	*Plan*	13
3.	Wesleyan School	*Plan*	14
4.	School as suggested by Education Department	*Elevation*	15
5.	Ditto	*Plan*	16
6.	Polytechnic School at Zurich	*View*	24
7.	School at the Hague	*Plan*	25
8.	School at Amsterdam	*Plan*	25
9.	American Model Primary School	*Plan*	29
10.	Newton Primary School	*View*	29
11.	Newton Primary School, Philadelphia, U.S.	*First Floor Plan*	30
12.	Capen Primary School	*View*	30
13–16.	Ditto	*Plans*	31
17.	Hollingsworth School, Philadelphia, U.S.	*View*	32
18.	Ditto	*Plan*	33
19.	The "Tasker" School, Philadelphia, U.S.	*View*	34
20.	Ditto	*First Floor Plan*	34
21.	Melon Street School, Pennsylvania, U.S.	*View*	35
22.	Ditto	*Plan*	37
23.	Wood Street School, Philadelphia, U.S.	*View*	38
24.	Ditto	*Plan*	38
25.	The "George M. Wharton" School, Philadelphia	*Plan*	39
26.	Girls' School, Boston, Massachusetts	*Ground Floor Plan*	40
27.	Ditto	*Basement Plan*	40
28.	Ditto	*Second Floor Plan*	41
29.	Ditto	*First Floor Plan*	41
30.	Ditto	*View*	42
31.	Lurgan Model School, County Armagh	*View*	51
32.	Ditto	*Plan*	51
33.	Cork National Schools	*Plan*	52

NO.			PAGE
34.	French Mixed School, S. Pardoux les Cars	View	57
35.	Ditto	First Floor Plan	58
36.	Ditto	Ground Plan	58
37.	Salle d'Asile	Block Plan	59
38.	Ditto	School-room Plan	59
39.	Salle d'Asile, at S. Etienne, Limoges	Elevation	60
40.	Ditto	Plan	61
41.	Collége Chaptal, Paris	Back Elevation	62
42.	Ditto	Princpl. Elevation	63
43.	Ditto	First Floor Plan	66
44.	Ditto	Ground Plan	67
45.	Royal Grammar School, Chemnitz	View	69
46.	Parish School in the Kurfurstenstrasse, Berlin	View	77
47.	Ditto	Second Floor Plan	78
48.	Ditto	First Floor Plan	78
49.	Ditto	Ground Plan	78
50.	Ditto	Basement Plan	79
51.	Ditto	Block Plan	79
52.	Frankfurter-Strasse School, Berlin	Plan	82
53.	Konig Wilhelm Gymnasium, Berlin	View	89
54.	Ditto	Second Floor Plan	91
55.	Ditto	First Floor Plan	91
56.	Ditto	Ground Plan	92
57.	Ditto	Basement Plan	92
58.	Gymnasium at Liegnitz	View	95
59.	Ditto	First Floor Plan	96
60.	Ditto	Ground Plan	97
61.	Gymnasium at Marburg	View	98
62.	Ditto	First Foor Plan	99
63.	Ditto	Ground Plan	99
64.	Gymnasium of S. Andrew at Hildesheim	View	101
65.	Ditto	First Floor Plan	102
66.	Ditto	Ground Floor Plan	103
67.	Sophien Realschule, Berlin	View	106
68.	Ditto	Second Floor Plan	106
69.	Ditto	First Floor Plan	107
70.	Ditto	Ground Plan	108
71.	Realschule, Halberstadt	View	109
72.	Ditto	First Floor Plan	110
73.	Ditto	Ground Plan	110
74.	Stadische Realschule, Cöln (Cologne)	View	111
75.	Ditto	Second Floor Plan	112

LIST OF WOODCUTS.

NO.			PAGE
76.	Stadische Realschule, Cöln (Cologne)	First Floor Plan	113
77.	Ditto	Ground Plan	114
78.	Höhere Burgerschule, Wiesbaden	View	115
79.	Ditto	First Floor Plan	116
80.	Ditto	Ground Plan	116
81.	Bezirkschule, Dresden	View	121
82.	Ditto	Second Floor Plan	122
83.	Ditto	First Floor Plan	122
84.	Ditto	Ground Plan	123
85.	Ditto	Block Plan	124
86.	Ditto (Turnhalle)	View	124
87.	Bezirkschule and Gemeindeschule, Dresden	View	125
88.	Ditto	Second Floor Plan	125
89.	Ditto	First Floor Plan	126
90.	Ditto	Ground Plan	126
91.	Bürgerschule, Dresden	View	127
92.	Ditto	Second Floor Plan	127
93.	Ditto	First Floor Plan	128
94.	Ditto	Ground Plan	128
95.	Kreuzschule, Dresden	View	131
96.	Ditto	Section	131
97.	Ditto	Second Floor Plan	132
98.	Ditto	First Floor Plan	133
99.	Ditto	Ground Plan	134
100.	Polytechnic School at Aix-la-Chapelle	View	136
101.	Ditto	Second Floor Plan	137
102.	Ditto	First Floor Plan	138
103.	Ditto	Ground Floor Plan	139
104.	Chemistry School, Aix-la-Chapelle	First Floor Plan	140
105.	Ditto	Ground Floor Plan	140
106.	Ditto	Basement Floor	141
107.	Gymnasium at Cottbus	View	146
108.	The Theory of Austrian Education	Diagram	149
109.	Handel's Academy, Vienna	View	151
110.	Imperial Gymnasium, Vienna	View	154
111.	Ditto	Second Floor Plan	155
112.	Ditto	First Floor Plan	155
113.	Ditto	Ground Floor Plan	155
114.	Ditto	Interior of Hall	156
115.	Desk for Graded Schools (No. 1)	Detail Drawing	172
116.	Ditto (No. 2)	Detail Drawing	173
117.	Ditto (No. 3)	Detail Drawing	173

LIST OF WOODCUTS.

NO.			PAGE
118.	Graded School of 210, showing the Dual Desk	Plan	174
119.	Infant School for 120	Plan	181
120.	Ditto for 170	Plan	182
121.	Ditto for 300	Plan	184
122.	Infant Gallery of Maximum Size	Section	186
123.	Ditto	Half Plan	187
124.	Ditto	Half Elevation	187
125.	French Gallery	Plan	188
126, 127.	Details of French Gallery	Section & Elevatn.	188
128, 129.	Desks for Infant Schools	Detail Drawing	190
130.	Uncovered School-room	View	191
131.	Playground. Home and Colonial Society	Plan	192
132.	Double Inclined Plane	Sketch	193
133.	Wooden Swing	Sketch	193
134.	Parallel Bars	Sketch	193
135.	Horizontal Bars	Sketch	194
136.	The Climbing Stand	Sketch	194
137.	Entrance Porch (Infants')	Plan	205
138.	The "Double" Staircase	Section	207
139–143.	Ditto	Plans	208
144–146.	The "Short-Flight" Staircase	Plans	209
147.	Lavatory and Cloak-Room combined	Section	210
148.	Ditto	Plan	210
149.	Interior of Lavatory	View	211
150.	Mechanism of Lavatory	Sketch	212
151.	Latrines	Section	215
152.	Ditto	Plan	215
153.	Urinals and Latrines	Interior Sketch	215
154.	Ditto	Plan	216
155.	Village Teachers' Houses, Derbyshire	Section	218
156.	Ditto	Elevation	219
157.	Ditto	First Floor Plan	219
158.	Ditto	Ground Plan	220
159.	Ditto	Section	220
160.	Ditto	Elevation	221
161.	Ditto	First Floor Plan	221
162.	Ditto	Ground Plan	222
163.	Caretaker's Rooms	Plan	223
164.	Method of opening Windows by Iron Rods	Sketch	225
165.	Sliding Partitions	Elevation & Plan	228
166.	Hanging of Sliding Partitions	Section	229
167.	Wood Block Flooring	Section	230

LIST OF WOODCUTS.

NO.			PAGE
168.	Plank Flooring laid zigzag	Section	230
169.	Milton Mount College	View	240
170.	Ditto	First Floor Plan	241
171.	Dormitory, showing sub-division into Cubicles	Interior View	242
172.	Ditto	Ground Plan	243
173.	Parish School in the Pankstrasse, Berlin	Plan	246
174.	Parish School in Naunyn-Strasse, Berlin	Plan	247
175.	Gymnastic Apparatus for Open-Air use	Front Elevation	250
176.	Ditto	End Elevation	251
177.	The Spring Board	Diagram	253
178.	Apparatus for Deep Jumping	Diagram	253
179.	The Swing Trees	Diagram	254
180.	Fixed Parallel Bars	Diagram	254
181.	Adjustable and Fixed Parallel Bars	Diagram	255
182.	Adjustable and Movable Parallel Bars	Diagram	256
183.	Gymnastic Hall at Hof	West Elevation	258
184.	Ditto	Long Section	258
185.	Ditto	Plan	259
186.	Ditto	North Elevation	261
187.	Ditto	Cross Section	261
188.	The Boyd Grate	View, Plan, Sectn.	276
189.	The Boyd School Grate	View	277
190.	The Longden Grate	Section	279
191.	Ditto	Plan	279
192.	The Gurney Stove	Sketch	280
193.	Ditto with fresh air	Section	280
194.	The Gurney Stove, placed below-ground	Section	281
195.	The Gill or "Studio" Stove	Sketch	282
196.	Warming and Ventilating Apparatus	Section	285
197.	Locality of First Board School in London	View	291
198.	Old Castle Street School	View	293
199.	Ditto	Second Floor Plan	294
200.	Ditto	First Floor Plan	294
201.	Ditto	Ground Plan	295
202.	Harwood Road School	View	297
203.	Ditto	Second Floor Plan	298
204.	Ditto	First Floor Plan	298
205.	Ditto	Ground Plan	299
206.	Jonson Street School, Stepney	View	301
207.	Ditto	Second Floor Plan	301
208.	Ditto	First Floor Plan	302
209.	Ditto	Mezzanine Plan	302

LIST OF WOODCUTS.

NO.			PAGE
210.	Jonson Street School	Ground Plan	303
211.	New North Street School	North Elevation	306
212.	Ditto	Second Floor Plan	307
213.	Ditto	First Floor Plan	307
214.	Ditto	Ground Plan	308
215.	Ditto	Basement Plan	309
216.	Ditto	West Elevation	309
217.	Winstanley Road School	East View	311
218.	Ditto	South View	312
219.	Ditto	Ground Plan	313
220.	Ditto	First Floor Plan	314
221.	Eagle Court, Clerkenwell	View	315
222.	Ditto	First Floor Plan	316
223.	Ditto	Ground Plan	316
224.	Angler's Gardens School	View	318
225.	Ditto	First Floor Plan	319
226.	Ditto	Plan	319
227.	Ditto	Ground Plan	320
228.	Wornington Road School	View	325
229.	Ditto	First Floor Plan	326
230.	Ditto	Ground Plan	327
231.	Aldenham Street School	View	329
232.	Ditto	Second Floor Plan	330
233.	Ditto	Ground Plan	331
234.	Orange Street School	View	333
235.	Ditto	Second Floor Plan	334
236.	Ditto	Ground Plan	335
237.	West Street School	First Floor Plan	337
238.	Ditto	Ground Plan	338
239.	Camden Street School	View	340
240.	Ditto	First Floor Plan	341
241.	Ditto	Ground Plan	342
242.	Mansfield Place School	View	341
243.	Ditto	Second Floor Plan	344
244.	Ditto	Ground Plan	345
245.	Haverstock Hill School	Block Plan	347
246.	Ditto	View	348
247.	Ditto	Plan	349
248.	Industrial School at Ardwick	Ground Plan	357
249.	A Gallery Lesson	View	359
250-252.	The Cologne School Desk	Sections	364
253.	Ditto	Section	365

LIST OF WOODCUTS.

xxiii

No.			PAGE
254.	Desks and Seats in a Gemeinde Schulhaus	Section	367
255.	Dutch Double Desks with Seats attached	Section	367
256.	American Double Desk with Single Seats		368
257.	Moss's Patent School-Board Desks		369
258.	The "Angel" Standard		369
259.	Alternative Design for Desk Standard		369
260.	The Sheffield Desk		370
261.	School Desks from the King William Gymnasium, Berlin		371
262.	Swedish Single Desk with Lifting Seat	Sketch	372
263.	American Single Desks and Seats		372
264.	American Double Desks with Seats attached		373
265.	Alternative Method of graduating Desks in Class	Section	374
266.	Ditto	Section	374
267.	Home and Colonial School Society's Desks	Sketch	375
268.	Double Class-room, showing dual arrangement of Desks	Interior View	376
269.	Drill. (1.)—"Return"	Sketch	377
270.	,, (2.)—"Slates"	Sketch	377
271.	,, (3.)—"Lift" (or "Raise")	Sketch	378
272.	,, (4.)—"Desks"	Sketch	378
273.	,, (5.)—"Stand"	Sketch	378
274.	,, (6.)—"Out"	Sketch	378
275.	Drawing Easel in Victoria School, Berlin	Sketch and Section	380
276.	Ditto	Plan	380
277.	Head Master's Desk		381
278.	Master's Chair		381
279.	Class-room of a Gemeinde School, Berlin		382
280.	Teacher's Desk and Blackboard from the King William Gymnasium, Berlin		382
281.	Head Mistress's Desk with flat Top		383
282.	Mistress's Work-table with folding Top		383
283.	Pupil Teacher's Desk		384
284.	Cupboard for Boys' School		385
285.	Ditto		385
286.	Mistress's Cupboard		385
287.	Cupboard for Girls' School		386
288.	School Fire-guard		387
289.	Hook for Caps, Bonnets, &c.		389
290.	Framed Easel with T Slide		390
291, 292.	Easels from the Royal Gymnasium, Chemnitz		390
293.	Portable Table, or "Object" Lesson Stand		391

LIST OF WOODCUTS.

NO.		PAGE
294.	Blackboards	391
295.	Blackboard swung on Pivots	392
296.	Map-hook	393
297.	Cabinet of Objects	394
298.	Models illustrative of Mechanical Powers	395
299.	Cabinet for Weights and Measures	396
300.	Abacus on Stand	397
301.	French "Compendium"	398
302.	Hammock for Infants used in the French Asiles Communaux	402
303.	Warming Apparatus	412
304.	Class-room Window	414
305.	Class-room Door	414
306.	Sunburner	415

SCHOOL ARCHITECTURE.

INTRODUCTORY.

The industry of those connected with, or interested in, the teaching of the young, has made us familiar with some of the most favourable conditions under which school organization, management, and discipline, have been commenced and maintained in different parts of the civilized world. Since the first establishment of Sunday Schools at Gloucester, in 1783, followed soon after by the introduction of the rival systems for poor children by Bell and Lancaster, works on education of more or less value have appeared from time to time. In more recent years, the number has greatly increased, and the subject has been discussed with growing interest as the various opinions of leading philanthropists and educationists have become known. Till, at length, the blue books of the Schools Inquiry Commission have given us instructive reports by Mr. Matthew Arnold, Mr. Fearon, and others of Her Majesty's Inspectors of Schools, on the state of education at home and abroad, and on the best means to be adopted for gradually raising the tone and extending the benefits of education in Great Britain.

The scholastic buildings have not received the same attention. Some of the writers on education give an occasional glance in this direction, but rarely with any serious intention of discussing many of the points of school arrangement. There is no complete handbook on planning and fitting-up school-houses sufficient as a guide to school-founders, school-boards, architects and others, showing the various arrangements which may be considered best for the health, comfort, and effective teaching of children, and setting forth how the different parts of the building should fit together so as to form one harmonious whole. Education itself, being the question of most pressing necessity, has naturally received the first attention.

The Elementary Education Act of 1870, passed by the personal exertions of the Right Hon. W. E. Forster and the concurrence of all political parties, has, by establishing School-Boards throughout the country, given a new impetus to popular education. It has provided not only the means so long wanted for erecting suitable buildings for Elementary Schools, but the power of compelling, if necessary, the attendance of children. The movement has already, in turn, rendered more than ever apparent our sad lack of Secondary Schools, and the painful truth that, in England, the educational era has only commenced.

An inquiry, therefore, into the best methods of building and furnishing school-houses cannot be deemed unnecessary. Its subject must rank as second in importance only to the question of the education itself. The establishment of schools for the intellectual culture of all classes of the community having become a matter of national importance, the buildings in which the great work is to be carried on should be carefully studied to ensure the perfection of convenience and fitness in every part. If popular education be worth its great price, its homes deserve something more than a passing thought. School-houses are henceforth to take rank as public buildings, and should be planned and built in a manner befitting their new dignity.

A large number of new edifices, intended for use as public Elementary Schools, are completed or in course of erection throughout the country, but it by no means follows that the plans adopted are always, or on all points, models of school contrivance. Public interest has been much more excited on the question of cost than on that of skilful planning. The importance of education is only beginning to be understood by the average Englishman: that of judicious arrangement of school-buildings lies some distance in the wake. It is precisely this aspect of the educational question which claims our attention to the exclusion of others. In spite of the temptation to roam free over some of the many subjects interesting to the educationist, the scope of our inquiry must be strictly limited to the building and fitting of school-houses in the best manner. The subject may be dry and uninviting to many, but its study has become a necessity.

The attention of all who are charged with the instruction of the young, whether their establishments be public or private, cannot be too earnestly directed to the connection between good schools and good school-houses, and to the great principle that, to make an edifice good for school purposes, it should be really built for children and their teachers. It may be necessary, from time to time, to touch on systems of teaching, so as to render apparent the reasons of the planning in particular cases, but no further. The plan of the school-building depends so much on the method of tuition that an acquaintance with the latter is of the first necessity to the school-architect. For years it has been common with architects to regard school planning as almost too simple and easy a branch of their art to deserve serious study, and published works on the subject could be quoted to show that, when studied, attention to external prettiness was regarded as the chief matter. It is, therefore, of some consequence to connect the plan of the building with the system of teaching to be followed. When this latter is clearly and definitely settled, architects will always be found able to produce buildings suited to it. But they must first understand in what manner their buildings are to be used.

In discussing some of the principal points which go to make a good school-house, we do not confine ourselves to any one kind or description of school nor to the buildings in any one country. The main interest of the question, at present, undoubtedly lies with the primary or elementary schools, and with the application of hints or ideas, derived from whatever source, to the elementary school-houses being or to be erected in England. The schools of Germany afford excellent plans, as well as architecture of a monumental kind, often suitable for English secondary schools. The practical details to be hereafter mentioned will apply, generally, as much to one kind of school as to another, being almost always common to all.

In England, hitherto, owing to the absence of a general system of public education adapted to the whole population, the only moneys derivable from the public purse towards the erection of school-houses have been the "grants in aid" made by the Education Department of the Privy Council, and which only bore a small proportion to the total cost. The remainder was derived from the subscriptions of those interested in the establishment of the new school—either Churchmen, Roman Catholics, or Dissenters of some denomination,* and the result has been, not only to necessitate the greatest economy in the school buildings, but so to contrive their plan as to render them occasionally useful for lectures, concerts, tea meetings, and the like, often to their great detriment regarded simply and solely as school-houses. The funds having been raised with great difficulty, every feature, the absence of which did not endanger the annual grant, was dispensed with, proper arrangements on all hands curtailed, class-rooms (except one solitary room for the youngest infants) abandoned, until the building consisted simply of one room for each sex and for infants, the plan being usually a parallelogram or L shape.

Now, however, education having emerged from the denominational stage and become a national question, the new buildings for school purposes should be planned solely with a view to convenient and effective teaching, and to proper sanitary

---

* Since the establishment of School Boards, all these kinds of schools have been called, for the sake of convenience or contrast, "Denominational or Voluntary Schools."

arrangements. Thus, the widths of the school and classrooms will be decided in relation to the number of benches and desks to be used. Both will be carefully proportioned as to length and height. The class-rooms will be so placed in connection with the schoolroom as to be economical in plan and easy of supervision on the part of the master or mistress. The windows will be ample, and so disposed as to throw the light in the right places, as well as to be useful for summer or occasional ventilation. The doors will be so arranged as to afford easy means for the dispersion of the school, and for access to the yards and playgrounds, without sacrifice of desk-room or other convenience. The fireplaces (or other method of warming) will be contrived so as to avoid roasting or rendering uncomfortable either teacher or children during the progress of a lesson; and the furniture and fittings will be suited to their respective purposes.

Hitherto the Education Department have had to contend with paucity of means on the part of school promoters, and their difficulty has chiefly been to bring up the buildings to the required mark, or nearly so. Now, however, the evil must surely be of a widely different and even opposite character, requiring the exercise of a firm restraining hand, so that the new erections shall be economically built and made fit for schools, if fit for nothing else. Formerly, the Department had to deal with clergymen and others usually possessing some knowledge of school arrangements, and simply lacking money. Now they are met by School Boards sometimes numbering gentlemen new to the work, and able, if they like, to squander large sums of money over their own theories of school-planning, however much these may be in opposition to the opinion of the best authorities.

It will thus be seen that (so far as we treat of Public Elementary Schools) our object is not to start new-fangled ideas which might appear directed to revolutionise the system in use in this country—a system which differs from any other, and which has grown through many years of care and studious effort—but rather to present it in the new light which is the necessary outcome of new legislation; to supply more copious reasons for, and explanations of, the points to which importance is attached; and to glean from other countries, as well as from many sources of observation and experience in our own, such further hints as may tend to develop still further the principles of English school-planning.

Throughout all the general principles and minor details which we may hereafter discuss, the subject of *hygiene* should be ever present, though perhaps never mentioned. To the school-architect, hygiene means the rules which should regulate the situation, construction, ventilation, warming, lighting, and furnishing of the building. To the school-manager and the schoolmaster it means the time, quantity and kind of work, exercise, and rest, which may be found most favourable to the health and development, physical and mental, of the children.

The school-architect should remember at the outset that he is building for children varying in age, sex, size, and studies, and therefore requiring different accommodation: for children engaged sometimes in study and sometimes in recreation; for children whose health and success in study require that they should be frequently and every day in the open air for exercise and recreation, and at all times supplied with pure air for respiration; for children who are to occupy it in the hot days of summer and the cold days of winter; and this for different parts of the day in positions which become wearisome if the shape and relative positions of the seats and desks have not been studied for comfort in every respect; and which may affect symmetry of form, quality of eyesight, and even duration of life; for children whose manners, morals, habits of order, cleanliness, and punctuality, temper, love of study and of the school, cannot fail to be in no inconsiderable degree affected by the attractive or repulsive situation, appearance, out-door convenience and in-door comfort, of the place where they are to spend a large part of the most impressionable period of their lives. This place, too, it should be borne in mind, is to be occupied by teachers whose own health and happiness are affected by most of the circumstances above alluded to, and whose best plans of order, classification, discipline, and recreation may be utterly baffled or greatly promoted by the manner in which their school-houses are constructed and furnished.

CHAPTER II.

EXISTING SCHOOLS.

Necessity of State Control over School Buildings—Want of Secondary Schools—Lancasterian System—Stow—Wesleyan Schools—The Education Department—Mixed System—Switzerland—Holland.

To understand fully the principle that a good school requires a good school-house, it is necessary to know something of existing schools in this country and to see how they compare with others. Wherever good teaching has been appraised at its full value, there should good buildings be found for the purpose. Where really first-rate education exists in inferior or unsuitable buildings, it will generally be found that better accommodation is contemplated, or that the education itself is not fully appreciated by those providing the funds. The importance of good early training is not immediately understood by any people. When once realized, proper buildings follow as a matter of course. The English nation as a whole has not grasped the importance of the Elementary Education Act, and the greater number of the new elementary school buildings will probably be completed before the real position of this popular boon is established.

Perhaps among a people having so much to say on the liberty of the subject and favouring freedom verging on license in all directions, it would seem rather hard to deny a person the right of teaching a few children in any place which might be selected. In Germany this is done systematically, and even the private schools are not exempt from State interference. There a certificate of competency (the result of examination) is far more stringently insisted on generally than it is, even yet, with us. The State appears to hold that an incompetent teacher is an injury to society, and both teachers and school-houses are under its control.

In Holland also a building cannot be used for school purposes, public or private, without an inspection and certificate of fitness. Any one conducting a school, in a building declared by the district inspector to be unsuitable, is liable, for the first offence, to a fine of from two to four guineas a week; and for a second, to a double fine and imprisonment of a week or a fortnight.

Apart from the necessities of the poorer population, for whom primarily the new Elementary Schools are intended, we find the condition of the children of the lower middle classes mainly educated in private schools, sometimes day-schools, sometimes boarding-schools.* Irrespective of Dame Schools and others of the lowest class, whether as to buildings or instruction, whose days are numbered and whose history deserves no record in an architectural work, this *better class* of private schools deserves attention. A large proportion of the buildings in which they are conducted are quite unfit for the purpose. What is the history of not a few of them? A gentleman — it may be a University man — sufficiently learned to give his pupils a good education, but utterly ignorant of sanitary science or the ordinary principles of domestic economy, finds in some back street or outlying country place an old house, left long without a tenant on account of its defective drains, bad water, damp walls, leaky roofs, smoky chimneys, or some of the other numerous diseases to which old houses of the badly-built sort are prematurely liable. The smallness of the rent reconciles him to such of the evils as he can himself see, while a little tinkering, and the application of some cheap paint and paper, effectually conceal the remainder. The largest room, perhaps with cubic air-space enough for half a dozen persons under ordinary domestic arrange-

---

* Known, since the advent of School Boards, as "Private Adventure Schools."

ments, and utterly innocent of any pretence to ventilation, is set apart for the school-room, furnished with benches and desks as uncomfortable as possible, and forthwith packed with a score, or more, of pupils for seven or eight hours a day! The bedrooms are crowded in like manner, and the result is that the children go home for the holidays looking pale and ill, or, as their parents fondly believe, very much "over-worked." This is no overdrawn picture. There are numerous schools of the kind, both in London and the provinces. They are under no Inspector's eye. They ask for no Government grant. At present they cannot be meddled with. But it is not too much to predict that the time will arrive when the use of such buildings for the purposes of education will not be tolerated. This time, however, is not likely to be until all classes, and all sects and parties, come to realize fully the benefits of education itself. The Duke of Newcastle's Commission ascertained in 1861, that these private—non-inspected—schools, were especially popular with the well-to-do working classes, from their supposed *gentility*, and were on the increase. The number of children they contained in 1861 was no less than 573,536. Some of them were good, but of the vast majority, whether as regards buildings, teachers, or methods of instruction, it would be difficult to speak too strongly. It is not, of course, pretended that we have now no good middle-class or secondary schools; but in a year or two hence, when Mr. Forster's Act shall have fulfilled the first part of its mission in the creation of suitable buildings for public elementary education, the English position will be that, although pride may be felt in the possession of such public schools as Eton, Harrow, Rugby, Westminster, Winchester, Charterhouse, Durham, and others—available chiefly for the children of wealthy parents—there will still remain a great gulf between the well-appointed elementary schools and these—a gulf which can only be bridged over by serious legislation in the direction of middle-class, practical, and technical schools. The children of the peer and the peasant will then have been provided for; but those of the middle-classes (of those classes which are sometimes regarded as the strength of the country) will yet, to a great extent, remain. These secondary schools will also be necessary as the promotion-schools for the holders of scholarships from elementary schools.

In dealing with the question of school-planning, and touching on the plans now or formerly in use in England, there is no attempt to describe the history, development, and working of the existing schools. That has already been completely accomplished by Mr. Bartley in his excellent work, "The Schools for the People." *

The English method of teaching, based on the employment of "pupil-teachers," is in use, more or less, in some other countries, but is not always developed in the same way. Large and wide school-rooms, common to the Lancasterian (now called British) Schools, are still in use both in England and America. The origin of the system was, in the expressed opinion of Lancaster himself, that with the aid of monitors, one master could conduct a school of even 1,000 children. A gradual change of opinion, however, came over the managers of this class of schools, and about the year 1850, the pupil-teacher system, first introduced in 1846, appears to have superseded that of the monitor. The large room, as shown in woodcut No. 1, is arranged to give an entire system of benches and desks facing in one direction towards the master. In this case

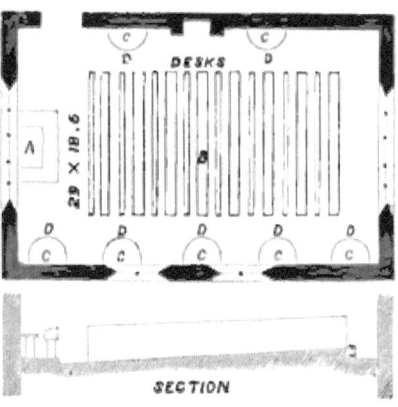

1.—LANCASTERIAN SCHOOL.

the whole school-room resembles nothing but itself, or a huge class-room based on the Continental model, but much exaggerated. It is, however, sometimes planned with benches and desks

* Bell and Daldy, 1871.

placed in groups down both walls, and the scholars then face each other from opposite sides of the room. Floor-space is left at the sides of the room to enable classes to be drawn out and receive lessons, standing in a semicircle, they being at D D, the monitors at C C, the teacher at A.

In Ireland and in America, though not hitherto in England, it has been a favourite plan to give each pupil a separate seat, although the desks themselves are made for a greater number, the gangways occurring generally at intervals of two, although (in Ireland) sometimes of three, and even of four. The advocates of wide and large school-rooms maintain that there is a greater power of control on the part of the teacher, and that, in spite of the great noise (and to an on-looker, the great confusion and babeldom) during lesson-time, there is a more than compensating element in the superior general intelligence promoted by the "sympathy of numbers," obtained by simultaneous teaching. These school-rooms, supposed to be obsolete in England, and so regarded by the Education Department, are advocated by the Leeds, Worcester, and other School Boards; and by some able schoolmasters and educationists.

On the contrary, the system in use among the whole German-speaking race, from Berlin to Vienna, turns on the theory that each class should be taught in a separate room by a separate and fully-qualified master, and that the same educational treatment should be maintained from the earliest to the last moment of school life. Two features, universal with us, a "pupil-teacher," and an infant-school gallery, are there wholly unknown to the system. The question of wide or narrow school-rooms, again, does not enter into their calculation at all, as the school-room is omitted altogether in favour of a collection of class-rooms. The contrast must not here be further developed, for the schools of Germany, studied and matured on a system so different from ours, and applied alike to every class in the country, from the highest to the lowest, are of sufficient importance to claim a separate notice.

Among existing English schools, those belonging to the Wesleyan Methodist Denomination deserve some notice.

Perhaps no others not directly connected with the Glasgow Normal Seminary are so directly the outcome of Mr. Stow's training system. When the tide of opinion set in strongly against the monitorial system, as advocated by Bell and Lancaster, Mr. Stow earnestly urged the employment of a greater number of properly qualified teachers, and condemned the improper employment of monitors—"the work of a master done by apprentices." But he never condemned the pupil-teacher system. In some sense, he is the author of it; for he invariably laid down the principle that mere teaching was insufficient—that a system of training was necessary. This was first adopted, in England, at the Cheltenham Training College, and is exactly what the English method now aims at. The Wesleyans have always held the opinions of Mr. Stow in reference to the sympathetic influence of large numbers on each unit of the number. The woodcut No. 2 shows a school on Mr. Stow's system, and it is equally the plan of a Wesleyan school. The chief feature is the large gallery, whereon a collective lesson may be given to a very large number — perhaps two-thirds of the whole school.

2.—SCHOOL ON STOW'S SYSTEM.

In the Wesleyan schools this is used as the theatre for the religious instruction which forms so vital an element in their system. The three points on which Mr. Stow chiefly laid stress, were, a competent teacher, a good gallery, and a well-furnished playground.

Woodcut No. 3 shows the plan of a Wesleyan school, which in turn is equally that of a school on Stow's system. It will be noticed that here the schoolroom is 30 feet wide, with a double system of desks, one set facing the other, while the large gallery rakes both, and, like the desks, is commanded by the teacher. The objection is, that the attention of the child is always liable to be diverted to those directly opposite and as full of play as himself. A general peculiarity noticeable in Wesleyan schools is that the

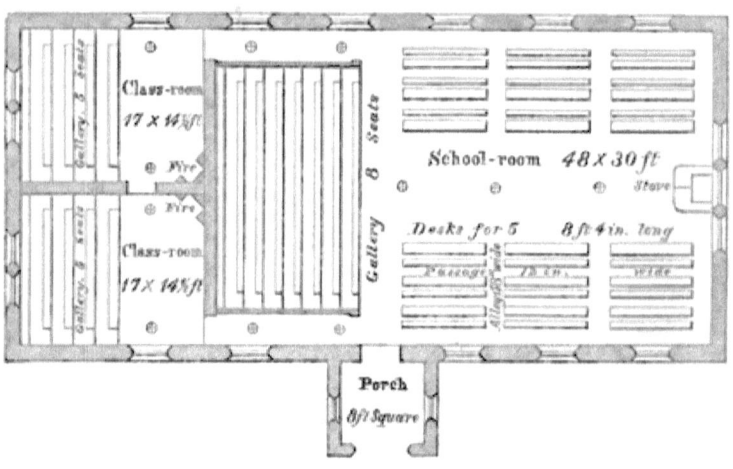

3.—WESLEYAN SCHOOL.

galleries are very large, and the class-rooms too small. In fact, the necessity of imparting religious instruction by means of a simultaneous gallery lesson to some extent governs the plan. The gallery, only occasionally in use, occupies much space in the school-room, and, unless economy be neglected, some of this space is obtained at the expense of the class-rooms. These schools are all, or nearly all, used as Sunday schools, and designed quite as much for that purpose as for the more secular education of the week-day. Wesleyan schools have, however, been in advance of the Church schools in providing more frequently a fair number of class-rooms, a want now pretty generally recognised. The foregoing examples refer to buildings contrived before the passing of the Elementary Education Act. The Wesleyan body do not now object to the arrangement of desks shown in woodcut No. 118, in the chapter on School Seats, especially if raised on steps one behind the other. It then approaches the gallery principle.

The work of the Education Department in reality dates from the year 1839, when an Order in Council created the "Committee of Council on Education," although the various educational establishments of the Government were not finally united under the new name until a second order in 1856. Their efforts have been so directed to the economy of school planning that an excellent foundation has been laid for the labour of School Boards and their architects. It is to them we owe official sanction to the pupil-teacher system, and also to the principle of arranging school desks *on one side only* of the school-room. The accompanying plan, taken from the well-known "Rules," is a sufficient reminder of the kind of school alluded to. The wide schools of the Lancasterian and British type are abandoned, equally with the class subdivision of Germany.

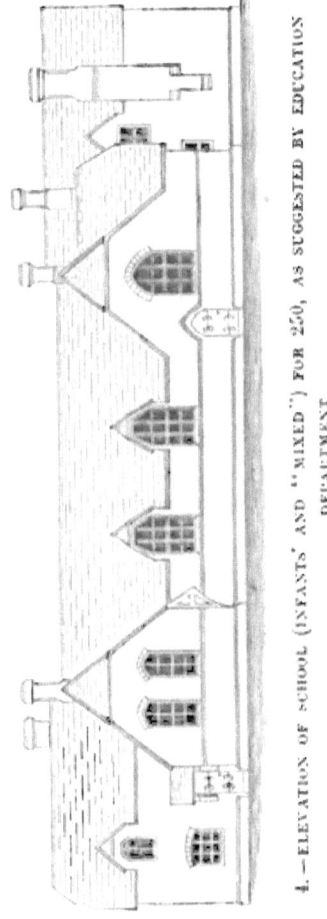

4.—ELEVATION OF SCHOOL (INFANTS' AND "MIXED") FOR 250, AS SUGGESTED BY EDUCATION DEPARTMENT.

Evidently the object has been to combine in one system the good points of several, and not to lose the use of the pupil-teacher working under supervision. The "sympathy of numbers" is not lost sight of, while separation of the classes is partly obtained by curtains. Economy of plan is insisted on, and no greater width of schoolroom allowed beyond what is strictly necessary to draw out the class in front of the groups of desks. To allow supervision on the part of the teacher,

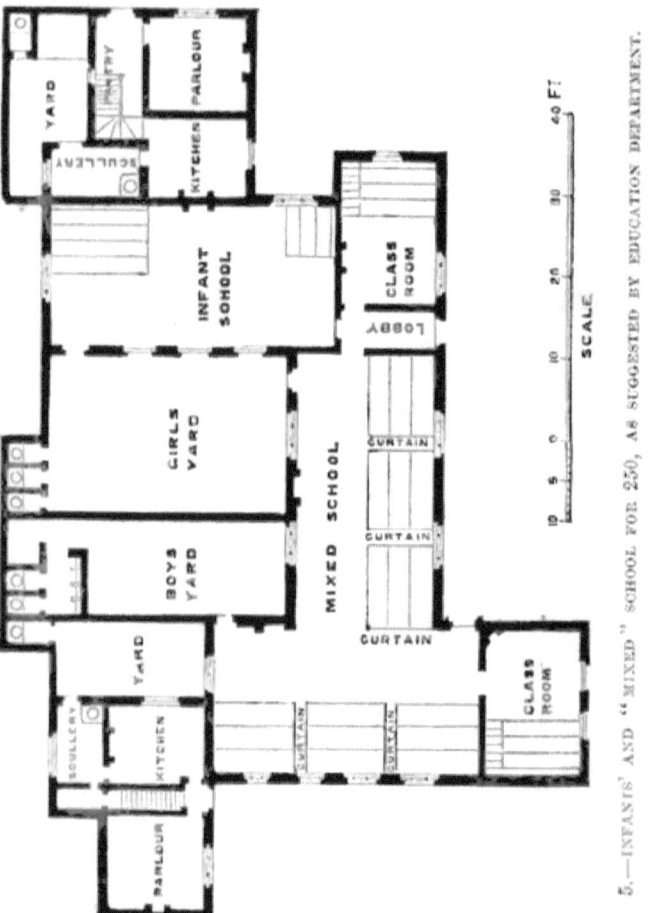

the use of long desks necessitated a space to be left between each row, and the number of rows was therefore limited (though not absolutely) to three. It is only now, however, that, by an immense improvement in the shape and mechanism of desks space is economised; five rows instead of three can be used both in school-rooms and class-rooms, and, instead of being limited to 20ft. as a maximum, the school-room may be economically built at 22ft. wide. While improvements of like nature, affecting the plan of the building, may be devised from time to time, the fact remains that what is now being done in school-planning is but the natural development of a system already laid down, stimulated by the presence of greater requirements.

On the other hand, and as throwing light on the plans issued by them, we must remember that the work of the Education Department has been mainly directed to the kind of school hitherto called "National," these being by far the most numerous, and belonging to the Church of England—the State Church. Mr. Bartley tells us that of those which received Government aid in 1869, the Church schools numbered no fewer than 808,364 children, as compared with 217,438 in Dissenting, and 67,768 in Roman Catholic buildings, or, nearly three times the number of all the rest put together. And not alone this: another fact is to be remembered. The assistance of the Department has been sought so much more frequently in the villages and smaller towns than in London and the great seats of industry (the latter often consisting of only one parish), that their plans and rules, as gradually developed, really apply rather to small than to large— to country rather than to town—schools, and always to an extremely limited condition of purse. The smallness of the school, too, led to the adoption of the "mixed" system, but only as an expedient, and not as a principle. The number of children being too few in the majority of cases to admit of the cost both of a master and a mistress, the boys and girls were taught together in the same school by a *master*. Wherever we find a Church school of large size, then the sexes are taught separately and in separate schools. In other words, the managers of this type of school have never really believed in the mixed system. They have, however, believed too much in, or carried too far, the pupil-teacher system, as proved by the fewness of class-rooms.

If any ground is to be gained by the study of existing schools, the defects as well as the merits of the system from which we chiefly develop must be fully discussed.

Mixed Schools—schools above the infant stage where boys and girls are taught together—have not been much in favour, partly because of the opinion that their adoption narrows the field for the employment of mistresses to Infant Schools only, and partly from other reasons. The objectors contend that few women can maintain discipline among the more turbulent male element in a Mixed School. The advocates of the system, on the other hand, point to America, where the whole of the primary education of the people is carried on in such schools principally conducted by mistresses. Practically, the mixed system among us has been regarded as a make-shift; and here in schools where it is in force we generally find a *master, not a mistress.* The late Mr. David Stow,* the eminent philanthropist of Glasgow, strongly advocated it, and his influence is still shown by the high favour in which the system is held in Scotland. The great majority of the Wesleyan schools adopt it, and, in some of the admirable establishments belonging to the Wesleyan body, may be seen classes of girls and boys trained together to the end of the course, through all the school grades, and even to the age of fourteen or fifteen. Among these schools, also, few of the teachers are women, except for the girls' needlework lessons, &c., as required by the Code. A greater number, however, may be anticipated, for the body are now encouraging, on principle, the employment of women teachers in Infant and Mixed Schools up to the age of twelve as a maximum. In this they set a good example.

The mixed system, if adopted, would affect our plans chiefly on minor details, such as the mode of access to the yards, the staircases, cloak-rooms, &c., and it is not proposed, therefore, to discuss it at length. It may be stated, however, that it is

---

* "The Training System." By David Stow. London: Longman, Brown, Green, & Longmans.

chiefly found among the English-speaking race. In America the system is in high favour, though by no means universal, and generally confined to the Primary Schools, which extend to the age of ten or eleven. On the continent of Europe it is found in Holland and Switzerland. In London the general opinion is somewhat against mixed schools, especially for the senior classes, although many old ones are in actual work.

Professor Huxley's Committee reported, in June, 1871, to the School Board for London, as follows:—

"It is universally agreed that Infant Schools may be mixed, not only without detriment, but with positive advantage to the children.

"With respect to Junior Schools, so much depends upon the previous training of the children, and upon local circumstances, that we do not think it advisable to lay down any general rule regarding them.

"On the other hand, while evidence has been brought before us tending to show that, under certain conditions, Senior Schools may be mixed, we are decidedly of opinion that the Senior Schools provided by the School Board for London should be separate."

The division of schools into Senior and Junior—so that in the lower the sexes may be mixed, and in the upper kept distinct—is only feasible in establishments of very large size. It is to be found in Scotland, but only in schools of 1200, 1500, or more. And the truth appears to be that, whether as a logical consequence, or in view of economical working, a graded school should either be entirely mixed or entirely separate. If the organization be for a school of 1500 children, then it is easy to mix the junior departments and to separate the seniors. If, however, the numbers be about one-half, as now generally preferred, then the senior departments become so small as to be better, and more economically, arranged in the form of class-rooms attached to the school.

We now come to the consideration of the central feature of the English system of teaching—that on which the planning of our *elementary* schools depends—the pupil-teacher system. Unless the school architect has some idea what this means, he cannot fail to be more or less in the dark as to his plans, and will often be unable to decide which of two or more possible arrangements is the best. It is diametrically opposed to the German system. And so is an English to a German plan. It has points of resemblance to the teaching systems of Holland and of Switzerland. And yet the school plans of the latter countries are quite unlike ours.

To those who wish to study the question more fully, we recommend a perusal of the recently published pamphlet * of Dr. Rigg. It would be difficult to mention any one more competent to speak with authority on the English sytem of teaching than the present Principal of the Wesleyan Training College at Westminster. He says:—"Men go hastily to Germany, and see a German school taught only by adult teachers, a teacher to each separate class; or they go to Edinburgh and see a similar plan in operation in a number of excellent schools; they see or hear besides that a similar plan is in operation in the expensively appointed and efficient schools which are among the shows of some towns in the United States, and they come to the conclusion that modern science is opposed to the employment of pupil-teachers, and requires that only adult teachers should have any charge of children. Whereas the fact is that, *for Elementary Schools*, the pupil-teacher system is the only truly scientific system; that the separate class and teacher system is the very old system, which has been superseded because of its ineffectiveness, and that where this system succeeds, it is under conditions altogether different from those which belong to the elementary schools of this nation. . . . It may be practicable in Germany, where thirty pounds a year in gross payment is good pay for a teacher, to have a public school manned only by grown teachers. In this country, where the teacher is paid six times

---

* "Primary Education in England." By the Rev. James H. Rigg, D.D. Hamilton, Adams, and Co.

the amount, this would not be possible. Even in Germany the result is that the classes under the charge of each teacher separately are far too large for thorough and efficient teaching at all points, and through all the school-time. No teacher can by himself, and without the aid at least of effective monitors, teach constantly a class of seventy or eighty children, although they may be nearly of the same age and standard of attainments. There are always great differences in children. The child who is quick in arithmetic is often slow at reading. The clever reader is often a dull calculator. Ability in grammar by no means implies cleverness in geography, and *vice versâ*. In particular, to teach reading and dictation to backward scholars requires minute and individual drill. Children to be well taught must often be broken up into small squads with a monitor or pupil-teacher over them. Although on certain subjects a teacher may be able to give capital and effective simultaneous lessons to a large class, yet for many purposes of eliciting thought, and correcting error, and supplying defect, he must be able to deal with each scholar apart, or as one of a much smaller section. Besides all which, for slate-work and paper-work, than which nothing can be more important, to be properly examined, it is certain that the teacher needs to have a comparatively small class. All the educational science of England fixes the maximum at from thirty to forty; sciolism, which phrases largely about Continental education and the German system, talks about teaching with ease and efficiency a class of eighty. By the pupil-teacher system a scientific form and organization has been given to that provision of pupil-aid in schools, the need of which is universally felt in public elementary education, and the rude and ready form of which, open to grave objection, and altogether imperfect and unsatisfactory, is found in the use of monitors—sometimes recognised as such, and sometimes extemporised, as necessity compels, and used without formal recognition."

This is all no doubt perfectly true. The writer evidently possesses the strongest faith in the system he advocates, and writes with vigour. For Elementary Schools,—for training children up to the age of twelve or thirteen,—the pupil-teacher, or English, system is not only the most economical, but probably the best. But its danger lies in the employment of too many pupil-teachers in one school—just as some other systems are faulty in not having any. The plan of the infant and mixed school, already given, cannot be regarded as other than a very early development of an English plan. Only one class-room is provided to the mixed school, and the number of pupil-teachers working with one master is evidently considerable. As the number of public Elementary Schools increases, some reduction will probably be made in the allowable number of pupil-teachers, and the number of class-rooms will be increased. Professor Huxley's Committee, already referred to, reported that the minimum number of teachers for a Junior or Senior School of 500 children should be 16—viz :—1 principal teacher, 4 assistant certificated teachers, and 11 pupil-teachers; and that the teaching staff should be increased by one assistant certificated teacher and three pupil-teachers for every additional 120 children. And no ideas of false economy should allow this proportion of pupil-teachers to be exceeded. The abuse of a system often proves its greatest condemnation.

Popular education in Switzerland is admitted on all hands to be effective, practical, and generally excellent. Perhaps it is, on the whole, better than in any other European country, except Holland. But, as each canton has its own laws, and the consequent organization, the system of instruction is not identical throughout. Formerly, the Swiss schools were carried on in the old châteaux, converted for school purposes as best they might be. Now, however, new and excellent school-buildings are to be found in most of the towns of the Confederation. The course of study usually embraces six years, commencing at the age of six, in the Primary Schools. These consist of two divisions—the Elementary and the Practical—to each of which are assigned three years of the course. This is, so far, unlike our system, and, as we proceed, it is still further unlike; for the Swiss possess what we have still to obtain—viz., a complete system of Cantonal Schools for those requiring further study to enable them to enter the University, or to acquire the scientific knowledge of an art or trade. These Secondary Schools are, therefore, divided into two sections, the Grammar School and the Technical or Trade School.

The Primary Schools we may continue to call, for the sake of convenience, Elementary Schools, although the upper department is called by a higher name, and really forms a good secondary school of popular kind. The plans present few features which, to the architect, are different from those of Germany. This is natural enough when we consider how much of the Republic may be regarded as German-speaking, and how many of their best teachers are selected from Germany. It is not necessary to give any special illustrations of their arrangements, as these will be sufficiently understood when we come to deal with the subject of German plans. Perhaps Zurich contains as many good examples of Swiss school buildings as are to be found in any one Canton, and, for our purpose, may be taken as the type of all. There are eleven school districts, wherein educational affairs are managed by a chain of different, yet connected, public authorities. Each commune, as well as each district, has its board or committee, whereon the teachers themselves—forming in Zurich a society of considerable influence—are not without a voice. To the higher of these bodies all school plans are submitted for approval, with power of appeal to the council of education. In the town itself the Grosmünster School (adjoining Zwingli's Church, containing 1000 children) is worthy of a visit. Another, in the Wolfbach, and under the same management, is entirely new, has corridors 15 feet wide, and is completely fitted in every respect. It accommodates 300 boys and 500 girls, and its cost of site, buildings (including the indispensable *Turn-halle*), and furniture, amounted to £25,000. It is entirely free. Near the Wolfbach is a large cantonal school planned with a central court, but badly. The site as usual, however, is fine. The playground, here open, is fitted with gymnastic appliances which are used with great zest by the boys in their play.

Those who wish to study fully the subject of Swiss education may read with advantage the report of Mr. Arnold on Secondary Schools in Switzerland, and note the system in operation. Among others are to be found the Elementary, the Secondary, the Cantonal, the Veterinary, the Agricultural, the Training, and the Polytechnic Schools; also the University. Certainly no visitor to Zurich should neglect an inspection of the famous Polytechnic School there. Whether in its arrangements,

6.—POLYTECHNIC SCHOOL AT ZURICH.

its teaching, its size, or the splendid site it occupies on a height overlooking the town, it must be regarded as a singularly commanding educational establishment. Yet its plan cannot be compared with some German models, and in point of architecture it is inferior. The defects of mass and skyline are aggravated by a very conspicuous position.

Holland deservedly stands high among those countries whose works we are studying, if an education diffused among all ranks, from the highest to the lowest, be any claim to distinction. Not only are the lower classes intelligent and well-informed, but the upper classes of the population generally have attained a high average of mental culture. In many respects the Holland of to-day may claim to be in the educational van. After the "Bewaar Schoolen," the Kinder Garten schools of Holland, come the Public Elementary Schools, which appear to be divided into four kinds, the "Armen Schoolen," the "Tusschen Schoolen," and the "Burger Schoolen" of two classes. But their plans start no new question for the English builder. Some of them consist, as in the country schools of the province of Utrecht, of rooms about 52 feet by 26 feet, intended for about 150 children, and bisected by a glass partition. As a rule, they are of one storey, though, where land is of increasing value this is not maintained. The two illustrations which we give of Dutch Schools have a certain resemblance to some English, though very old, plans. No. 7 represents a large square school at the Hague, where the chief difference consists in the position of the classes in the room, and not in the room itself, the classes being placed *dos-à-dos* instead of *vis-à-vis*. The monitors, at *a a* in the Dutch School, would, in one of our exploded Lancasterian examples, be placed down the centre, and they, instead of the children, would be back to back. No. 8 is the plan of a school at Amsterdam

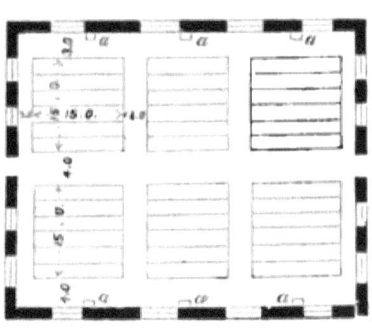

7.—SCHOOL AT THE HAGUE.

8.—SCHOOL AT AMSTERDAM.

for 500 children. Having only one line of desks, it slightly resembles a school planned according to the rules of the Education Department. The great length in proportion to the width is a defect to a more serious extent than is usual with our old schools. Indeed, the wide school-room (No. 7) would, of the two, be found to be the best for purposes of teaching. Both these types of school-rooms may now be considered out of date, so far as their application to new buildings is concerned.

The lessons to be learned from Switzerland and Holland—the two European countries where primary education has reached the highest excellence—should not be overlooked. In both instances the schools are good and cheap, the youngest children are well attended to, and the sexes are taught together in the same classes. In the two last particulars, the practice of Holland and Switzerland differs from that of Germany and Austria. Compulsory attendance everywhere prevails, and religious instruction is given in all the schools. It is so common to find in England the delusion that teaching in Board Schools must necessarily be of a purely secular character to the exclusion of religious subjects, that the custom prevailing in other countries which possess a national or state system of education should be quoted in disproof. In France, Italy, Germany, Switzerland, and Holland the systems are not only national and religious, but denominational. It should be added that the denominations are not incessantly squabbling over their 'isms, as though each worshipped a separate and rival deity.

CHAPTER III.

AMERICA.

Coloured — Alphabet — Primary — Normal — Grammar — High Schools — Specimens — Girls' School at Boston — General.

No people make more determined efforts to obtain information on the subject of schools and schoolhouses from all available sources than those of the United States. The general movement in favour of education is regarded with a deep interest, and in every civilized country the American representative is on the watch to report to Washington the facts concerning any progress which may have been made. As one out of many instances of this vigilance, it may be mentioned that the main points of the English Elementary Education Act of 1871, were known and discussed among American educationists before the average Englishman had grasped them. The enthusiasm of the nation on the subject of schools, generally, may well be a source of pride.

Turning our attention, therefore, to the New World, we should naturally expect to glean many valuable hints from among the numerous examples of recently erected school buildings. The American text-book for school plans has been for many years Barnard's excellent and well-illustrated work. There is no work in the English language on the same subject so complete. But many of the plans are now regarded as old-fashioned; and it is to the annual reports of the School Boards in the several States of the Union that we must look for the latest information. Generally they show a decided opinion in favour either of the dual or single arrangement of desks, never a preference for the long-length system which has so long been common with us. The plans of the buildings themselves, though sometimes sufficiently ordinary, are sometimes highly ingenious.

The American school-manager has one difficulty more than encounters us in England. The increase of population, by reason of the continual immigration from nearly all other parts of the world, is so great that, year by year, it has always been very difficult to develop education at an equally rapid pace. As an illustration of this, it may be mentioned that, according to the last report of the Superintendent of the Schools of New York, there are, of the old log school-house, the first kind of school-building (if building it can be called) ever attempted in the New World, no less than one hundred and twenty still remaining in the Empire State alone. But now that the black man is emancipated from slavery and possesses the same rights of citizenship as his former masters, he too must be educated. And in towns where the lowest stratum of the population consists of Irish and people of colour, it is natural to expect to find both in the same public school. Experience has shown, however, that to teach them together involves more fighting and general confusion than is desirable every day of the week. They will no more work peaceably in the same school-room than oil and water will mix. "Coloured" schools are, therefore, frequently provided.

In America there are many schools on the Kinder Garten model, but, generally, the first step is the "*Alphabet*" School, sometimes (as in the Southern States) called the "Infant" School. These, for the poorer class, are public. The better classes, there as elsewhere, learn the alphabet either at home or in a private school.

We find next the "*Primary*" School, which corresponds to the Junior Graded portion of an English Elementary School. In this class of school the mixed system is universal, but is applied oddly. For, although in the class-rooms the sexes are mixed, in the schoolroom they are commonly placed on different sides. And it is a form of punishment to place a single naughty child among a whole class of the opposite sex. The woodcut (No. 9) shows a "Model" Primary School such as we see attached to a Training College as the type of a country school. Like the Lancasterian plan, No. 1, given at page 11, it is simply a huge class-room. The raised gallery so familiar to English Infant Schools, is never used here, any more than in one of the class-rooms of a Prussian school. The teacher's seat is placed on a raised plat-

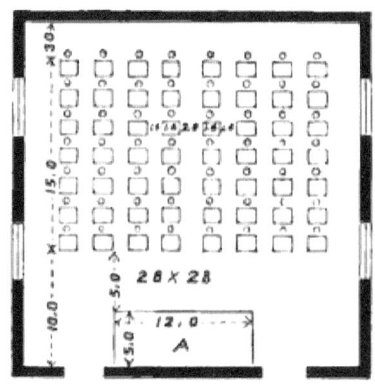

9.—AMERICAN MODEL PRIMARY SCHOOL.

form, and the children sit on separate seats fixed upon a level floor, and with a gangway dividing each from the next. In towns,

10.—NEWTON PRIMARY SCHOOL.

the primary schools are commonly of three, four, or more storeys, the youngest children being placed on the lowest floor, and so on upwards, according to age.

The Newton Primary School, Ludlow Street, Philadelphia (woodcut No. 11), is so planned that, without abandoning the use of the large schoolroom, the entire area is convertible into six class-rooms by means of partitions sliding horizontally. Thus, at pleasure, the whole can be used as one large room, or can be sub-divided into a series of rooms for separate classes.

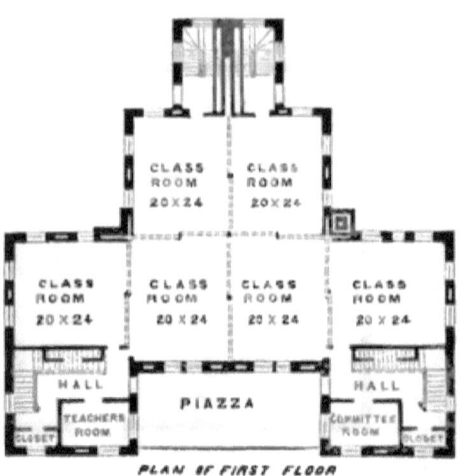

11.—NEWTON PRIMARY SCHOOL, PHILADELPHIA, PENNSYLVANIA, U.S.

12.—CAPEN PRIMARY SCHOOL.

 When in their sub-divided condition do they

thoroughly exclude sound? Do they not continually get out of gear and cause interruption of work until set right?

The Capen Primary School, at Boston, Massachusetts, consists of girls' and boys' schools, each of three single rooms 28 feet square, on three separate floors, and is a fair illustration of a common type of plan. Underneath the schoolhouse are playgrounds, or playrooms for the sexes protected by their position from stormy weather. This last arrangement is one which has found favour in many instances among the plans of the new Elementary Schools of London. There are two staircases, in this case, a second not having been,

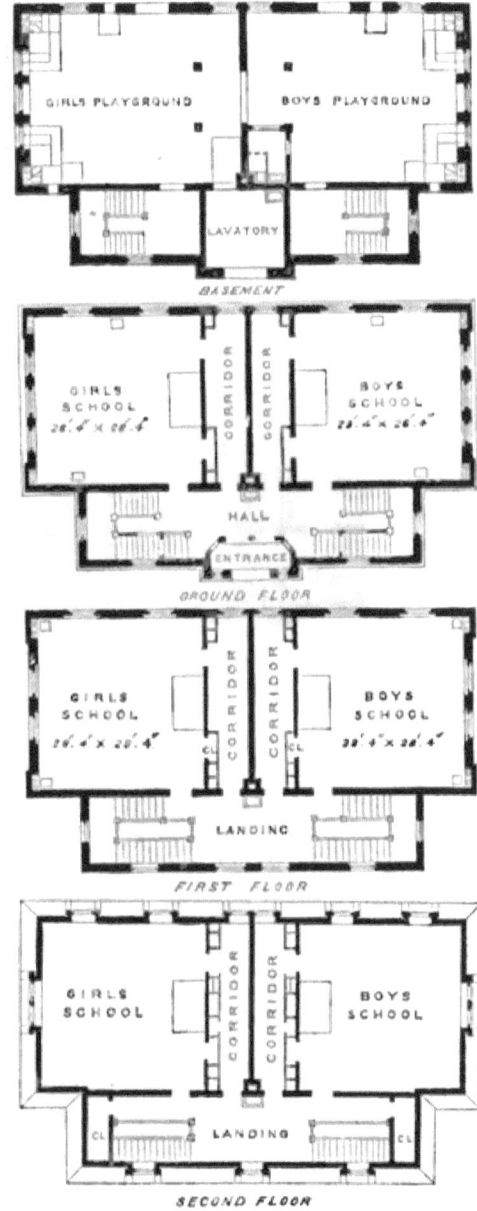

13, 14, 15, 16.—PLANS OF CAPEN PRIMARY SCHOOL.

previously, a common provision in Boston Primary Schools. The external architecture is of brick with heads and cills of granite, and the cost was as follows, viz., land 6145 dollars, building 34,716 dollars, and furniture 2075 dollars, Total 42,937 dollars. Children are not admitted to Primary Schools without passing an examination, and, at least, knowing the alphabet. A large proportion of the entire number of children in the United States

17.—HOLLINGSWORTH SCHOOL, LOCUST STREET, PHILADELPHIA, U.S.

never get beyond these Primary Schools. In the large town of Philadelphia it is placed at 53 per cent., or more than half. The education thus amounts to little more than the three r's.

The "*middle*" school is a secondary school intermediate between the Primary and the higher schools, and forming a stepping-stone to the latter, just as these in their turn form the approach to the University. In this kind of school, as in all others above the Primary, it is common to find (especially in New England) military drill, including the use of the rifle compulsory under the municipal regulations. Government provides the rifles, and drill takes place once or twice a week.

Among the higher schools are the "*Normal*" Schools for the higher education of women and for training mistresses;

"*Grammar Schools*" for both sexes; "*High*" Schools wherein boys are trained for commercial life, the foreign languages taught being principally French and German; and "*Latin*" schools, where the classics form a principal element in the preparation for college.

To select more plans of American schools. The Hollingsworth School, Philadelphia, (woodcut No. 18) provides on one floor a group of four separate class-rooms, and also five other class-rooms, which can all be thrown into one by means of partitions sliding horizontally into hollow walls specially provided. No class-room, when the partitions are drawn out, is converted into a passage-room, but each can be entered separately from the landing. The plan involves a serious expense in the item of staircases, and the entire arrangement is sacrificed to the desire for sliding partitions. The proper lighting of the rooms is utterly ruined. And the question naturally presents itself, "Are moveable partitions of such paramount importance?"

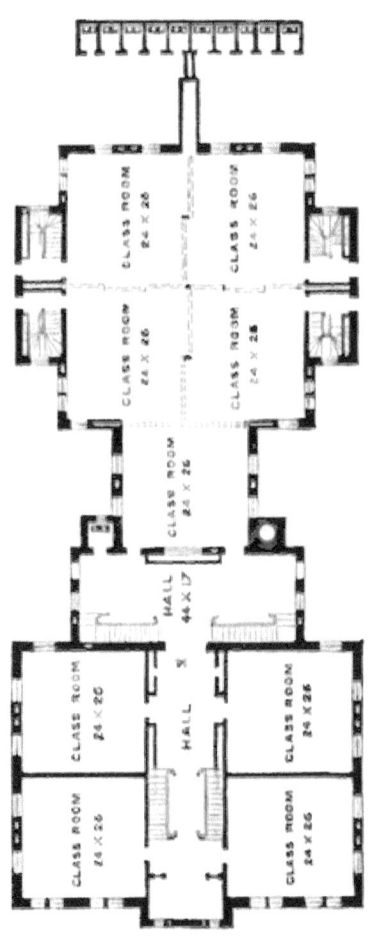

18.—PLAN OF HOLLINGSWORTH SCHOOL.

In the Tasker School (woodcut No. 19) the partitions do not slide horizontally, and the plan would be thoroughly German, if only the lighting were always on the left of the children.

The Melon Street School (woodcut No. 22) gives six class-

19.—THE "TASKER" SCHOOL, PHILADELPHIA, U.S.

rooms available for being thrown into one, in the same manner as the Hollingsworth School. The plan is extremely interesting to the English eye at the present moment, as combining the advantages of a hall of assembly with those of the separate class-room system. But it is impossible to believe that, with the rough usage common to schools, so vast a system of moveable partitions is not liable to constant disorder, and consequent interference with the working of the school.

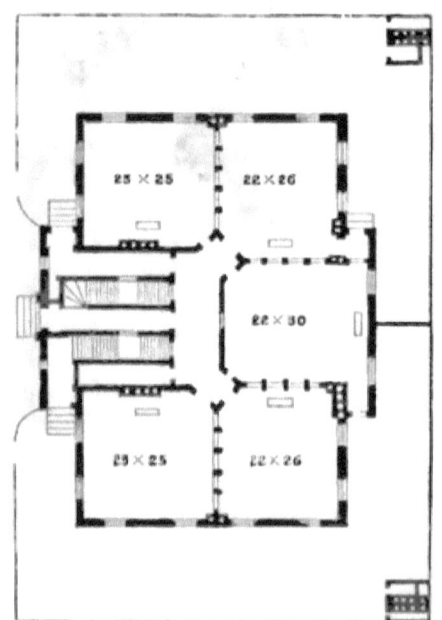

20.—PLAN OF FIRST FLOOR "TASKER" SCHOOL.

The Wood Street School (woodcut No. 24), on a totally different general plan, gives five rooms similarly arranged. The exterior view (No. 23), presents a characteristic specimen of school architecture in the United States.

21.—MELON STREET SCHOOL, PENNSYLVANIA, U.S.

The great number of convertible plans shows a remarkable development of planning in New England, and indicates the current of opinion on a system which has been much discussed in our own country, and not hitherto with much favour.

The George M. Wharton School (woodcut No. 25) differs from any of the preceding, but in principle more nearly approaches the Tasker School (woodcut No. 30). The plan is little more than a parallelogram, subdivided into seven classrooms, no provision being made for horizontal sliding divisions. The lighting is not carefully attended to, and thorough ventilation for summer is impossible. No exterior view is given, for the architectural character is that of a detached, well-built warehouse.

The architectural designs as set forth in the woodcuts are, one and all, extremely plain, not particularly school-like in character, and of no special English interest, except as shewing what American Schools are like. Philadelphia, the largest town in Pennsylvania, is divided into 20 school divisions or sections. Its population is said to be increasing at the rate of 20,000 per annum. And its expenditure on school buildings in 1871 was 184,842 dollars.

The report on the St. Louis public schools for 1871 describes the constant difficulty experienced in maintaining a sufficient supply of proper schoolhouses, even when the cost is defrayed out of the public purse of the town, or from accumulated real estate. The following table shows at a glance the enormous increase in population, and the development of the common school system. Although, apparently, of local interest only, in reality it represents what is occurring very generally through the country.

Year.	Population.	Enrolment in Public Schools.	Revenue from Real Estate (in dollars).	Percentage of entire population in Public Schools.
1841	20,826	350	4,200	1·2
1851	83,439	2,427	14,220	2·9
1861	163,783	13,380	33,497	8·1
1871	325,000	31,087	53,221	9·6

A considerable proof of the appreciation of good education is given in the condemnation of the practice of using unfit hired buildings for school purposes. Every year some of these become necessary from the rapid increase of numbers, but it is held that every year steps should be taken to provide proper school accommodation for ascertained wants. So great has been the difficulty in supplying schoolhouses in number equal to the demand that some years ago it had become necessary to forbid the admission of all pupils under seven years of age. The working of this rule has not been found beneficial to education, and the age of five is now recommended. The new school-buildings are constructed with external walls of much greater thickness than is common to private structures of similar size and height, the object being to rest the weight of all floors and roofs thereon, and thus to avoid the use of objectionable iron columns in the rooms and to secure the utmost freedom in arranging moveable partitions without constructional difficulty. It is admitted that the problem of warming and ventilation has been perplexing, and is only partially or approximately solved, the invention of a *self-regulating* apparatus for warming being among the unsupplied wants.

The High and Normal School-house for Girls in Boston (woodcut No. 30), erected in 1870, is said to be the largest, most costly, and most substantial schoolhouse of recent erection in the United States, and should, therefore, claim a more detailed description, although it may not be all that the English public would look for. It is located between Tremont and Shawmut Avenue, and has frontages to Newton and Pembroke Streets.

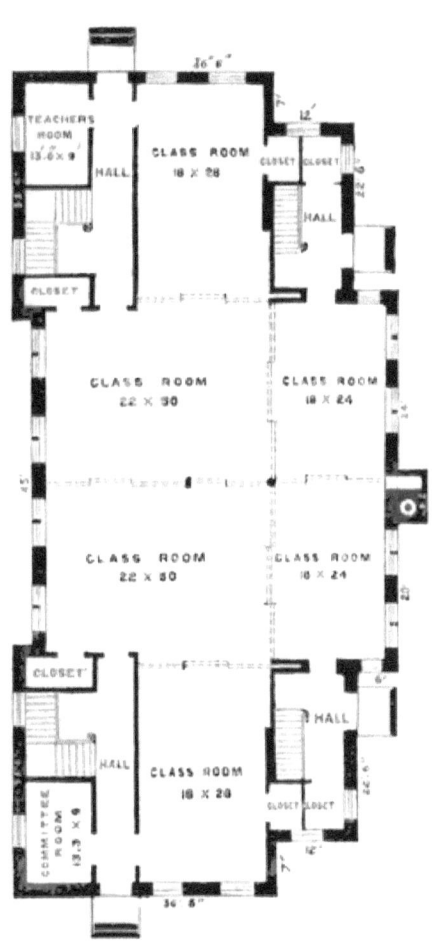

22.—PLAN OF MELON STREET SCHOOL.

The site itself has a frontage of 200ft., and a depth of 154ft. The building measures 144ft. frontage, and is 131ft. deep. The model school department is placed in the basement, and the open

23.—WOOD STREET SCHOOL, PHILADELPHIA, U.S.

play-ground or yard attached thereto is 154ft. by 54ft. The space unoccupied by building is paved with bricks, at a level of about 3ft. 6in. below the street, and the basement story is thus about 16in. above the level of the play-grounds. As so many English authorities urge that no schoolhouse should exceed two stories in height, it may be remarked that this building is, including the rooms provided in the high roof, no less than five stories high.

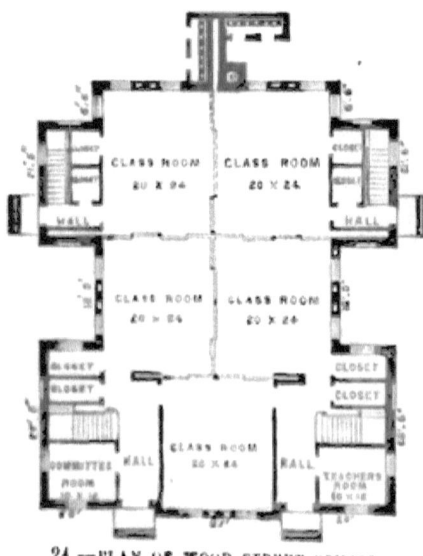

24.—PLAN OF WOOD STREET SCHOOL, PHILADELPHIA, U.S.

On plan, a corridor, 12ft. wide, extends across the building from the middle of the Newton Street (or south-westerly) side to the middle of the Pembroke Street (or north-easterly) side; and at right angles to this corridor, in the middle of the building, is a hall 22ft. by 77ft., at each end of which are two rooms 30ft. long, which, together with the hall, occupy the whole length of the building.

The hall and corridor divide the building into four equal sections or quarters, which are subdivided as follows :—At the left of the entrance at the Newton Street side is a waiting or reception-room, 16ft. by 22ft. Beyond the waiting room is a passage leading from the corridor to the master's room 14ft. by 27ft., which is connected with the waiting-room, and also with the advanced class-room, 45ft. by 30ft. The inner portion of this westerly part is occupied by

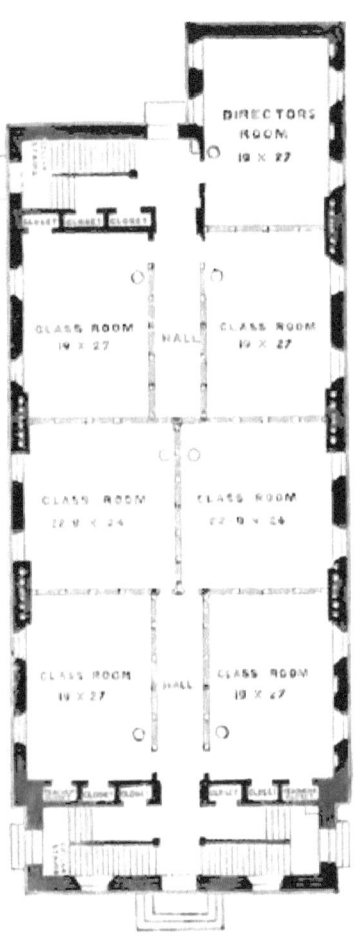

25.—THE "GEORGE M. WHARTON" SCHOOL, PHILADELPHIA, U.S.

a passage leading from the hall to the advanced class-room and the master's room; a staircase leading to the basement story, a cloak-room for the advanced class, master's closets and air-flues.

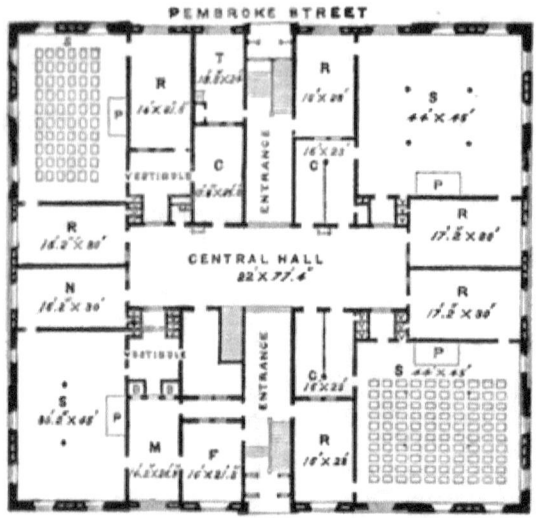

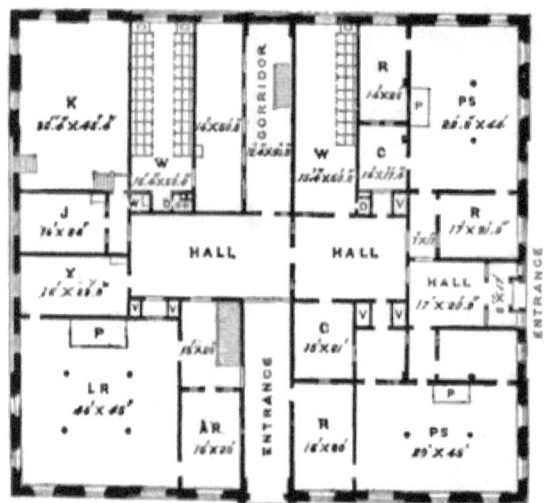

Reference.

B. Airflues.
C. Cloak Rooms.
F. Waiting Room.
J. Janitor's Room.
K. Boiler Room.
M. Teacher's room.
N. Library.

R. Recitation Room.
L. R. Chemical Lecture Room.
P. S. Class Rooms.
T. Mistress' Dressing Room.
W. Latrines.
X. Laboratory.

26, 27.—GIRLS' SCHOOL, BOSTON, MASSACHUSETTS.

CHAP. III.]   AMERICA.   41

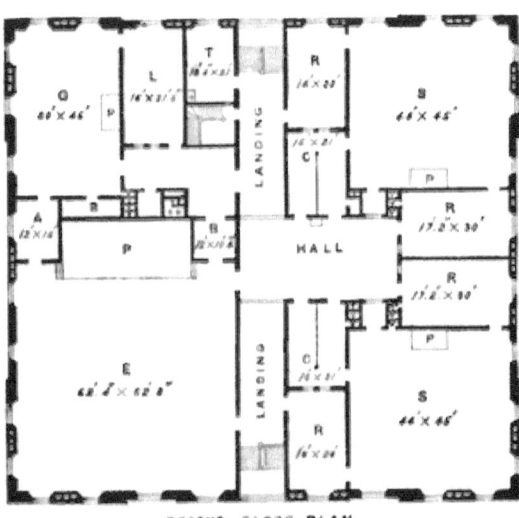

SECOND FLOOR PLAN.

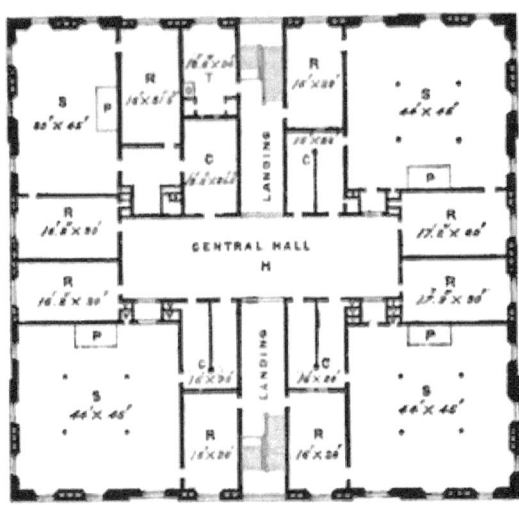

FIRST FLOOR PLAN.

Reference.

C. Cloak Room.
E. Assembly Hall.
G. Drawing Class Room.
H. Central Hall.
L. Apparatus Room.
R. Recitation Rooms.
S. Class room.
T. Mistress' Dressing Room.

28, 29.—GIRLS' SCHOOL, BOSTON, MASSACHUSETTS.

At the end of the central hall, and occupying the middle portion of the north-westerly side, are two rooms, each 16ft. by 30ft., one of which was designed for a library, while the other is a recitation-room. In the northerly corner is a class-room, 30ft. by 45ft.; at the south-easterly side of the class-room is a recitation-room, 16ft. by 32ft., between the inner end of which and the central hall is a large brick foul-air

30.—GIRLS' SCHOOL, BOSTON.

shaft and chimney; also a passage leading from the class-room, recitation-room and cloak-room.

At the right of the entrance on the Pembroke Street side is a dressing-room, 14ft. by 24ft., for female teachers, and a cloak-room, 14ft. by 25ft. At the left a recitation-room, 16ft. by 28ft., a class-room, 44ft. by 45ft., and another recitation-room, 17ft. by 30ft. The remainder of this portion is occupied by a cloak-room, 16ft. by 21ft., and by ventilating shafts. Through the middle of the latter is a passage leading from the hall to the class-rooms.

The southerly quarter has the same amount of accommodation, and is arranged precisely like the easterly quarter last described. And, further, the same arrangement is carried through the three stories above the basement in the south-easterly quarter of the building, and the westerly quarter of the second story—that is to say, two class-rooms, two recitation-rooms, and a cloak-room in each case. The northerly quarter of the second story contains a class-room, 30ft. by 45ft., with two recitation-rooms, cloak-room, teachers' dressing-room, &c., as in the northerly quarter of the first storey.

The westerly quarter of the third storey is devoted to an assembly-hall, about 62ft. by 74ft. 6in. In the northerly quarter of the third storey a room for drawing, 30ft. by 45ft., a cabinet for apparatus, 16ft. by 32ft., also a teacher's room, cloak-room, &c., as in the same quarter of the stories below. In the westerly corner of the basement story is the chemical lecture-room, 44ft. by 45ft., fitted with all the necessary apparatus and appliances. On the northerly side, and adjoining the lecture-room, is a laboratory, 16ft. by 30ft. Adjoining the inner end of the cabinet is a passage and staircase, leading to the storey above. In the northerly corner is the boiler-room, 30ft. by 45ft., in which are three boilers, each 3ft. 6in. diameter, and 16ft. long, which supply the steam for heating the building. The room for coals occupies the space between the outside of the building and the line of the street. At the southerly side of the boiler-room is a room for the janitors, 16ft. by 24ft. On the easterly side of the boiler-room are the water-closets, 22 in number, for the High and Normal Departments, occupying a space about 30ft. wide and 50ft. long. The remainder of the space in the north-westerly half of the building is occupied by the central hall, and a staircase at the Pembroke Street end of the corridor.

The whole of the south-easterly half of the basement is devoted to a model school, accommodating about 150 primary, and the same number of grammar-school pupils. The entrance, cloak-rooms, water-closets, and other accommodation for this department are separate and distinct from those of the other departments. This portion consists of a large class-room, 30ft.

by 45ft. in each of the two corners, and connected therewith are in each case two smaller rooms, 16ft. by 25ft. The remainder of the space is devoted to cloak-rooms, water-closets, hall, and passages.

The number of pupils accommodated in the High and Normal Departments is measured by the seating capacity of the seven large class-rooms, each having 100 single desks, and the three smaller class-rooms, each having 75 single desks, making a total of 925 in these departments, which, with the 300 in the model school, makes a grand total of 1,225 pupils.

The internal finishings throughout are of pine, painted, grained, and varnished. The floors are of southern pine, and the "trimmings" of the staircases are of black walnut.

The heating is by steam-boilers placed in the basement, and all the rooms are heated by hot air radiated through the apparatus connected therewith. The halls and corridors are heated by direct radiation.

Electric bells and speaking-tubes place the master's room in communication with all the principal rooms. Water is carried to each floor at two places in the central hall.

The external design is simple, and presents a fair type of American school architecture. The basement is faced with a light-coloured Maine granite. Above this is pressed brickwork with freestone Nova Scotia dressings. On the face of the arch-stones the name of the school is cut in large raised letters. The main cornice is of wood, with copper spout, and is ornamented by a Mansard roof, having cast-iron cresting or "snow guards." On the middle of the flat portion of the roof stands an octagonal turret 30ft. in diameter, and intended to be used as an observatory. The main extracting-shaft of the ventilation passes up through the centre of this, and terminates in an octagonal cupola 13ft. in diameter, having on each face outlets 8ft. high by 3ft. wide. In the case of this school, the cost has been: Of site, 60,206 dollars; of building, 234,563 dollars; and of furniture, 15,947 dollars. Total cost, 310,717 dollars.

A study of trans-Atlantic schoolhouses, as set forth in the numerous published reports, leads generally to the conclusion that those of New England are the best, and yet do not present perfect models for English imitation. The general plans are often (as already shown) highly ingenious, the sites ample, and the expenditure munificently liberal, but in a vast country engaged in a constant struggle to meet efficiently the enormous annual increase of population by a free education, it would be too much to expect to find the subject of our inquiry reduced to a science. As in England, there is much critical investigation and discussion of education itself, but no trace that some of the vital points affecting the buildings (and, therefore, indirectly the education), such as the proper amount, distribution, and kind of light, the necessity of "through"—or summer—ventilation, the most wholesome, efficient, and economical kind of artificial ventilation, and others, have, as yet, been sufficiently tackled at close quarters or in the careful manner common to Germany. Their nomenclature differs from ours without any particular advantage; for instance, the school-porter is a "janitor," the airflues are "ventiducts," and the class-rooms are "recitation-rooms." The general conclusion is that American primary schools are founded chiefly on those of Ireland, while others, especially the higher schools, are based on the German model. It is to this latter circumstance we owe the fact that their class-rooms are frequently called "schoolrooms," while, if there be one, the examination-hall (the German *Aula*) is called the hall.

While observing carefully and critically what has been and is being done, not only in England, but throughout the civilised world for education, the Americans do not conceal their preference for the German system as a whole, and draw the broad conclusion that to a superior system of education the Prussians owed their success over the French in war. Recollecting how far behind England has been in the race, and that Prussia had adopted the principle of compulsion in reference to primary schools more than a century ago, no Englishman can wonder at, or complain of, American preference for German school-plans. It is noticeable that their system is sometimes called by themselves the "English-American," and it is probable that the new impetus given to the English system, which must steadily grow and develope from day to day, will not be entirely lost upon the future school-building of the United States. Already the Infant School system, hitherto very much neglected, is becoming better and more general. The raised gallery is beginning to be used therein. Instead of the single seat and desk used in the Primary Schools, some of the teachers now advocate in those schools the use of the dual arrangement, common enough in the higher schools of America, and universally adopted in the elementary Board Schools of London. Other developments may be expected. School-planners may always look with interest and expectation to a people remarkable for mechanical skill, ingenuity, and inventiveness, and proud of their popular education.

The student of educational systems rather than school architecture may study with advantage the work (published since the above was in type) on "National Education," by J. H. Rigg, D.D.; Strahan and Co., London.

CHAPTER IV.

SCOTLAND AND IRELAND.

Religious Instruction—Compulsory Attendance—Educational Aspect—Buildings in the two Countries.

THE policy which directs the education of a people, like the policy which determines the action of a Government, while exercised for the general good by the light of a superior intelligence, must be largely influenced by national idiosyncrasies, and by the necessity of complete harmony with popular convictions. A species of legislation striven for by one people as for a national blessing, may be strenuously resisted by another, and looked upon in the light of an unmixed evil. This is especially so with reference to new laws on the subject of education; a system of public instruction proved by the evidence of a high success to be right for one race, may yet be found eminently unfit for another.

Scotland and Ireland present a curious contrast considered in relation to schools, and therefore to schoolhouses. The different ideas prevalent in the two countries on the subject of education may be expected to influence the form of plan in greater degree in time to come than in the past, in proportion as school-planning becomes more skilful, and the buildings bear greater evidence of more definite intention. Both nations agree on the general principle that religious instruction should be given in their public schools. The application of the principle is so very different as to leave, in detail, little of practical agreement. Scotland,—ever in the van in modern times where education is concerned,—tells us that, among the little of human wisdom which has been acquired in the course of ages, may be counted the certainty that a knowledge of reading, writing, and arithmetic has been proved to be for the benefit of mankind and should, therefore, be extended to each individual. Ireland, on the other hand, avers that the spiritual interests of a child are of far more importance than any merely secular knowledge, that, without the inculcation of the Roman Catholic faith, all human teaching is worse than useless, and that the traditions of the Church are even older than the Bible itself.

The Scottish system does not present, perhaps, under the head of religious teaching, a sufficient difference from that of England to warrant the expectation of novel forms of school buildings. The Irish system, if developed under the wishes of a large section of the people, and made to include religious teaching according to the Roman Catholic faith, would probably present some new features.

On the question of the character, amount, and even the presence of religious teaching in public schools, several varieties or degrees of difference are found in different countries. Ireland and France insist on the teaching of Roman Catholic dogma, and in neither country, at present, are all the children brought within the scope of education. In Germany, the religious instruction forms part of the regular school course, like reading or writing, and is usually given by rote without much interest on the part of the teacher. In England, a limited time is devoted to it, the Bible is regarded as the text book, and dogmatic or sectarian teaching is prohibited. In Scotland, religious instruction is left by law to the option of the various School-boards, and generally includes the use of the Bible and the Shorter Catechism. In Australia an extreme course has been adopted, for the new Education Act of that country absolutely prohibits religious teaching in public elementary schools during ordinary school-hours.

The subject of the application of compulsion by law to attendance at school presents also many marked differences of opinion or practice. Germany adopted direct force so long ago that its application is now a matter of course with the people. In England the use of compulsion is optional with the School Boards, who may decide whether it is desirable in the circumstances of their own districts. In Scotland it is compulsory: the Elementary Education Act (Scotland) of 1872 was framed to reach and educate the hitherto neglected residuum of the population, and the Boards have no choice but to compel the children to attend school. In Ireland, as in France, — two Roman Catholic countries,—it does not exist in any direct form.

In Ireland sectarian differences run high, and the spirit of compromise is almost unknown. The Roman Catholics refuse "the Bible without note or comment" as a common ground of agreement, exhibit a crucifix and images of saints in every schoolroom, open school each day with characteristic ceremonial, and inculcate doctrine at every opportunity during the lessons. The Church Education Society (Protestant) is equally determined to maintain its own programme of religious teaching. And both these rival religious bodies object to their schools being placed under the Board of National Education, however moderate and even neutral its character. Among these burning jealousies, and in a country where partizanship on one side or the other is as much the rule as though it were an article of faith, the schools established directly by the Board not infrequently expire from sheer inanition. Irish parents who prefer thoroughly good education to religious opinion are so few in number that a sufficient gathering of children in average attendance to justify the expense of a school cannot be maintained, the schools dwindle and have to be closed, and the work of education as carried on by the Board is seriously impeded. It would seem that any attempt to secure to Ireland the system in force in England, might, from the preponderance of the Roman Catholic element, be regarded by many in the light of concurrent endowment aggravated by its educational disguise. In the face of strongly marked antipathies between different religious communions, an amicable settlement on a common, or common sense, basis, even on matters like the multiplication table, though not absolutely hopeless, will be arrived at with no slight difficulty. The wider the differences of opinion and belief, the narrower the possible field of united action. The only safety for the cause of education would appear to lie in vigorously setting aside or ignoring the differences. To the on-looker, an absolutely secular education would appear to be the most probable, if not the only solution. At the same time, the opinion of some well-informed Irish Protestants is known to be in favour of a plan by which Government might have the power to insist,—firstly, on the suitability of the school premises, secondly, on the efficiency of the teacher, thirdly, on compulsory attendance, and fourthly, on an inspection of results in the *secular* portion of the education. This plan leaves the religious instruction entirely with the clerical authority, be it Catholic or Protestant. By some process or other, the warfare of creeds in Ireland should be prevented from retarding the establishment of a national system of public instruction.

It should not be forgotten that the neglect of education in Ireland is a matter of comparatively modern times, and may, perhaps, not be entirely untraceable to centuries of English rule out of harmony with the convictions of the great bulk of the population. Those conversant with ancient authorities quote the writings of William of Malmesbury and the Venerable Bede to show, that as early as the seventh century, or long before Scotland (or England either) had much pretence to education, Ireland was already famous through Europe as a seat of learning. The forty literary institutions of Borrisdale anciently existing, and the colleges of Lismore and Mayo, the latter of which is said to have contained, at one time, no less than two thousand English students alone, are some evidence in support of this happy recollection of Irish history. The excellent foundation formerly existing, was destroyed, or diverted from its natural channel, even from the time of Henry the Eighth downwards, by a series of laws of mistaken kindness. It is some consolation to reflect that a nation which has once been foremost in education, may not impossibly, with suitable opportunities, again assume a leading position in the same department.

The schools of Ireland afford examples of skilful management

31.—LURGAN MODEL SCHOOL, CO. ARMAGH.

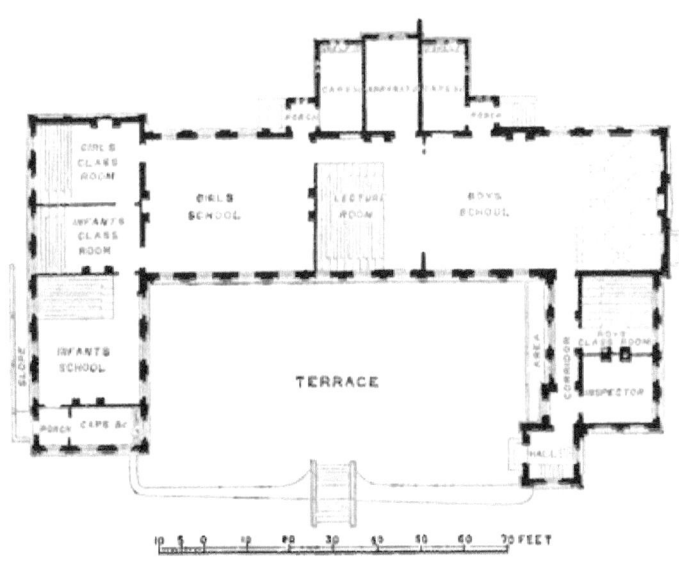

32.—PLAN OF LURGAN MODEL SCHOOL.

rather than of architectural planning. Like other things, school-houses cannot be perfect models if their cost is curtailed too rigidly on necessary points. And this is stated to have been

commonly the case in every part of the sister country. There
are not wanting excellent specimens of the school-house, but the
type is usually very simple and well-known. The Model National
Schools in Dublin, will repay a visit if only to observe the
admirable discipline and system. The Lancasterian principle of
a large and noisy school-room may not be the visitor's *beau-idéal*,
but the noise is that of work, and is under the most complete
control. Mr. Stow's training system is here well understood,
though not adopted precisely as the author in his time advocated.
In the play-grounds, all the children play together without any
separation, broad marks in the pavement indicate the quarter set

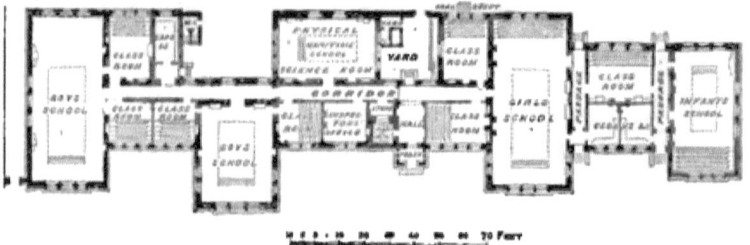

33.—CORK NATIONAL SCHOOLS.

apart for each department, and training and discipline supply the
place of high brick walls. The class for needlework, where poor
children are taught to mend their own rags, and to make clothing
of different kinds, is, in principle and in practice one of the most
admirable things of the kind in the United Kingdom, and
deserves the especial study of all ladies who wish to be useful in
connection with School-boards. The Model National Schools
at Lurgan, Co. Armagh, (woodcuts No. 31 and 32) and also the
National Schools at Cork, (No. 33) present, among others, good
specimens both of buildings and school organization carried out
under the auspices of the Irish National Board of Education.
Should "justice to Ireland" ever proceed so far as to compass
the enactment of a really suitable educational law applicable to
all, its efficient and equal administration can hardly fail to give us

in course of time further good specimens of elementary and other schools.

The higher schools of Scotland are, in the main, based on the plan adopted in Germany, and the buildings consist of a series of separate class-rooms. The best specimens, some of considerable size, are to be found in Edinburgh. The influence of Mr. Stow has been powerful in respect of the lower schools, and his ideas, in modified form, are in much favour in Scotland. The new Act for Scotland, the operation of which is already creating a further provision of elementary schools, will be of immense service, and the "Juvenile" schools will probably almost all adopt the mixed system. Hitherto, the primary education of the Scottish people has been in great measure carried on, with other of more advanced kind, in the parochial schools. Now, the primary and secondary schools are likely to take their natural place as separate, though allied, institutions. The new elementary schools will probably resemble in their teaching those of England, and the school-houses may be expected in their plans to resemble also those of this country, subject to the differences of arrangement involved by the adoption of the mixed system, and the consequent division into senior and junior departments. Another educational phase, already begun, may also be expected to proceed further. The rich school endowments of Scotland will by degrees be applied with greater effect to public instruction in that country, in the same manner as in England, although, in both cases, the process of development may be of a somewhat deliberate character.

CHAPTER V.

FRANCE.

General—Varieties of Schools—Want of Elementary Education—Statistics—
Ecoles Mixte—The Crèche—Salles d'asile—The Collége Chaptal at Paris.

A COMPARATIVE inquiry, however slight, into the state of school architecture, speedily shows that the main source of inspiration for future development in England is not to be found in France. This is more especially true of buildings for primary education. Yet the study of French schools, higher and lower, furnishes occasional information which may well be applied to our own works. Before the year 1789, France had been well supplied with establishments for higher and classical education, grammar schools and colleges, some of which, with their endowments and the religious or lay congregations under which they flourished, were destroyed between 1789 and 1794. The wars of Napoleon, succeeding the social disorder which marked the later years of the eighteenth century, were not favourable to the development of public instruction or to the cause of education, and little progress was made. The country is said to have been as well supplied with grammar schools and colleges in 1789 as in 1849. Ordinances, passed between 1791 and 1794, may have laid the foundation of Normal and Polytechnic Schools, but Primary Education (under which is classed in France the elementary and superior, the communal and private schools) remained in the most deplorable state until the law of 1833, framed by M. Guizot on the advice of M. Victor Cousin, compelled the establishment of schools in every commune by the joint action of the Commune the Department, and the State. The higher schools are good, and, in their teaching are not very different from those of Germany, while the principles of planning are also similar. Their variety and number must excite admiration. Among French public schools are to be found almost all kinds of establishments calculated to train the mind before entering on any special career. There are schools of agriculture, horticulture, astronomy, architecture, engineering, drawing, modelling, and chemistry; preparatory-technical and technical schools, trade schools, naval schools, schools for teaching the theory and practice of clock and watch-making, schools of mines, and many kinds of workmen's and foremen's schools—in some cases held in the evening. The higher schools, art and trade schools, in their completeness and system leave little to be desired. They probably formed the models for some of those in other countries, belonging to a later period, which have been erected on an increased scale of cost, fitted with new improvements, and which manifest a greater regard for sanitary science. Some are famous through the world. Successive rulers of France have always encouraged public instruction of the higher, and of the technical kind. The results of the latter are conspicuously manifest in more than one art and trade wherein French taste, skill, and technical knowledge sometimes competes with and excels that of our own country.

The Commission on Technical Instruction appointed by Imperial decree in 1863 produced, in the form of a blue book, a report so valuable that in 1869 it was reprinted for the use of the House of Commons. In its concluding pages, this work points out the weak part of the national educational system in France, viz., *the want of elementary instruction*. The heads of large industrial establishments are quoted as pointing out the necessity of a degree of general preparatory instruction, proportioned to the extent of professional or industrial education which is intended to be its complement, so as to enable each individual to follow his career with success. Some among the most eminent are said to declare plainly that the deplorable, and far too general, absence of primary instruction among even the most intelligent workmen formed one of the greatest and most lamentable obstacles to the development of their faculties and the progress of industry.

The great educational inquiry in France, of which this blue book was one of the results, produced reports from the school-inspectors showing the necessity of constructing new schoolhouses at an estimated expenditure of 275 million francs. Efforts have not been wanting on the part of Government to ameliorate this condition of things, and new schools, the result of a more energetic application of the law of 1833, are in course of erection both in Paris and the Departments. The direction of endeavour is to diffuse elementary education as widely as possible *without the exercise of compulsion*, and may arise either from a fear of "religious" agitation, or a belief that the absence of direct force is better suited to the genius of the people. The result is that education never reaches a considerable number among the lower stratum of the people,—a fact which contrasts unfavourably with Switzerland, Holland, and all Germany.

The statistics recently published by the Minister of Public Instruction, show that out of a total State expenditure in the year 1865 amounting to 19,918,121 francs, 6,863,100 francs was devoted to primary schools, 3,141,000 francs to secondary instruction, lycées, communal colleges, &c., while the cost of superior schools, normal, literary and scientific establishments, faculties, the Institute, &c., amounted to 7,493,071 francs. The remainder represents the cost of the staff and *matériel* of the central administration, and of the general service of public instruction, inspectors, academy administration, &c. At the present time it is said that out of the 35,000 Communes comprised in the Empire, about 1,000 have no schools, while some 10,000 have school-buildings either unfit for their purpose, or of inferior kind.

It remains to speculate whether the school-buildings in actual existence for primary education are well filled, and on this point we have no information. The most serious fact is that in France the law does not enforce attendance at school. There is no method of securing regular daily instruction but through the influence of the priest; and, if parents either object to the religious aspect of the teaching given, or desire to avoid sending

34.—FRENCH MIXED SCHOOL, S. PARDOUX LES CARS.

their children to school, the latter are allowed to *gaminer* about the streets.

Later French law, enacted in 1850 and 1867, requires every Commune to provide a suitable site for a primary school and teacher's house. In populations under 500, the schools may be mixed. Above that number, separate provision for boys and girls must be made. The mixed system is accepted only as a necessity of the rural districts, never as a principle, and may be said not really to exist in France. Where compelled, by the smallness of population to adopt an *école mixte* the expedients used to keep the sexes as separate as possible are amusing. Those found in the school at S. Pardoux les Cars (woodcut No. 36), indicate the economical origin of the system. The stove is placed in the centre of the room, which is large enough for two classes. A partition equally divides the space. The boys are in one class, the girls in the other, isolated by the partition which meets the platform midway. Thus the one teacher is able to supervise both classes on strictly separate principles, and the stove warms both divisions of the room equally. The arrangement carries out in an ingenious manner the instruction of the Government Circular requiring separation of sexes in mixed schools by a barrier not less than one and a half metres high.

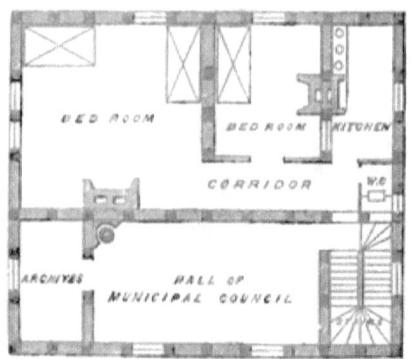

35.—S. PARDOUX LES CARS. FIRST FLOOR.

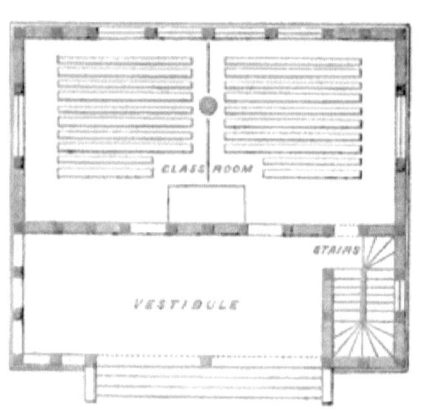

36.—S. PARDOUX LES CARS. GROUND PLAN.

The education of girls and boys together, only possible where a very high kind of moral training is exercised, and which forms one of the noblest features of the American, the Swiss, the Dutch, the Wesleyan, and the Scottish systems, appears to be regarded as practically impossible in France. In villages and rural districts, the combination in one group of the *Mairie* and *Justice de Paix* with the *École*, is not uncommon, and the present instance furnishes an example of this union of scholastic and municipal purposes. The rooms required on the first floor (woodcut No. 35) for the teacher's residence and for the *Mairie*, together occupy a space sufficient for the provision, underneath, of the school-room already described, and of a large vestibule or covered playground.

When all pretence of the use of a mixed system is gone, and elementary schools are planned separately for boys and girls, the specimens to be found in France present little of special interest or novelty.

The species of day-nursery organized for the very young children of poor parents, and known by the name of the *Crèche*, for another reason need not here be discussed. It can hardly be classed among educational establishments.

The *Salle d'Asile*, or hall of asylum, is the French version of an infant school, and presents certain arrangements not common elsewhere. It is usually carried on by the Sisters of some one of the religious orders. The covered play-

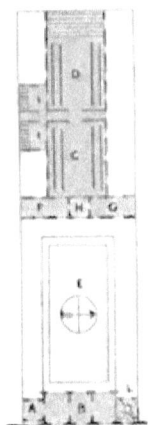

Reference.
A. Entrance.
B. Caretaker.
C. Play-room.
D. School-room.
E. Playground.
F. Bath-room.
G. Basket-room.
H. Fuel.
I. I. Latrines.
L. Staircase to Mistress's house.

37.— BLOCK-PLAN OF SALLE D'ASILE.

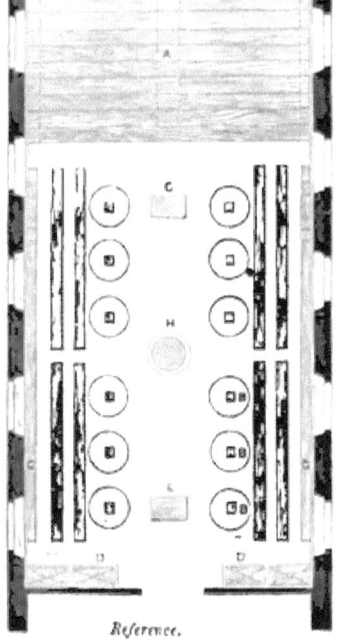

Reference.
A. Gallery.
B. Stands for examples.
C. Black-board or Compendium.
D. D. Hammocks or Couches.
E. Table.
F. Benches.
G. G. Stalls or Arm-chairs.
H. Stove.

38.— SCHOOL ROOM OF SALLE D'ASILE.
FOR 200 CHILDREN.

ground is considered indispensable, the class-room is not. The programme comprises a school-room, a covered playground or play-room, an open play-ground, a basket-room, a small kitchen, and a bath-room. A Mistress's room is invariably provided in close contiguity, and not entered from the school-room. A selection from the school studies of M. Uchard shows (woodcut No. 37) the general plan suggested for such a building, and (No. 38) the internal arrangements suitable for a school-room to contain 200 children. On the block-plan, the Caretaker, or the Mistress, has a residence next to the street, immediately behind is the open playground, then comes the covered play-ground, and lastly the school-room. The larger plan (No. 38) represents the portion marked D on the Block Plan, comprising the school-room, which bears a striking resemblance in its main features to the plan of an English Wesleyan school.

The *Salle d'Asile* at S. Etienne, near Limoges (woodcuts 39 and 40), is an instance of an actual school on the above theory.

39.—ELEVATION OF SALLE D'ASILE, AT S. ETIENNE, LIMOGES.

The end view (No. 39) shows the Mistress's house of two stories contained in the central portion, the upper floor being devoted to bedrooms. The Mistress's house is usually placed towards the front, and through its spacious entrance and corridor the children reach the *préau couvert*, or covered playground, fitted with seats next the walls, and warmed in winter by a stove.

Unlike the English covered playground, the *préau couvert* is really a warmed room. The former is usually open to the external air on one side. In the latter the poorer children have a provision equal in comfort to any play-room to be found in the houses of the rich. The internal arrangements of the schoolroom suggest the old-fashioned type of plan already described at pages 13 and 14 (woodcuts 2 and 3), but the seats placed at each side of the room have no desks attached, and the gallery is of a size to accommodate *the whole* of the children. Just inside, and on each side of the door, is either a cot (*lit de repos*) or a hammock swung or folded up, as shown in the concluding chapter. Here also the separate system is maintained. The boys are seated one side of the schoolroom and the girls on the other. When all are on the gallery, separation of the sexes is effected by a very wide central gangway, undesirable for teaching purposes, painted on the steps in black. A specimen of the kind of gallery used in a *Salle d'Asile* in the Rue des Ursulines will be found in

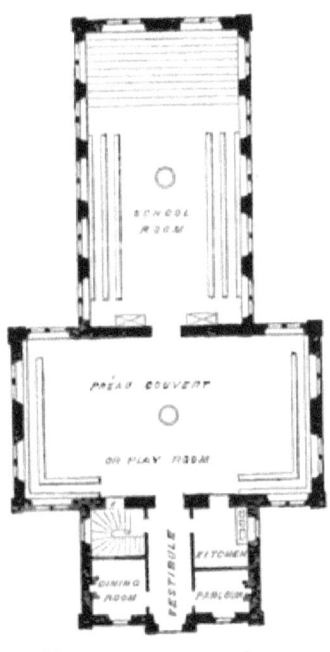

40.—PLAN OF SALLE D'ASILE, S. ETIENNE, LIMOGES.

the chapter on Infant Schools. The lavatory is usually placed in the covered playground, or in a place specially provided, like the basket-room or bath-room. The basket-room, in some cases, forms a part of the hall serving as dining-room for the children. The lavatory and resting-cots are sometimes crowded into the same space. Occasionally, the covered play-ground is fitted with shelves on which to place the dinner-baskets. The fire for cooking the children's food is sometimes in the bath-room, sometimes

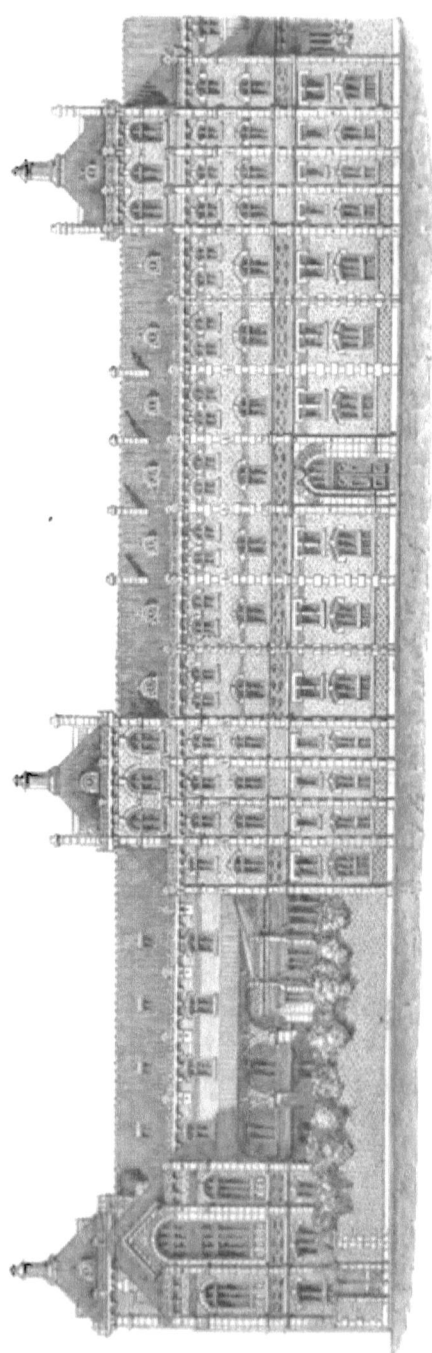

41.—BACK ELEVATION OF THE COLLÉGE CHAPTAL, PARIS.

in a separate room. These various notes indicate some of the methods adopted in France for maintaining the physical powers of young children, who require food more frequently than adults, and for persuading an attendance at school where compulsion has not been applied. The study of the history of education in France, though in itself sufficiently interesting, does not lead us to practical conclusions tending to the improvement of school-planning in our own country, sufficient to warrant a prolonged pursuit. Results of greater importance may be obtained from the same amount of investigation carried on among the schools of Germany. Those who wish to study the national

CHAP. V.]

aspects of the teaching, rather than the school-buildings, of France (and also of Holland and Switzerland) may consult with advantage the works of Mr. Matthew Arnold* and others. From among a very large number of published school-plans, both French and Belgian, we select one example as perhaps likely to be useful.

The Collége Chaptal at Paris, erected in 1865, presents an instructive and recent example of a French school comprising in one group, an elementary, a middle, and an upper school department. It is situated on the Boulevard des Batignolles

* "The Popular Education of France," by Matthew Arnold, M.A. Longmans, 1861.

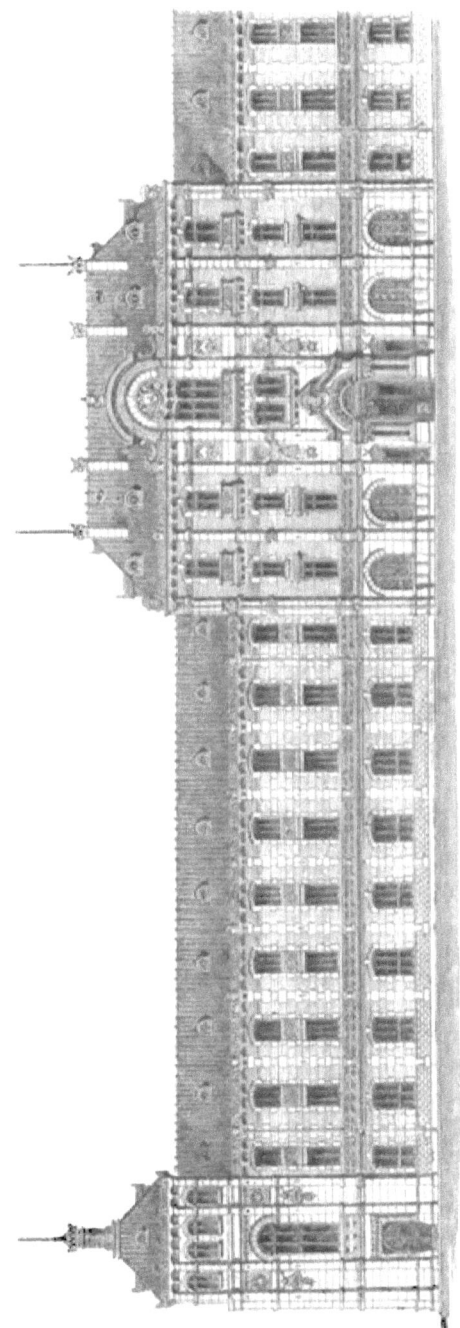

42.—PRINCIPAL ELEVATION OF THE COLLÈGE CHAPTAL, PARIS.

and Rue de Rome, Eighth Arrondissement. An account, accompanied by drawings, and published in the Transactions of the Intime Club by M. E. Train, the architect of the building, enables us to give such description and illustrations, as tend to throw light on the plan and architectural thought exhibited, and also to some extent on the theory of education which has moulded the internal arrangements. In principle the school is similar to a German *Gymnasium*, consisting of two divisions, and with a Preparatory School attached. The provision of "studies," or small class-rooms, only half the size of the usual German class-room, marks a difference. The school buildings also include provision for boarding and lodging, with the attendant arrangements. The accommodation is for 1000 students, of whom 600 are boarders, and 400 are day scholars, divided as follows : viz.,—

	Boarders.	Day Scholars.	Total.
The Preparatory School, or Little College . . .	250	200	450
The Middle School, or Middle College	200	150	350
The Upper School, or Great College	150	50	200
	600	400	1000

Each division of the school has an ample playground with covered sheds. A large Drawing Class room is provided, divided into three sections, and available, when undivided, for use as a hall for music. Also a handsome chapel, capable of holding 500 or 600 worshippers, and a library with reading and working rooms. There is a lodging for the Almoner. The Gymnastic-hall is situated near the Little College. An Infirmary for the sick students is provided at the southern end of the main block of buildings. It contains a large room for patients, a dispensary, a room for bandages and herbs, a convalescent room, and a bath room. Also, three separate rooms for patients suffering from disorders of an infectious nature. The doctor's residence is contiguous; and, to complete the arrangements of this miniature hospital, there is a small garden devoted to the sick. Each main staircase communicates with all the floors. Water cisterns are provided in every part of the building. The lighting of the principal rooms, corridors, and staircases is by gas. The building is described under four divisions; (1), the administration; (2), the schools; (3), the management; and (4), the sleeping-apartments.

1. *Buildings for the Administration.*—These contain all the services for control. An easy covered communication is provided for families on the side of the principal entrance, and for students on the side of the college buildings. There are cloak rooms, latrines, waiting-rooms, and a porter's lodge. The Directors' private room is on the first floor, with ante-chamber, and Secretary's room. There is an archive room in communication with the Director's room, and available also as studies, for the use of the Council of Administration, for public receptions, and for examinations. The Prefect-general's private room is near the Secretary's room.

For the management of the establishment, there is provided a manager's office with ante-chamber, a strong-room, a writing-clerk's room, a storeroom for books and papers, and another for dresses, clothes, and shoes. There are also residences for the Prefect, for the Under-Prefect, and for the Director of the College.

2. *The School-buildings.*—(*a*) The Little College has nine studies (*études*) each, for 30 to 35 pupils. Eight classes for 50 each. Two large lecture-rooms or amphitheatres for 150 to 200 pupils, easily accessible from the street. Eight Masters' rooms. Eight Servants' rooms. The private room of the Under-Prefect is on the ground floor. Attached to the Little College or Preparatory School are, besides, two studies for younger pupils and one extra class-room. (*b*) The Middle School or Middle College contains eight studies, each for 25 to 30 pupils. Nine classes for forty each. Two amphitheatres easily accessible from the street. Seven Masters' rooms. Seven Servants' rooms. The General Inspector's private room is on the ground floor. And he,

F

also, is provided with a residence. (c) The Upper School, or Great College. Here there are six studies, each for 25 to 30

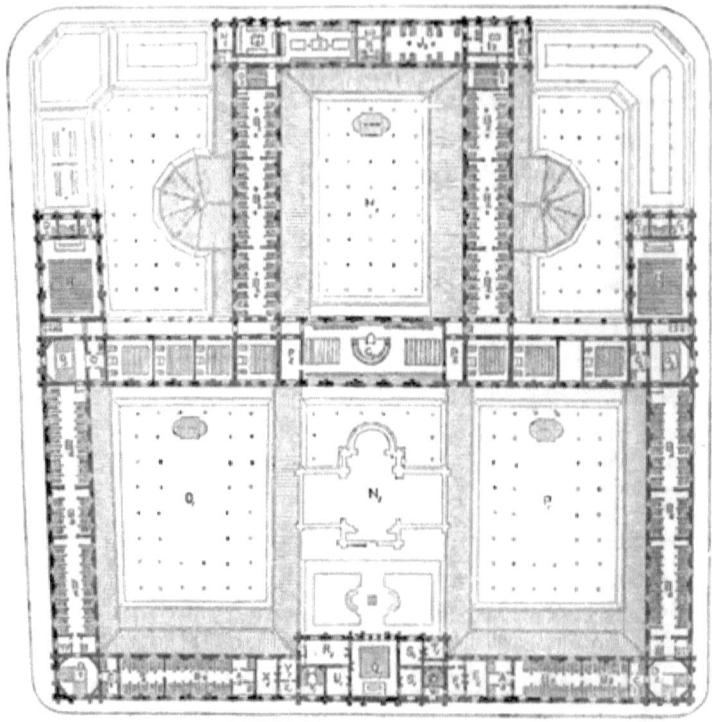

43.—THE COLLÉGE CHAPTAL, PARIS. PLAN OF FIRST FLOOR.

Reference.

A. 2. Masters' Rooms.
B. 2. Dormitories.
C. 2. Masters' Rooms.
D. 2. Staircase to Dormitories.
E. 2. Steward's Room.
F. 2. Strong Room.
H. 2. Lecture Room.
I. 2. Refectory.
J. 2. Sister (Infirmary).
K. 2. Convalescents' Room.
L. 2. Library.

M. 1. Upper School.
M. 2. Library.
N. 1. Chapel.
N. 2. Kitchen.
O. 1. Preparatory School.
O. 2. Bath Room.
P. 1. Middle School.
P. 2. Drawing Class Room.
Q. 1. Committee Room.
R. 1. Directors' Room.
S. 1. Writing Clerks.

students. Four classes for 50 each. The amphitheatres or lecture rooms, serve also for museums. That for physical science contains

200 students, and that for chemistry the same number. There is a laboratory for the Professor and one student. A room for

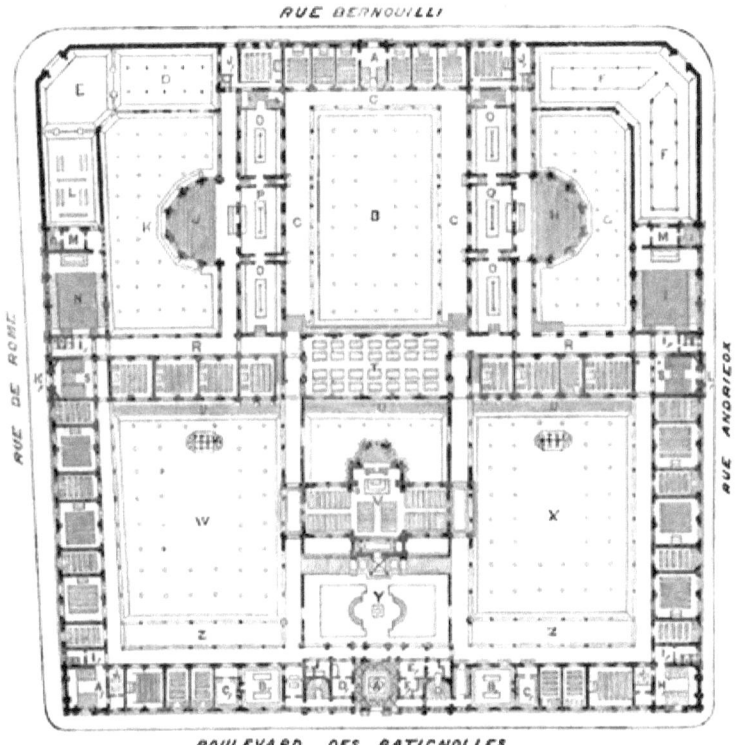

44.—THE COLLÉGE CHAPTAL, PARIS. GROUND PLAN.

Reference.

A.	Vestibule.	F. 1.	Cloak Room.
A. 1.	Staircase of Preparatory School.	G.	Second of Middle Court School.
B.	Court of Upper School.	G. 1.	Superintendent's Room.
B. 1.	Parlour.	H.	Lecture Room for Chemistry.
C.	Shelter.	H. 1.	Staircase of Middle School.
C. 1.	Directors' Room.	I.	Lecture Room of Middle School.
D.	Service Court.	J.	Lecture-room for Physical Science.
D. 1.	Porter's Lodge.	K.	Second Court of Preparatory School.
E.	Kitchen Courts.	L.	Gymnastic Room.
E. 1.	Teachers' Dining Room.	L. 1.	Entrance.
F.	Infirmary Garden.	M.	Teachers' Room.

natural history specimens. Five Masters' rooms. Five Servants' rooms. For each of the three divisions, physics, chemistry, and natural history, there are six rooms for reciting lessons, for examinations, and for other purposes.

3. *The Management Department.*—This comprises the arrangements for the physical comfort of the pupils. Two refectories, each capable of accommodating 300 boys, are planned in close proximity to a large kitchen, with a bakery having as far as possible a north aspect, coal cellars, and oil cellars. There are also in this department, eight Servants' rooms, a scullery, larder, bread stores, meat stores, laundry, and the Steward's office. A special entrance from the street into the court, easily accessible from the kitchen, is provided. There are cellars for wood, coal, wine, vinegar, and oil. A Steward's room (*sommellerie*) is also described as near the refectory. There is a dancing-room; a fencing room; two rooms for foot-baths; a house for the Steward close to his work, an immense linen-room with a working-room (*ouvroir*) attached; a large cloak room; a room for brushing clothes; a store for clothes and for trying-on; and two boot-cleaning rooms.

4. *The Sleeping Apartments.*—The dormitories each contain from 25 to 30 beds, with wash-basin for each pupil. Latrines and Servants' rooms are placed in contiguity. The dormitories are never less than 13 feet in height.

45.—THE ROYAL GRAMMAR SCHOOL, CHEMNITZ.

CHAPTER VI.

GERMANY.

Principles of School System—Classification of Schools—Prussia—Elementary Schools—Description of ordinary Parish School-house—Grammar Schools—King William Gymnasium in Berlin—Cottbus—Liegnitz - Marburg—Hildesheim—Commercial Schools—Sophien—Realschule in Berlin—Halberstadt—Cologne—Höhere Burgerschule at Wiesbaden—Schools of Saxony—District, Parish, and Burgher Schools at Dresden—Königlische's Gymnasium at Chemnitz—Kreuzschule at Dresden—Polytechnic and Chemistry Schools at Aix-la-Chapelle—Education of Women—Defects in German system—General excellence.

THE educated German compares favourably with the Englishman of corresponding rank when tested by his acquirements. If trained simply as a gentleman, and without reference to any profession, he usually knows with some accuracy besides his own language, Latin, Greek, some modern language (as English or French), History ancient and modern—with a thoroughness seldom taught among us—Geography and some elementary mathematics and physics. He has been taught singing as part of his school training, and has probably pursued from choice the study of some musical instrument, he dances, rides, and is intelligent and courteous. His physical powers have been developed by gymnastics scientifically taught. If intended for some trade, art, or profession, he knows less or more of special subjects,—Greek, Latin, mathematics, drawing, physical science, &c. according to the requirements of the career in view. Not great on the box seat, nor in leaping a five-barred gate, his riding or driving may not be perfect. On the other hand, his knowledge of such modern language as he has professed to learn, is excellent, and in particular, his acquaintance with geography forbids him to speak of going "down the Rhine" when describing his route to Switzerland—an error by which he is apt to gauge the primary education of the British tourist. It may not be that his accomplishments enable him to adorn society in a greater degree than his English rival. They are certainly more useful, and prepare him better for the competition of life. This is simply the result of his education.

When we turn to those of lower condition, no parallel can be drawn. The middle-class German has an enormous advantage over the middle-class Englishman, in the possession of a matured educational system which enables him to acquire, step by step, the knowledge necessary for advancement in after life.

And, when the advantages afforded by the schools for technical instruction are considered; that is to say, those found in schools for teaching the theory (and, in part, the practice) of architecture, construction, civil and mechanical engineering, chemistry, agriculture, metallurgy, mining, and other arts and sciences, we can only wonder that the Englishman has been able, even in some degree, to hold his own against the competition fostered by every educational advantage carefully applied to economical results. When the future career of the young German has been decided upon, his course is quite simple, and there is no difficulty in defining the direction of his studies and the steps to be taken for the development of his faculties and the acquisition of the knowledge necessary for any business, art, trade, or profession. The schools are his patrimony, the real sources of the wealth and progress of the nation and the individual.

From the buildings in which such a system of education is carried on, useful hints for future schoolhouses to be erected in England are likely to be derived.

On the continent, Prussia may be said to have taken the lead in education during the present century, and even as far back as the date of her adoption of compulsory primary education in 1763. Immediately on the completion of the seven years' war with Austria, in which the resources of Prussia were quite exhausted, Frederick the Great gave to his poor, but victorious country, the boon of education. It is natural, therefore, to find among neighbouring states, some imitation of the leading system extending even to questions of building. In point of architectural planning, the following is sufficiently close. The methods of arranging and imparting instruction,—the same generally throughout the German empire, Austria, and Switzerland, differ only on minor points in different states. The school-plans, with a strong family resemblance, differ in the same way. The differences are due more to the varying opinions of the authorities on subordinate questions, and to the varying skill of architects, than to any conflict on the general method or principle of education. Considered as specimens of planning, we find ourselves dealing with one leading idea in different schools of the same kind. The first point which strikes the visitor to German schools is the uniformity with which one system of teaching is applied alike to all the children from the youngest to the eldest. The second is the uniformity with which one system or principle of planning is applied alike to different kinds of scholastic buildings. Herein lies the origin of much of the difference between a continental and an English schoolhouse. The system of public instruction is almost, if not quite, as military in spirit as that which governs the army, and the buildings do not escape the *régime*. If Berlin may be described as a vast barracks, German schools may equally be classed as a series of small barracks. In degree, the rules are, perhaps, less rigid than those of France, where any freedom to the teacher, such as the choice of books, &c., is never allowed, and where every day, the same lesson is being given at the same moment from the same book to the same kind of class in all the different schools of the land.

In England, compulsory attendance begins at five years of age, and children are allowed to attend from the age of three. So far as being public schools is concerned, our Infant Schools now rank equally with Eton or Harrow. In Germany infants are not recognized. Philanthropy or private charity has organized institutions on the Kindergarten principle, where those under six years of age can be amused or occupied, and perhaps taught some little, as in the younger part of our Infant Departments. These, however, form no part of the school system. Government looks on them, it is true, with a favourable eye, and even inspects their buildings, which are mostly hired, not specially erected for the purpose. Considered as public schools, they do not exist. Sanitary law does not allow them to be conducted in improper premises; but their teaching is not recognised as part of the national system.

At the age of six, a German boy goes to an elementary school. Theoretically, he goes under compulsion,—practically, of his own pleasure, for the German parent no more thinks of depriving his child of tuition than of breakfast. The usage of years has converted it into a habit. And the habit has come to be regarded as a privilege. The appearance of the school is very different from that of an English model. There is no general school-room. No raised gallery where the child can receive "simultaneous instruction." No breaking the business to him gradually. There is a series of class-rooms entered from a wide corridor. He is placed in one of these, fitted with benches and desks precisely similar to, but smaller than, those used by boys twice his age, and there he commences that intellectual drill which is continued till the age of fourteen. Such a system must give the dull boy a better chance, for the most awkward recruit will make a tolerable soldier if drilled regularly, and among others, for a sufficiently long time. It can hardly fail to raise the masses of a nation. On the other hand the tendency to destroy individuality of character must be ranked as a loss.

The most important fact to be noted, is, that the government of the country takes charge of every child's education whether in public or private schools, by endeavouring always to secure, first, good teachers: secondly, good school-houses: and, lastly, satisfactory results. The private establishment only differs from the public in being managed by an individual for his own profit, and in its public examinations being conducted by the principal and his assistants instead of by the Government Inspector, the Provincial Board, or the High Council of Education. No school can be commenced without a licence from the authorities as to the suitability of the premises. None but certificated teachers are allowed to be employed. The subjects taught are under strict regulation. The main examinations must be thorough, and, as a rule, of a public nature. All this refers not only to those for elementary instruction. It is true of every school in the land forming part of the national system, although not always followed out in precisely the same way. The low type of private elementary school which Mr. Forster's Act will surely, in time, eradicate from English soil, the "Classical and Commercial Academy," as well as the "Establishment for young ladies," would not be tolerated in Germany for a week. This is said without prejudice to many honourable exceptions which exist as the result of private enterprise.

The application of similar principles to English schools of every kind is of the gravest possible moment. Below the Universities, every kind, from those of famous history downwards, might be subject to government control, either by periodical inspections or other form of verification. Some guarantee to the nation is required as to the thoroughness and high quality of the education imparted, and as to the suitability of the buildings used. A Rugby under a Dr. Arnold could be none the worse for such an inspection, while hundreds of inferior schools would be benefited or abolished by its universal application. If it should be found impracticable to adopt universal inspection, at least the fitness of teachers and premises might be controlled by the licence, or some other efficient method. If it is inexpedient or unnecessary to test the *results* of education in all establishments, at least the safety of the *process* to mind and body might be ensured.

A consideration of the comparative value of school-buildings abroad, as embodying ideas applicable to England, leads to the conclusion that, though isolated specimens of fine character may be found elsewhere, those of Germany—especially of Saxony and Prussia—are, on the whole, the best. Evidence of careful thought and desire to perfect every arrangement and every point of detail occurs here most frequently. Usefulness, rather than show, is the principal characteristic. Economy is everywhere studied, yet proper or needful provisions are seldom lacking on the ground of expense. Education is always treated as a great question.

The German empire now numbers about 6000 Elementary Schools, containing altogether about 6,000,000 pupils. These figures show an average of 150 children under actual tuition for every thousand persons in the population, and are irrespective of the large numbers attending other primary schools placed under the wing of higher schools of different kinds. Among Elementary Schools (*Elementarschulen*) are included the Parish School (*Gemeindeschule*), and the People's (*Volkschule*). In Saxony the Elementarschulen are classified under the heads of District (*Bezirk*) and Parish or Poor (*Gemeinde*) Schools, the latter being to some extent free. The studies in these primary schools are not specially designed as a preparation for those of secondary kind, and therefore, in connection with one of the latter we nearly always find a Preparatory School (*Vorschule*) of two or more classes under the same roof, and managed by the same Director, to which children whose education is intended to be of a higher kind may be sent, and from which they are passed on, at the age of ten, to the higher school without preliminary examination.

Among Secondary and Higher Schools we find—

1. The Burgher (*Burger*) School, forming the favourite education of the lower middle classes, the future tradesman or merchant, or the official of lower degree. How efficient these are, and how generally popular, is shown by the rapid increase in their number in comparison with others during the five years between 1863 and 1868 in the eight old provinces of Prussia, as set forth in the following table: viz.:—

	In 1863.	In 1868.
Gymnasien and Pro-gymnasien	173	199
Realschulen	65	68
Bürger Schulen	21	41

2. Realschulen are of three kinds, (*a*) The Realschule of the first rank, the course of which is nine years, as in the Gymnasium, and wherein the studies differ from the latter mainly, in replacing, after a certain stage, dead languages by those in modern use. (*b*). The Realschule of the second rank, in which it is allowable to shorten the duration or alter the direction of the studies, and which stands to the Realschule of first rank as the Pro-Gymnasium to the Gymnasium. (*c*). The Higher Burgher School (*Hohere Burgerschule*), which is the least classical of the higher schools.

3. The Practical School (*Gewerbeschule*) which supplies a training in science, or art, or both, preparatory to the more complete scientific theory of the Polytechnic School.

4. The Gymnasium, forming the classical school preparatory to the University.

5. The Polytechnic or Scientific University.

The Higher schools of Germany are said to contain about 178,000 students. There are others, not mentioned above, forming connecting links between two kinds by combining to some extent the instruction of both, that for instance connecting the Realschule with the Gymnasium and called the Real-gymnasium. There are, besides, 20 universities with 624 professors of higher or lower rank, and nearly 14,000 students. The Polytechnic Schools contain between 4000 and 5000 youths preparing for professions.

As, in Switzerland, it is practically sufficient to take note of the schools in the Canton of Zurich, so in Germany an outline of the system of Prussia as set forth in the buildings of the capital, will be found most useful. In this slight description of the schools of Berlin, the endeavour is rather to present a simple picture of the state of school architecture as it arises from the system of teaching in a manner likely to be generally intelligible, than to confound the reader with a record of the many phases of difference occurring in the theory and practice of schools otherwise and mainly alike.

In Berlin, the schools are either under Royal patronage, or belong to single parishes and municipalities under the supervision of the School Board of the province (Brandenburg), or are under the direct control of the municipal authorities.

To the latter class belong the Higher, the Parochial, the Private Schools and Seminaries, and the Poor Schools.

The Gymnasien, the Real-schulen, and the Gewerbe-schulen (not being Royal Schools), are controlled by the magistrate. The remainder, such as the Girls', the Town, the Poor, the Parochial, and the Private schools, are administered by the Municipal School-Commission specially selected for the work.

School-houses erected by the municipality are very numerous and rapidly increasing in number. Year by year, new specimens spring up in different quarters of Berlin, as though here also efficient elementary education were a new thing. They are now built of common brick with a facing of first-rate red brick added afterwards. Till the permanent facing has been added, the brickwork exactly resembles in appearance that of Italy where it is intended for the reception of a final facing of sheets of marble. This method of executing brickwork is entirely unknown among us. Its advantage lies in producing finer effects by an appearance of finer workmanship. The school-plans continually approach nearer to perfection in the various minutiæ of arrangement and fittings, and it may be added that the quality of the education itself also steadily rises if the increased range of subjects taught is any criterion. The rapid increase of population

46.—PARISH SCHOOL IN THE KURFÜRSTENSTRASSE, BERLIN.

in the capital, since the close of the Franco-Prussian war and the consequent unification of the Empire, may lead us to expect still greater progress. The municipal Elementar-schulen usually consist of a Boys' and Girls' department under one director or head master. Each department has six class-rooms, with generally a reserve class-room, and a playground, the latter being, in the case of boys, fitted with gymnastic apparatus. Where expense is not to be very strictly considered, the addition is thought desirable of an Examination Hall (*Aula*), a drawing-class room, one or two small rooms for teachers, a residence for the master and another for the

78 SCHOOL ARCHITECTURE. [CHAP. VI.

caretaker. The presence of these latter is very rare, and some of them may be regarded as the exception. The drawing-class room

47.—PLAN OF SECOND FLOOR.

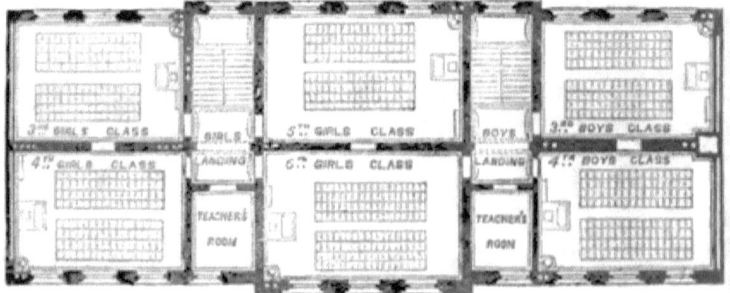

48.—PLAN OF FIRST FLOOR.

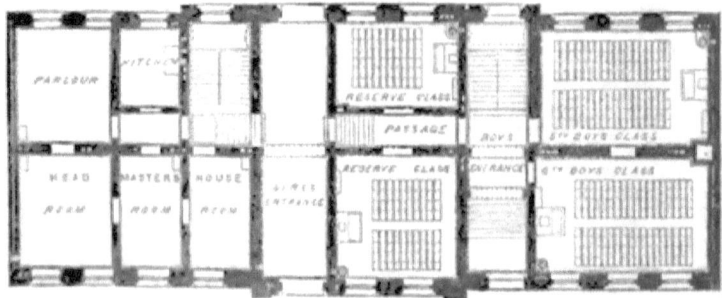

49.—GROUND PLAN.

in elementary schools is a mere theory. The Aula is seldom found, its cost being great, and its use little. As a specimen of

CHAP. VI.] PARISH SCHOOL-HOUSE IN BERLIN.

this class of school-house, exceptionally possessing an Aula, and

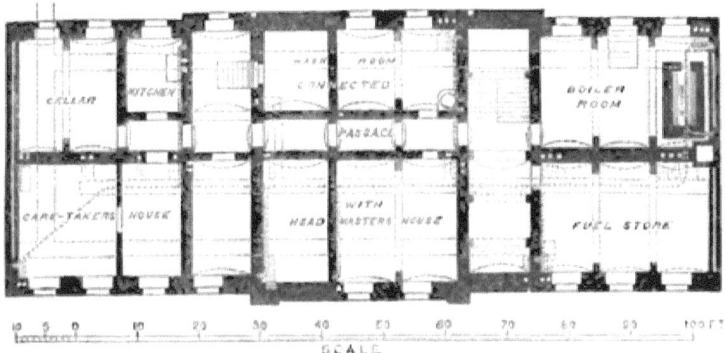

50.—BASEMENT PLAN.

forming apparently an advance upon the ordinary kind in Berlin, we give from the work of Herr Gerstenberg (woodcuts Nos. 46 to 51), plans and illustrations of the Parish School (*Gemeindeschule*) in the Kurfürstenstrasse.

The house of the Director or Head Master, commonly at the top of the building, is here on the ground floor, with the central portion of the basement allotted to it. The reserve class rooms, usually placed on an upper floor, are also on the ground floor, and planned so as to be used either by girls or boys, as required. The class-rooms for girls and boys, and the entrances and playgrounds, are perfectly separate, though both sexes are taught by men. There is access on each floor, for the teachers, from one school to the other; but on the first floor this is obtained through one of the class-rooms belonging to the

51.—BLOCK PLAN, SHOWING PLAYGROUND.

girls' school, and, on the second floor, through the examination hall. There are no lavatories, bonnet, cap, or cloak-rooms, but each class-room has ample provision of hat pegs arranged in a single line against the walls on two sides, respectively opposite the windows and the teachers' platform. The school comprises six class-rooms for boys, six for girls, two reserve classes, teachers' rooms, and the examination hall. The internal roof space is not utilized. The external design is of brick, and forms an average specimen of the kind of architecture followed in the poorest Berlin schools.

The importance attached in the United States to the provision of a general Hall of Assembly has already been shown in a previous chapter, and the sacrifices in point of convenience and cost made by the adoption of sliding partitions with the sole object of securing this advantage, have been alluded to. The Hall of Assembly there spoken of should not be confounded, either in idea or in reference to its use, with the German Examination Hall. The two things are totally different. America (like England) prefers in elementary schools the simultaneous method of teaching in full extent, and provides a room of sufficient size for assembling all the children *every day*, or whenever desired, for collective lessons, singing, or addresses. Germany adopts the separate or class system for every kind of school, high or low, and in no part of the course approaches the opposite method. The gallery raised tier above tier, which formed the favourite feature of Mr. Stow's teaching, and which is now invariably seen, though of modified size, in our Infant Schools, does not exist in any part of Germany except in the lecture halls of the highest schools. Even the Kindergartens are without it. In consequence of the public nature of the main examinations, a convenient room of large size is required in which these can be held. This room, hall, or aula, being used only some *three or four times a year*—viz., at the great examinations, or on the occasion of a fête like the Emperor's birthday—is usually located on the highest floor. The plain truth is, that in its building-wants the separate or class system has little or nothing in common with the simultaneous. Writers on education may argue that the two systems are identical, as both involve the principle of teaching a class rather than an individual. To the architect, and to the subject of school planning, they are very different. The theoretical programme of every German schoolhouse includes the aula, and the highest kinds are never without it, although practically it is hardly ever found in the people's or poor schools. Nowhere in Germany is it really a "hall of assembly," as in America. Those who advocate so expensive an addition to the cost of an English Public Elementary School may be perfectly right in their opinion, but they cannot quote the practice of the Fatherland with any show of reason. In doing so, they compare things essentially unlike pertaining to two different systems of training.

Gemeindeschulen vary in their arrangements, sometimes as suggested by the peculiarities of site, and at others according to the kind of children to be taught. The system of teaching the sexes separately, of course implies not only separate school arrangements, but separate playgrounds and staircases. The last are usually made extremely wide, and consist of very short flights. A single staircase thus occupies so much space that, although two are generally provided, it is no uncommon thing to find only one for the use both of girls and boys. The municipal school in the Frankfurter Strasse (woodcut No. 52) is an example (also from Gerstenberg) of this feature and of the more common kind of people's school which lacks the aula. The plan is of extreme simplicity—the simplest of all German schools —consisting of little more than four class-rooms on each floor divided into two groups by the intervening passage and staircase. That in the Schmidt Strasse, otherwise similar, has two staircases side by side. When the size of the school requires six or more class-rooms on one floor, examples having only one staircase become very rare. There is not always the provision of a gymnastic room (*turnhalle*), as in the schools respectively in Naunyn Strasse and Wartenburg Strasse; but the outdoor playground (*turn-*

platz) is seldom without its complement of gymnastic appliances. The aula is far more commonly omitted than the turnhalle, the newer schools being considered incomplete without the latter. The boys' playground is made much larger than that for girls generally, but even this rule is not invariable. Symmetry of arrangement is a constant weakness among continental school-

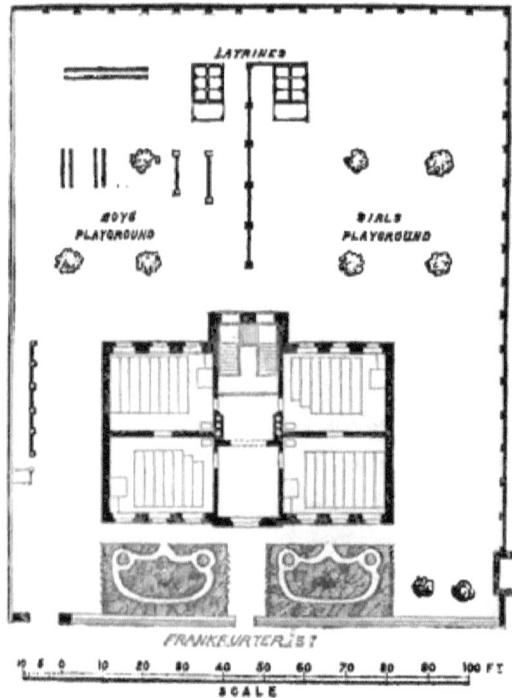

52.—PLAN OF FRANKFURTER-STRASSE SCHOOL, BERLIN, SHOWING PLAYGROUNDS.

planners, sometimes pushed so far as to interfere with the practical advantages of the plan.

The schoolhouses are mostly of three stories, the lowest being for the youngest children, and the highest for the eldest. When the site is very confined, care is taken to avoid the crowding of rooms on the ground floor, by which the size of all the rooms would be reduced and their usefulness impaired, but to place those which cannot be obtained on the ground floor without injury to health, on higher stories. The now obsolete official rule provided at least six superficial feet to each child—as compared with the eight feet of our Education Department. The new schools provide much more. In Berlin, as in London, the minimum allowance is always exceeded in a well-planned school, and is now really nine to ten feet, and, in the higher schools, even twelve square feet per head, according to the number to be instructed, and to the age of the pupils. This is considered essential for the sake of health.

The mixed system may, perhaps, be found under exceptional circumstances, and in country districts where necessity compels, but no new schools are organized on its principles in the large towns. The co-education of the sexes forms no part of the German educational faith. This may to some extent arise from the decidedly military objects constantly kept in view. Culture of the higher sort is not entirely confined to the stronger sex, although for the weaker a domestic education is much preferred. Every young woman is expected to study the kitchen and the economy of house-keeping, just as every young man is compelled to use the rifle and to understand the drill necessary for war.

The position of the building on the site is very much settled in reference to the subject of light. Dr. Wiese, the Minister of Education having charge of Secondary Schools in Prussia, says, speaking of these, "The best is an entirely open place. When possible, the class-rooms should look to the east, as there is then little trouble with the sun. If an easterly aspect be impracticable, the south or west should be chosen, and the sun as well as the wind kept out in the best possible way. A north aspect is only suitable to a room for drawing." "In the arrangement of the class-room, the seat of the teacher must be placed where he can overlook the whole at once. At his side he must have table, maps, &c. On the other hand, the pupils must have, before their eyes, the teacher in a convenient direction. The breadth of the room is dependent on this, and, still more, on the fact that the light from the window must be strong enough for those sitting against the opposite wall. The light should be admitted from the left side, so that the right hand may not cast a shadow. If light be admitted from two sides, the result is not so good, a glare on the drawing or writing rendering it difficult to see. Windows placed in front of the child would allow light to come directly into his eyes and be hurtful. Placed at the back of the class, they are admissible, and in a very large class, advantageous. As a rule it may be laid down that the scholar (if his eye be in its normal condition) can read writing on the wall at a distance of 27 feet. The length of the class-room, therefore, might be 30 feet. The breadth should not be more than 21 feet, because then pillars or complicated construction are required. The former occupy space, and prevent a free view of the teacher. The latter increases the cost. Even the old-fashioned method of supporting the tiebeam of the roof by an iron pillar only three or four inches in diameter is undesirable. A large class-room should be rather oblong in shape, and not less than thirteen feet in clear height."

In massing the desks for a class, it is considered that about three square feet should be allotted to each child, otherwise they will be too close together for purposes of health. The maximum number of children allowable to one class has been fixed at sixty, and this is only exceeded, even in elementary schools, in those rare instances where the exigencies of planning have necessitated a room of larger size. It is authoritatively affirmed that German experience shows this number to be the very greatest which can be efficiently taught by one teacher. In the higher schools it is much less, being commonly fifty for the lower and forty for the higher classes.

In arranging a new schoolhouse, the first condition is, that in every class-room the windows shall be on one side only of the room, viz., to the left of the children. No other windows are permitted. This rigid rule as to the admission of light is the very foundation of German school-planning, and completely governs the arrangements, the classes being grouped "end-on" or "side-on," according to circumstances of lighting. The form of the room is determined by the power of lighting well the desk furthest from the window-wall. This cannot be done at a greater distance than one and a half times the height of the room. To this may, of course, be added the width of the gangway beyond, varying from three to four and a half feet. The size of the room, in the other direction, is limited by the power of the teacher's voice, and never exceeds thirty feet. Between these two limitations, the German architect shapes his class-room for sixty pupils, or a smaller number.

Each room has (or, in theory, ought to have) two doors, one leading to the corridor or landing, the other to the adjoining room. The first door is of ordinary size. The second is larger (about 8' 3" × 4' 3"), made double and folding,—double, to stop the sound which would otherwise be heard from one class-room to another,—folding, to render more easy the management of both classes by one master in case of the absence of the other. The arrangement is also thought valuable for rapidly changing the air of the rooms during a few minutes' interval between lessons.

Windows are always placed high in the wall, and are (in theory at least) of different sizes according to the amount of lighting-power required. In a shallow room, of thirty feet frontage, four windows measuring about 7' 6" × 3' 6" are recommended in the long window-wall. In a deep room with a short wall where only three windows can be obtained, there should be nine or nine and a half feet by four and a half or five feet. Blinds of unbleached linen or green twilled stuff are preferred to all windows except those facing the north. White or too opaque blinds are condemned. Where the class-room unavoidably looks towards the street the windows are made double, to prevent sound. The favourite plan is to place them all, if possible, next a court or playground at the back. The temperature is regulated by a thermometer in each room, and is kept as nearly as possible at 16° Réaumur. The heating apparatus is never allowed to get cold during the whole winter. The German certainly knows how to warm his school, but his chief principle is to husband the heat and to admit fresh air as sparingly as possible. His ventilation is usually defective and the rooms are close, and, towards the close of the lessons, unwholesome. He points with pride to the thermometer on the wall which has not risen very greatly since work commenced, but he forgets that, although the thermometer is a test of temperature and may be appealed to in cases of over-heating, it cannot ascertain the depths or gauge the quantity of atmospheric impurity.

The old system of warming by means of separate stoves made of Dutch tiles is no longer applied to new schools.

The warming apparatus of an ordinary Gemeindeschule for 800 children costs altogether from 600*l.* to 800*l.*, and differs little from that in use in England. In Berlin, English engineers have possession of much of the work. To the casual observer, the stove may still appear to hold sway, for it often appears to stand in the corner of the room as usual. It is now of cylindrical form and is filled with coils of hot-water pipes usually 4 inches in diameter. By means of an ordinary cock accessible in the corridor, the heating of any one room can be cut off from the general system when no longer needed. The best position for the heating-coil is not in the corner next the door where it is usually found, but against the window-wall. For the class-room of either a primary or secondary school it is there least in the way, and most valuable both for warming and ventilation. The flues for extracting vitiated air are usually ten inches square, and lead from each room to one central shaft three feet square, which is carried out to about five feet above the roof. The extraction almost always lacks artificial motive power and is therefore always feeble. The real ventilation avowedly consists in opening the window after the scholars have been dismissed.

The foregoing remarks on some of the points which may be said to guide the German architect in his plans for elementary schools, apply with still greater force to the case of secondary and higher schools. All being organized on the same first principle —that of class division,—it is natural to find the most complete specimens in cases where considerations of cost have not been of the same importance, where the rank of the school has rendered desirable something more than the barest necessaries, and where every detail has received the highest kind of care and thought. As to the ultimate building-result and architectural meaning of the system, it is better to study representatives of the higher than of the lower kind of schools. And, as to an ordinary sort unaffected in plan and arrangement by the requirements of special studies and most likely to be useful to the English reader, we turn first to the Gymnasium.

The German "Gymnasium" is not, as might at first appear to the English mind, an arena for athletic exercises after the manner of the ancient Greeks. It is a public secondary school, where pupils receive a classical education. Here the mind is placed in training as the first object, though the body be not entirely neglected. It is, in fact, the German grammar-school.

The education given, as in the French secondary schools, is purely classical—that is to say, there is no professional turn given to the studies, and no preparation for future apprenticeship to any trade or business. This is the reason why drawing, if included at all, is only taught in the lower classes, and why little importance is attached to instruction in physical science. Those who, after leaving it, wish to carry their studies farther, pass either to the polytechnic school or the university; as a rule, to the latter. There is frequently a preparatory school attached, pupils not being admitted to the Gymnasium proper before the age of nine. By extraordinary diligence and rare ability they may complete the course of six classes in the same number of years; but, as the examinations are strict, very few do so. The average course may be taken as nine years, one for each of the three lowest classes, and two for each of the three highest. The maximum number permitted in one class is fifty. When the number grows beyond this, the class is divided into two.

Berlin has four royal and four municipal grammar-schools. Of these, perhaps the most interesting to the schoolmaster is the Friedrich-Wilhelm Gymnasium, whether from its ancient history, its education, or the flourishing state of its finances, it being the only higher school in Prussia, except one, where the expenditure is less than the receipts from pupils. But for architectural excellence we should select either the König-Wilhelm Gymnasium in Belle Vue Strasse, or the Cölnisches Gymnasium in Inselstrasse. In the case of the former, a personal inspection of the building is supported by the perusal of a description published at Berlin, which enables us to understand the theory of planning held by the authorities, as well as the mind of the architect in developing it into bricks and mortar.

The Cölnisches Gymnasium has an important architectural advantage over the other, in being executed in red brick and terra-cotta, which give to its poorer design and defective sky-line, a value in point of colour ever denied to the most masterly conception when carried out in cement, stucco, or other inferior material. It derives a distinctive name, not from any scholastic peculiarity, but from locality. The town formerly consisted of two as distinct as London and Westminster, one being Berlin proper, the other Cöln, the river Spree forming the division. The latter town is now, of course, completely blended with, and lost in, the former.

In the capital of the German empire, architects have long been given over to a Classic mind. This may have arisen partly from education, preference, or fashion, but is chiefly due to the absence of any good building-stone in the neighbourhood, and the consequent expense of carriage from immense distances. Cusps, tracery, crockets, and finials would be supremely ridiculous when neatly executed in cement, the usual material for a façade. The architecture is nevertheless excellent. The influence of Schinkel is everywhere seen, even in the works of less distinguished architects, and is marked by a beneficial effect on art, although sometimes acting as a check to originality. Those who would

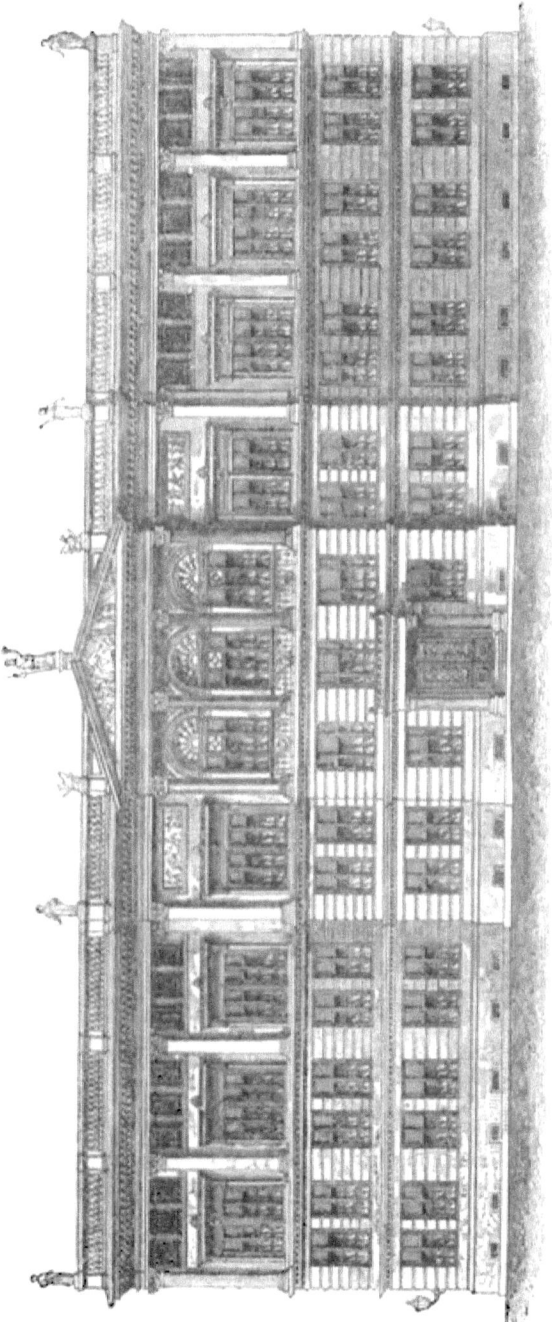

53.—KÖNIG-WILHELM GYMNASIUM, BERLIN.

gladden their eyes, too long accustomed to the "builders' Classic" prevalent in London, may see in Berlin how the same things, when in the hands of skilful architects, assume the shapes of scholarly, refined work, and attain a beauty often nearly allied to the picturesque, in spite of the use of an inferior material scarcely the result of choice. The cement has a hard, clean appearance, and never assumes the dirty, abject condition which is its sure fate in the smoky atmosphere of our own capital. This may be partly due to the practice of carefully saturating the work with linseed oil as soon as it has become dry. Brick architecture has sometimes been attempted, and is rapidly rising into favour. One building by Schinkel, in brick and terra-cotta, possesses details of rare beauty. The new town-hall is also of red brick, and must be termed Gothic in idea, though a civil character is sought by the use of round arches, instead of pointed, in the principal features. The German architect, well schooled in what may be termed the lower branches of his art, produces continually buildings carefully thought out in reference to their purpose, and not devoid of originality. Such work stands apart from the slavish copyism of bygone styles, without descending to the level of another sort belonging to another school of somewhat lawless character, which mistakes ignorance for originality, and mere comic lines for artistic effects. Lacking the complete and final sense of power and mastery over building material, it yet seldom rises beyond mediocrity except when treated classically. This Pointed town-hall may be fairly contrasted with the St. Pancras Hotel in London, where a civil character is obtained in a stately brick building with a free and bold use of the pointed arch throughout.

In spite of the unfavourable situation of Berlin in a flat, sandy waste, the population, even before the Franco-German war, had been considerably on the increase. The abolition of the rather heavy fine imposed on strangers before being allowed to settle in the capital has added materially to the influx; and now the gates of the town no longer represent its real boundaries.

CHAP. VI.] THE KÖNIG WILHELM GYMNASIUM. 91

On the side of the Potsdam and Anhalt gates a spreading population, chiefly of the better classes, long ago required the

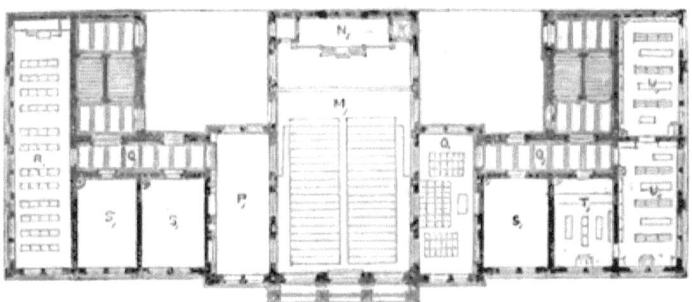

54.—PLAN OF SECOND FLOOR.
Reference.

M.	Aula.	R.	Drawing Class.
N.	Dais.	S. S.	Reserve Class-rooms.
O.	Song School.	T.	Scholars' Library.
P.	Ante Room.	U. U.	School Library.
Q. Q.	Corridors.		

establishment of a higher school, and in 1856 a Pro-gymnasium was opened under the patronage of the King. The rapid growth

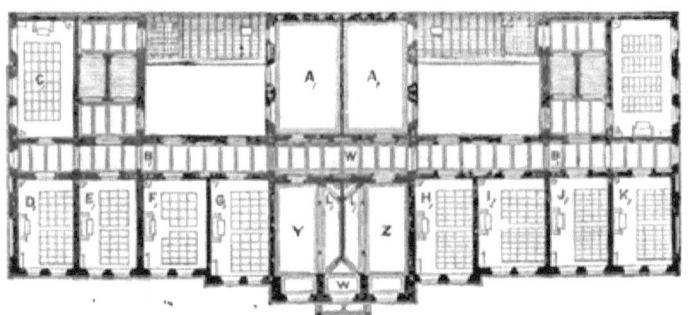

55.—PLAN OF FIRST FLOOR.
Reference.

A. A.	Museum.	I.	Quinta (a).
B. B.	Corridor.	J.	Sexta (a).
C.	Prima (a).	K.	,, (b).
D.	Secunda (b).	L. L.	Teachers' Lavatories.
E.	,, (a).	W.	Lobby.
F.	Tertia (a).	Y.	Directors' Room.
G.	Prima (b).	Z.	Conference and Teachers' Room.
H.	Quinta (b).		

of the neighbourhood soon rendered necessary a new building, which, in turn, was opened Oct. 24th, 1865, under the title of the König-Wilhelm Gymnasium, or King William Grammar School

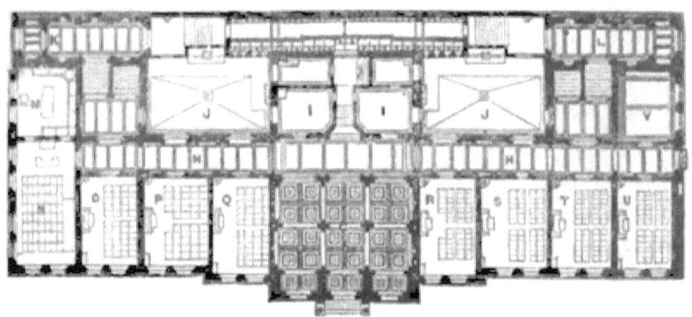

56.—GROUND PLAN.

Reference.

G. Entrance Hall.
H. H. Corridors.
I. Caretaker's house.
J. J. Courts.
K. Boys' Entrance to Gymnasium.
L. ,, ,, ,, Vorschule.
M. Physical Apparatus.
N. ,, Lecture-room.

O. Quarta (a).
P. Tertia (b).
Q. Quarta (b).
R. Third Class (b).
S. Second Class (a).
T. Third Class (a).
U. Second Class (b).
V. Treasurer.

(woodcuts 53 to 57). The approach to the site is not at present

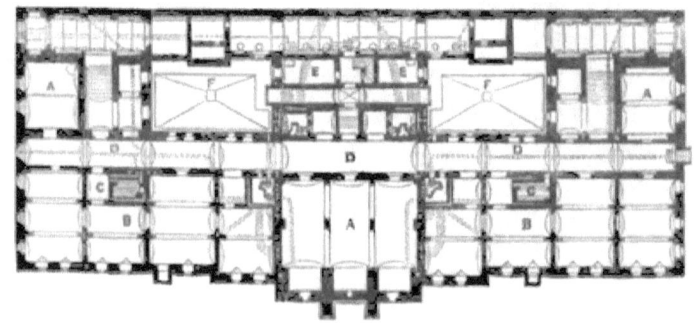

57.—BASEMENT PLAN.

Reference.

A. Cellar.
B. Cellar space.
C. Boiler.

D. Corridor.
E. E. Part of Caretaker's house.
F. F. Courts.

particularly good, but a new one, directly opposite the entrance is in contemplation. The building is set back a considerable distance from the street, and is surrounded by gardens. It is thus quiet for purposes of study. The site measures 453ft. by 216ft. There are two playgrounds, fitted with gymnastic apparatus, one for the elementary or preparatory school, the other for the gymnasium. The gymnastic hall, intended to be 107ft. by 46ft. 1in., has not yet been built.

The programme of accommodation to be provided was as follows, viz. :—

1. A hall for 600 persons.
2. An adjoining room, for visitors' room and for committees.
3. Twenty class-rooms—viz. :—
   (*a*) Six for the elementary or preparatory school, each grouped into two sections of 50 pupils.
   (*b*) Four for sexta and quinta (the two lowest classes of the gymnasium proper), and their sub-divisions of 50 pupils each.
   (*c*) Two for quarta of 40 each.
   (*d*) Six for under and upper tertia, secunda, and prima, of 40 each.
   (*e*) Two reserve classes.
4. A drawing class-room.
5. An inspectors' room.
6. Two rooms for physical science and apparatus.
7. A room for gymnastics.
8. A gymnastic apparatus-room (in direct communication with No. 7).
9. An apparatus-room for geographical and natural history department.
10. A committee-room (*vide* No. 2).
11. A master's-room.
12. A directors'-room (available also for the deposit of archives).
13. A masters' library.
14. A pupils' library (to contain also the study-books of the poorer pupils).

15. Two rooms for the solitary confinement of pupils under punishment.

16. Urinals and water-closets.

17. Caretaker's apartments.

This "programme," or theory of the building, is, in itself, an instructive outline. The kind of school is laid down in skeleton, and the method of discipline indicated pretty plainly. According to it, 20 instruction-rooms were required for 900 pupils. The architect's actual building (including reserve class-rooms devoted to special subjects) gives accommodation for 960 pupils in the same number of rooms.

The building has been erected in a very complete and costly manner throughout, with the intention of making it the most complete example in Germany. It is a kind of school useful for comparison with our own grammar-schools. The methods of planning shown in this and others cannot fail to be carefully studied at some future time when a demand for secondary schools arises in England, as the natural consequence of the new elementary schools now in progress.

The most important architectural feature of the interior—that of the great hall—is marked externally by a projecting portico of four Corinthian columns of artificial stone, which support an entablature and parapet. A balustraded parapet of the same material, 5 ft. 7 in. high, runs round both the main building and the wings, and is ornamented by six statues representing the Sciences. On the pediment of the main front stands the national emblem of Borussia, 9 ft. 3 in. high. All the statues are executed in terra-cotta. The professional reader, who may wish more closely to study the various technical points of this fine school, will find on referring to the end (Appendix A) a full description translated, in somewhat condensed form, from the German of Herr Gerstenberg.

From the copious and admirable work of Dr. Wiese, directed perhaps rather to the instruction of the educationist than that of the school builder, yet full of reliable information on questions affecting public instruction in Germany—some further examples of secondary schools are chosen.

The gymnasium at Cottbus, erected in 1867, presents in its exterior (woodcut No. 107, p. 146) a fair specimen of plain Renaissance treatment, unfortunately executed in cement. The plan is very simple. A fine portico leads to corridors, from which the class-rooms are entered, at each end of these the staircases, polygonal in form, project. The Ground Floor is used entirely for the gymnasium proper, but the First Floor provides rooms for students intended for professions rather than the University, and thus embraces in its work part of the course of a Realschule. The position of the aula is easily identified, as possessing the round-arched windows in the centre of the façade on the first floor. It is used in common by both schools.

The gymnasium at Liegnitz, also of the date of 1867, is one of

58.—GYMNASIUM AT LIEGNITZ.

the smaller grammar-schools in which the prominent position of the aula on the first floor (woodcut No. 58) and its height extending through two ordinary stories, appear intended to mark the importance of the examinations to be therein held. The mere fact of the presence of the aula, wherever placed in the building, occurring commonly in the higher schools of a nation so economical as the German, also points in this direction. In visiting many schools of this class, and seeing the hall often richly decorated and handsomely furnished, the inquiry "how often is it used" was always met by the reply "three or four times a year," or, "Not more than six times a year." Occasionally an instance occurs

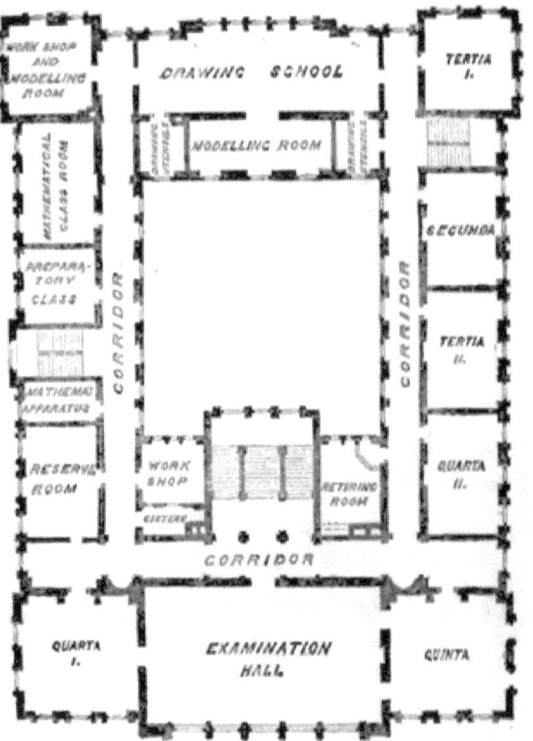

59.—GYMNASIUM, LIEGNITZ. PLAN OF FIRST FLOOR.

where the aula is also used for music and singing, but more commonly separate rooms are provided for music and drawing, in schools where these subjects are taught. The English mind, accustomed to the sight of extravagance, finds it hard to understand this kind of economy. The explanation is that the separate class method is carefully and logically followed, and the results of the system in its higher conditions, as proved by the examinations, are considered to be the crowning test, assuming in the eyes of the authorities quite a national importance. The common absence of the aula from the elementary schools is explained by the fact that the parents of the poorer children do not take so much interest in the examinations, and that, if there is to be no

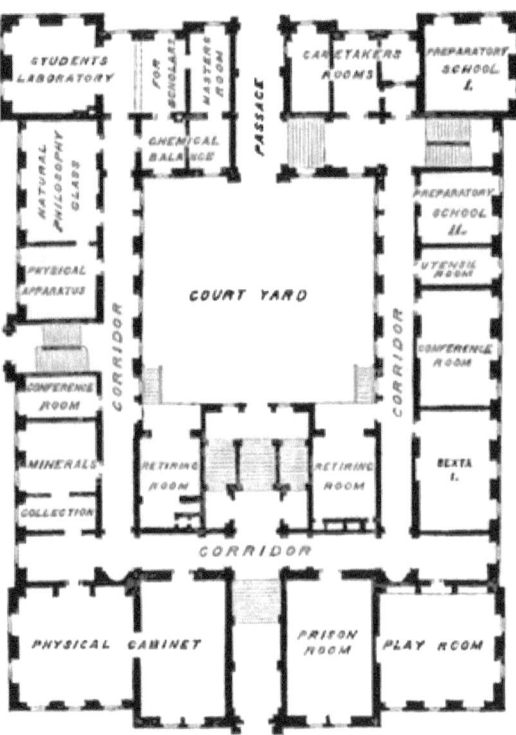

60.—GYMNASIUM, LIEGNITZ. GROUND PLAN.

audience, the examiners can as easily do their work from room to room as in a large hall.

The plan of the gymnasium at Marburg (woodcuts Nos. 62 and 63), like those of the König Wilhelm and the Cölnisches gymnasien in Berlin, shows a double arrangement of rooms in depth, and a main longitudinal corridor parallel with the principal front. Unlike those examples, the corridor is lighted only from the two ends except so far as any assistance is derived, in the middle of its length, from the windows of the main staircase. The principal entrance, placed as usual in the centre of the façade, leads directly into this staircase—which is of magnificent size —and communicates immediately with the corridor. On the left, this latter terminates in a small staircase which leads to the Directors' house located on the second floor. At this end of the building there is a projection at the back forming a wing wherein

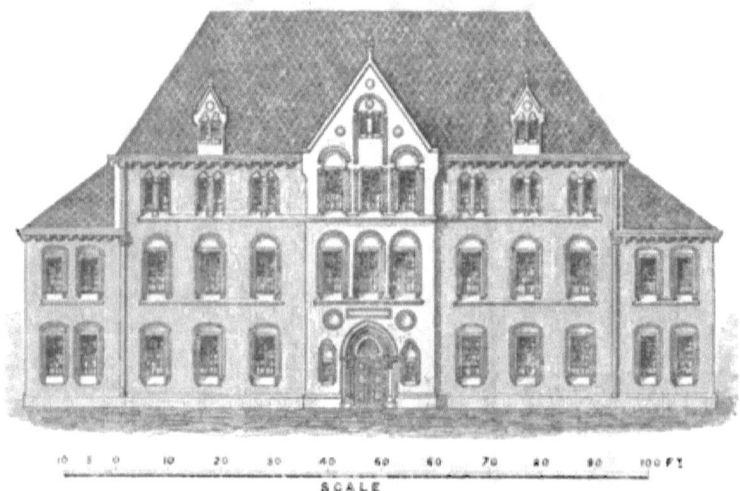

61.—GYMNASIUM AT MARBURG.

is placed, on the first floor, the spacious aula. The more usual position is in the front, where it commonly appears as the central feature of the external architectural design. The unusual variety in the size and shape of the rooms may here be noticed. Also the presence of two rooms instead of one (*carcer*) for the confinement of refractory boys. The rule of admitting light from one side only of the class-rooms is strictly followed, although other windows have been permitted to the aula. The external design (woodcut No. 61), although sufficiently simple, presents an odd compound of Gothic forms used in free intermixture,—arches of round, pointed, segmental, and trefoil shape being all found in the same front. Some of the ornamentation is equally odd in

CHAP. VI.] GYMNASIUM AT MARBURG. 99

detail, that for example to the window-jambs which cannot be

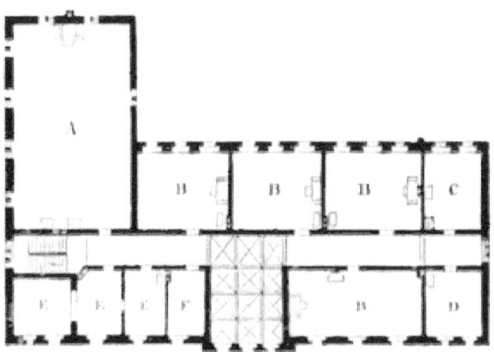

62.—PLAN OF FIRST FLOOR.
Reference.

A. Aula.
B. Upper Classes.
C. Conference Room.
D. Physical Apparatus.
E. E. Library.
F. Store Room.

63.—GROUND PLAN.
Reference.

A. Vestibule and Main Staircase.
B. Corridor.
C. Staircase to Directors' House.
D. D. Caretaker's House.
E. ,, Office.
F. G. Apparatus Rooms.
H. Natural History Collection.
J. J. Lower Classes.
K. K. Places of Confinement.
L. Caretaker's W.C.
M. Teachers' ,,
N. Latrines.

shewn on the small scale of the woodcut. A slight projection is made in the front wall in order to obtain the architectural feature of a central gable. The rest of the general outline is formed by depressing the two extremities, and omitting some rooms on the top floor, thus producing a broken effect. The use of steep roofs, relieved by two dormer windows flanking the central gable, is effective, but we miss the feature of the chimney-shaft which commonly plays so important a part in any grouping of the kind. Classic architecture has been chiefly affected for the classical schools of Germany, but this example adopts some of the main features of northern Gothic art, without securing the use of real materials in its construction. Artificial warming, now universal among German schools, requires only one smoke-flue, which is carefully kept out of sight by being placed away from the principal front.

The gymnasium of S. Andrew at Hildesheim, built in 1869, is a specimen of a remarkably rich building devoted to school purposes. In its exterior (woodcut No. 64) the purely decorative element, as distinct from the constructional, is probably more predominant than in the majority of German schoolhouses, and expresses itself by abundance of turrets, buttresses, crockets, finials, and other ornamental features familiar to Gothic art. The central portion of the principal front is profusely ornamented, with sunk tracery and patterns, until, rising in shapes of polygonal towers, capped with conical roof, the whole terminates in a forest of gablets and pinnacles. No one can deny the architectural propriety involved in the devotion of a greater amount of decoration to the highest classical colleges of the land than to the ordinary parish schools; but the present instance suggests either the possession of rich endowments, or the receipt of unusually large fees from pupils. The plan (woodcuts No. 65 and 66) differs in several particulars from any previously discussed, and marks the work of an architect not deficient in ideas. In point of depth, the whole of the building consists of single rooms; but in the front portion, and in one wing, the corridor is placed to the back and the rooms to the front; while in the other wing the

CHAP. VI.] GERMANY. 101

treatment is reversed, the corridor being placed to the exterior,

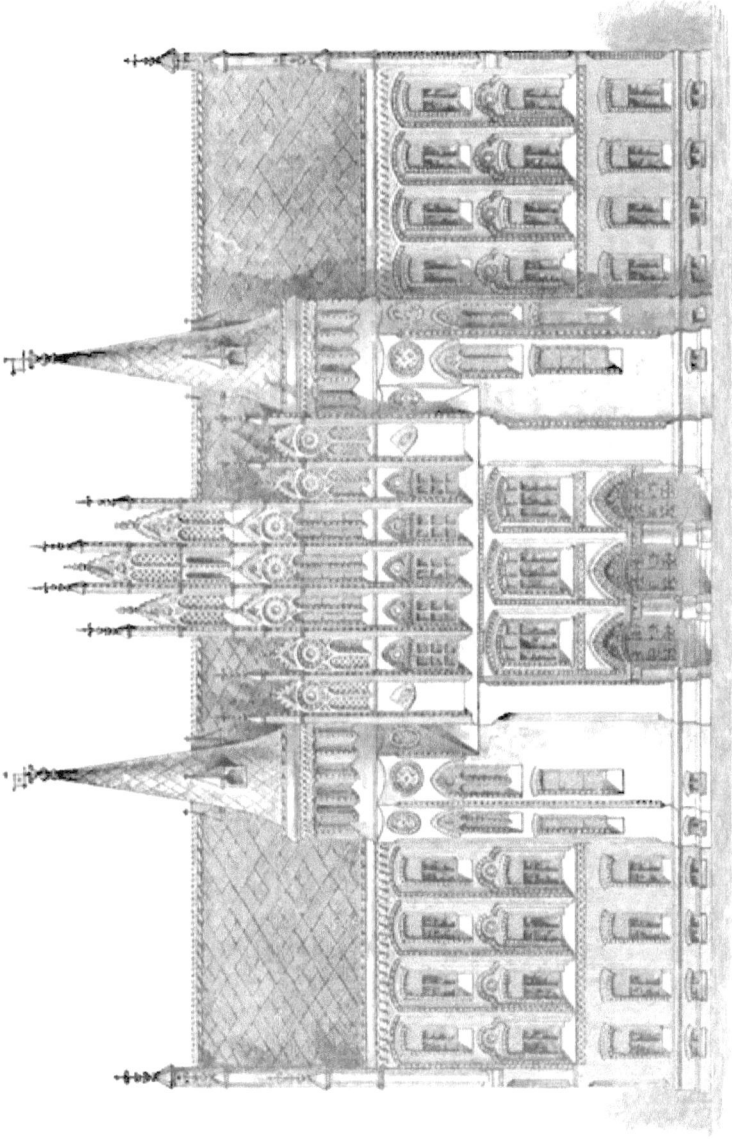

64.—GYMNASIUM OF S. ANDREW AT HILDESHEIM.

and the class-rooms towards the court. Evidently this unusual sacrifice of symmetry of plan has only arisen from the real or supposed necessity of lighting well the various rooms with particular aspects, and furnishes more proof—if more were needed—of the great importance attached to the question of good light by the skilful school-architect in Germany. Another feature, unusual in a German school intended for one sex, presents itself in the pro-

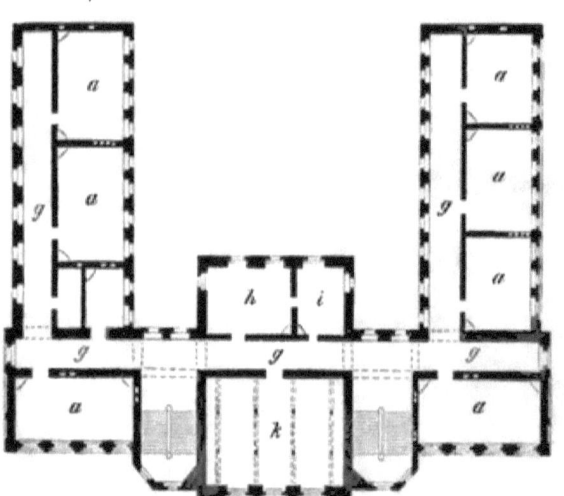

65.—PLAN OF FIRST FLOOR.

vision of two main staircases of considerable size, only separated in position by the width of the entrance hall. Ideas of symmetry may have had some share in this liberal provision, for the staircases both lead out of the same large entrance-hall, and are not approached separately from the outside. The two staircases were also, probably, designed to enable the class-rooms in the wings to be cleared more quickly and with less confusion than could have been effected by one staircase.

In these, the highest classical schools forming the preparation for the university, as well as in others, we mark the absence of certain accessories, always provided in an English school, and the presence of other provisions of a costly nature seldom, if ever, found among us. A little observation reveals the fact, that the minor features always present in an English and absent in a German school, are those on which personal cleanliness, comfort, and health in some measure depend. A very little more shews that the provisions lacking to English, and liberally supplied to German academical life, are those on which education itself, in its completer form, in no slight degree depends for excellence.

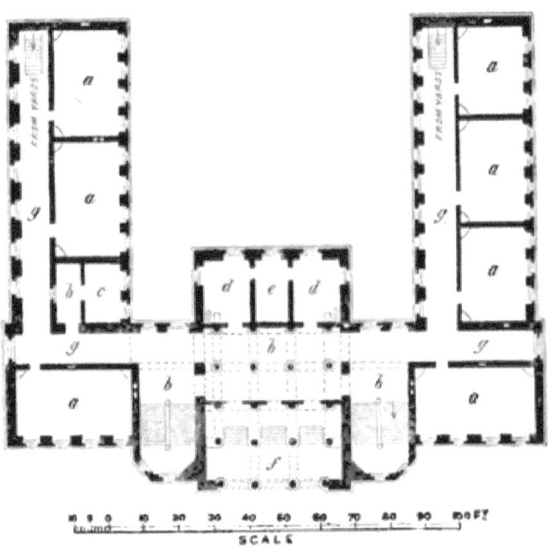

66.—GYMNASIUM OF S. ANDREW, HILDESHEIM. GROUND-FLOOR PLAN.

Reference.

a, a. Class Rooms.
b, b. Vestibule and Lobby.
c. Place of Confinement.
d. } Caretaker's Rooms.
e.
f. Entrance Hall.
g. Corridor.
h. Conference Room.
i. Directors' Room.
k. Library.

The gymnasium, for example, has no better provision as to lavatories and cap-rooms than the elementar-schule. Neither, in fact, has any. In both cases the single class-room forms a complete little school by itself, with separate teacher, separate appliances (of the more ordinary sort), and with the caps belonging to the boys composing the class hung round the walls of the room itself. The best authorities in Germany, including Dr. Wiese, all condemn this last practice, yet it is invariably followed. An admirable little contrivance, with sponge, water, and towel, is often found conveniently placed near the black-board, for the use of the teacher after drawing with chalk; but there is never any provision for the cleanliness of the children. They are expected to come to school clean, and to remain so. In speaking of this common deficiency to one learned director, and asking where, in his splendid school, the boys washed their hands, he burst into laughter, and, saying "There!" pointed to the pump in the middle of the courtyard. "Cleanliness is next to godliness," says the proverb, but the German schoolmaster usually holds the opinion that the observances connected with both, however commendable, should be kept away from his school premises.

The number, completeness, and value of the appliances and provisions of all kinds, bearing on the subject of education itself, present a remarkable contrast, and as they are found commonly in the gymnasium of a German town of very ordinary size, cannot fail to strike an Englishman with astonishment. At first sight, he supposes the series of classrooms to represent the only necessary part of the system. Then he finds the aula, already spoken of. Afterwards the separate director's-room, the conference-room for other teachers, the special large rooms provided for the extensive and costly natural history collections and physical apparatus, and for the ample library. When he, at length, discovers in the playground a large separate hall for no other purpose than the practice of gymnastics, he is apt to conclude that he is among an extravagant or thoughtless set of people. It is precisely because of the careful, painstaking and thorough manner in which the German works out, what he has to do, and because of his well-known habits of economy, that his adoption of such things after years of thought and experiment, together with his scale of expenditure in their production, deserves the serious attention of other nations.

Realschulen may be described as commercial schools for the middle classes. An exact translation of the term is difficult in few words, because in England there are no schools of quite the same nature. The German who wishes to give his son a superior education fitting him to enter one of the higher professions, or preparing him for the study of any of the five faculties of the University—Philology, Philosophy, Jurisprudence, Medicine, or Theology—would select, not a Realschule, but, a Gymnasium; not a Commercial, but, a Grammar school. The Realschule is rather the school forming the preparation for those professions which, in England, would remain after excluding law and medicine, although it is also applicable to mercantile pursuits. In describing it generally as a "commercial" school, let no one imagine that it bears the slightest resemblance to some of the commercial and scholastic impostures carried on in England under that name, and for which the nation is only responsible in that it has not yet interfered for their suppression. It is a school really excellent for its purpose and worthy of imitation. Another kind of description would call it the Grammar School for those not intended to pass to the University. The highest kind of Realschule affords, in fact, an education quite equal to that of the Gymnasium, and only different in giving a different direction, during the last stages of the course, to the plan of the studies. Architecturally, it becomes of little importance to trace the differences in the three kinds of Realschulen, because the mere length of the course cannot affect the shape or proportion of the rooms in the school-house, although it may easily alter their number. When the curriculum demands rooms of a totally different shape and use, as every Realschule does when compared with a Gymnasium or Gemeindeschule, then variation in principle of plan steps in. Or, when it is so shortened or altered that the higher classes become almost extinguished, modification of arrangement becomes necessary, and difference in general theory arises from the altered circumstances.

The Sophien Realschule and the Sophien Gymnasium in Berlin form a group of fine schools built on the same site and placed under one management for the more complete and perfect organization of the studies pursued, the pupils intended for the university being in the latter, and those intended for business

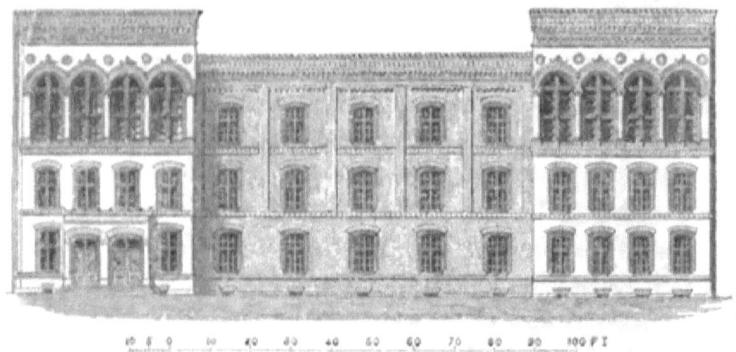

67.—SOPHIEN REALSCHULE, BERLIN.

or professions of the lower kind being in the former establish-

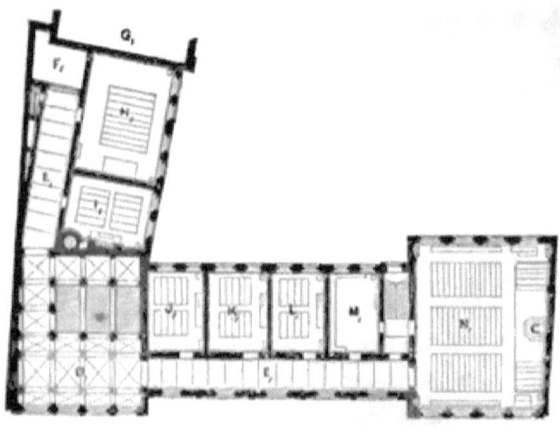

68.—PLAN OF SECOND FLOOR.

Reference.

D. 1. Landing.
E. 1. Corridors.
F. 1. Well for Light.
G. 1. Aula of adjoining Gymnasium.
H. 1. Singing Class.
I. 1. Secunda (B).
J. 1. Quinta (B).
K. 1. Quarta (B).
L. 1. Tertia (B).
M. 1. Ante Room.
N. Aula.

ment. The gymnasium usually has a course of six classes commencing with *sexta* as the lowest and terminating with *prima* as the highest. It is found that boys attending the Realschule never complete so long a course, but proceed to the practical study of some mercantile or professional pursuit on attaining a certain age, and the curriculum has therefore been shortened to enable them to obtain the most useful education for their purpose in a limited time. In the Sophien Realschule there are only five classes, and the highest *(prima)* is single while the lower are all

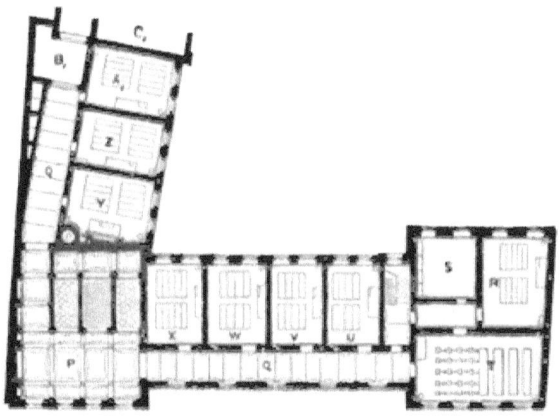

69.—PLAN OF FIRST FLOOR.

Reference.

P.	Landing.	W.	Quinta (A).
Q. Q.	Corridors.	X.	Quarta (A).
R.	Preparatory School (3rd Class).	Y.	Preparatory School (1st Class).
S.	Modelling Room.	Z.	Prima (A).
T.	Drawing Class.	A. 1.	Preparatory School (2nd Class).
U.	Secunda (A).	B. 1.	Well for Light.
V.	Tertia (A).	C. 1.	Adjoining Gymnasium.

in duplicate and have consequently two classrooms. The number of specimens of gymnasien already given render it unnecessary to reproduce plans of the Sophien Gymnasium, which immediately adjoins the Realschule. The scope and meaning of the latter is sufficiently explained by the woodcuts taken from the work of Herr Gerstenberg (Nos. 68, 69 and 70) which shew the position and use of every room in the building, together with the arrangement of the furniture. It will be seen that, on the ground floor, special rooms for the study of chemistry and physics not usually found in the gymnasium or grammar school are here provided in the Realschule or commercial school. The aula is at one end of the L shaped building. In point of external design (woodcut No. 67) the wing in which it occurs is balanced by the opposite wing containing the principal staircase. The singular bend given, on plan, to the projecting wing at the back appears to have arisen from a desire to bring the walls of the Realschule

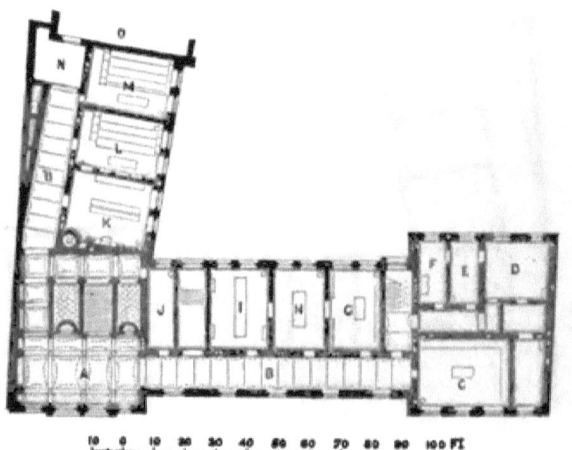

70.—SOPHIEN REALSCHULE, BERLIN. GROUND PLAN.

Reference.

A. Entrance Hall.
B. B. Corridors.
C. Natural History Collection.
D. Caretaker's Living Room.
E. ,, Bed Room.
F. ,, Kitchen.
G. Library.
H. Reading Room.
I. Conference Room.
J. Muniment Room.
K. Chemical Laboratory.
L. ,, Lecture Room.
M. Physical ,, ,,
N. Court for Light.
O. Adjoining Gymnasium.

and of the Gymnasium into one plane on the side towards the play-grounds. The buildings are executed of red brick, in the admirable manner (so far as workmanship is concerned) known to Berlin architects. Although at present used in school edifices with little of artistic skill or power, this re-introduction of brick as a material may perhaps lead to a more general use of that legitimate kind of building which honestly shows its material instead of hiding it under the disguise of plaster, of which the Prussians have such excellent precedents and suggestive examples in the ancient brickwork of Pomerania still remaining and lying close to their hand.

The exterior of the small Realschule at Halberstadt (woodcut No. 71), erected in 1865, does not appear to be finished in plaster

71.—ELEVATION OF REALSCHULE, HALBERSTADT.

or stucco. Dealing with Gothic forms, however simple, the German architect usually prefers a wall-face which shows its natural building-material without covering—seldom the case when a trabeated style is adopted, especially if good stone of large size be not easily obtainable. The portico is architecturally treated as a groined *porte-cochère*, its projection from the main wall being sufficient for the introduction, at each end, of an arch equal in size to one of the three in front, and for groining satisfactory in point of size if not in nature of material. The reason of this treatment, unusually imposing among school-buildings, is better seen by reference to the plan of the first floor (woodcut No. 72) where sufficient space for the aula is obtained by extending it over the portico. The somewhat rich Gothic effect sought in the design of the central portion of the façade is not supported harmoniously in the remainder, where the insertion of buttresses

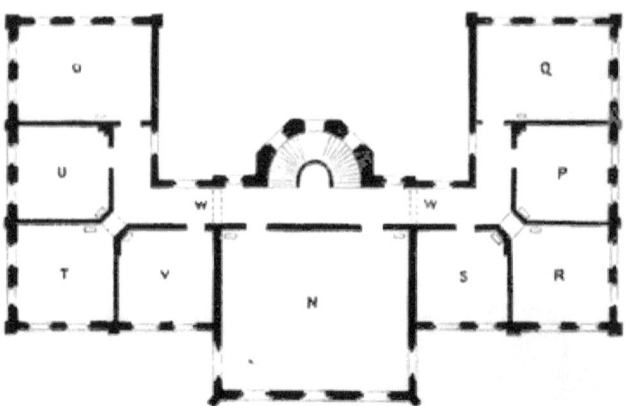

74.—PLAN OF FIRST FLOOR.

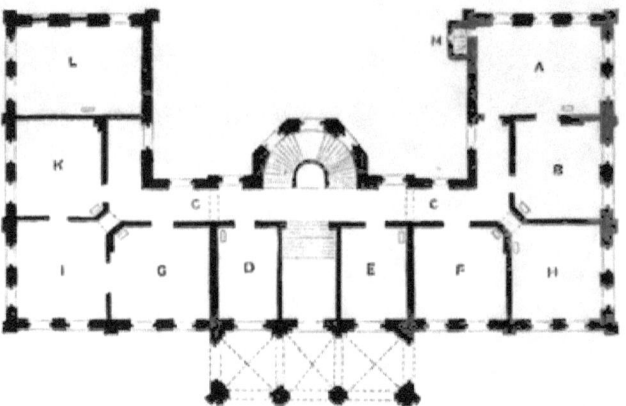

73.—REALSCHULE OF HALBERSTADT. GROUND PLAN.

Reference.

A. Lecture Room for Physics (with Staircase (M) leading to Chemical Laboratory in Basement).
B. Physical Apparatus.
C. C. } Corridors.
W. W. }
D. Directors' Room.
E. Conference Room and Library.
F. Sixth Class (1).
G. ,,   ,,   (2).
H. Fifth ,,
I. Second ,,

K. Third Class.
L. Fifth ,,
M. Small Staircase.
N. Aula.
O. Drawing Class.
P. First Class.
Q. Second ,,
R. Third ,, (1).
S. ,,   ,,   (2).
T. Fourth ,, (1).
U. ,,   ,,   (2).
V. Fifth ,, (1).

chiefly marks the style. The plan is pretty and symmetrical.

On the ground floor, (woodcut No. 73) after passing through the principal entrance or hall, the staircase lies opposite,—circular internally and semi-octagonal externally. Wings, running back some distance at right angles to the front, enclose the playground on two of its sides. As the arrangement does not comprise rooms on both sides of the corridor, it is not necessary for purposes of light to carry the latter through to the limits of the building. The rule of lighting the rooms from one side only is here departed from in the case of all the corner rooms, apparently for purposes of external architectural appearance.

The Realschule at Cologne (woodcut No. 74) is a fair example

74.—ELEVATION OF STADISCHE REALSCHULE, COLN (COLOGNE).

of an ordinary commercial school embodying present German ideas, both externally and internally. In consequence of its singular site it is L shaped on plan (Nos. 75, 76, and 77). The corridor, lighted from a yard, is on the long side, and the classrooms look towards the play-ground. The staircase is treated as an adjunct at one part of the wing. The aula has its usual place on the top floor of the building. There are nine classes each having two rooms. About one per cent. of the 600 scholars in attendance are free, having been drafted, by merit, from the

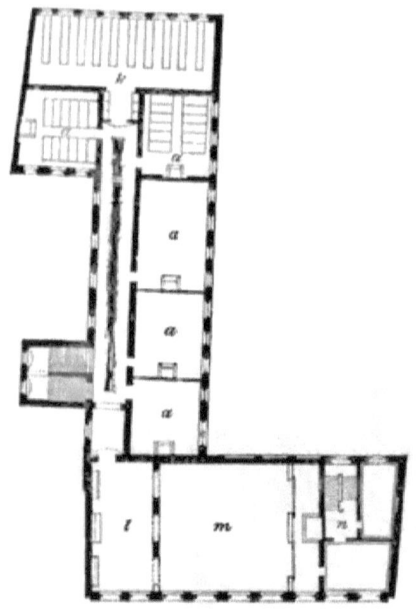

75.—PLAN OF SECOND FLOOR.

Reference.

a. a. Class Rooms.
k. Drawing Class.
l. } Aula (gallery over l).
m. }
n. Upper part of Director's House.

elementary schools. No difference of treatment is applied to the free scholars, as sometimes occurs in the elementary schools. They work and play together like the King's scholars with the other pupils in an English Grammar school. The warming is by six heating apparatus placed in the basement. The air is heated by contact with fire-brick in four cases, and with iron in two others. The air, thus warmed, is admitted to each room, on the side opposite to the windows, by an opening measuring $14\frac{1}{2}$ inches by $11\frac{1}{2}$ inches, made at a height of 4' 6" from the floor, and fitted with iron valves. It is supposed to rise on being brought among colder and heavier air, gradually to move across the room, to descend, and finally to make its exit by an opening of the same size, placed in the same wall but at some distance off,

76.—PLAN OF FIRST FLOOR.
Reference.

a. a. Class Rooms.
d. d. Directors' Rooms.
n″ ,, Salon.
e. e. Physical Science Collection.
f. Physical and Chemical Lecture Room.
g. Laboratory.
h. Private do.
i. Dispensary.

and at a height of 15 inches from the ground. The theory is that the *admission* should just clear the boys' heads when sitting at their work, and that the *extraction* should aim at removing the stratum of carbonic acid gas lying next the floor. The practice is to open the window when fresh air is desired. In Germany this occurs seldom. Like all those erected from the designs of Herr Raschdoff when Town Architect of Cologne, this schoolhouse shows skilful planning. It may not present in its exterior the highest artistic skill, yet is not without dignity. A study

I

of the second floor plan (woodcut No. 75) will shew why the two windows at each end of the principal front are differently treated from the five central windows forming the main light of

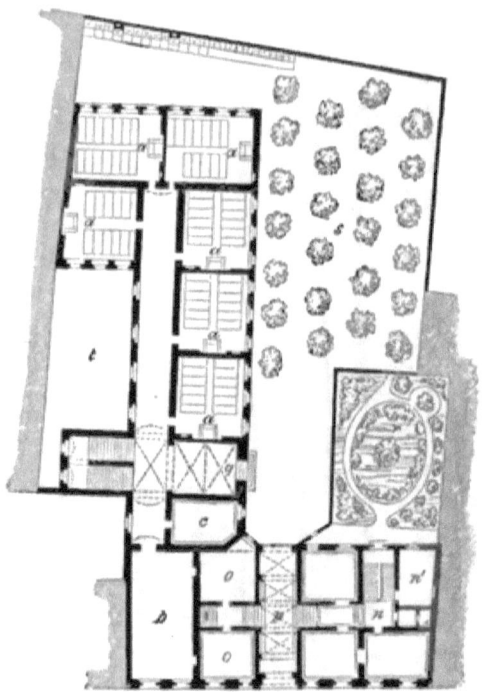

77.—STADISCHE REALSCHULE, COLOGNE. GROUND PLAN.

Reference.

a, a. Class Rooms.
b. Conference Room.
c. Teachers' Room.
n. Director's House (also above).
o, o. Caretaker's House.

p. Carriage Entrance.
q. School Entrance.
r. Director's Private Gardens.
s. Playground.
t. Court.

the aula. The building stands quite up to the street line with houses adjoining on both sides. It is executed in brick of different colours and, in style, is an adaptation of Rhenish mediæval architecture to modern ideas.

Among Realschulen, that of the third rank known as the Höhere Burgerschule has already been noticed as forming the favourite school for the education of the tradesman, the official of low degree—a numerous class in Germany—the small merchant, and others requiring a commercial rather than a classical education. In its building requirements it differs hardly at all from the gymnasium of corresponding size, but the rooms are sometimes differently used because of the change given to the direction of the studies. That at Wiesbaden erected in 1868 (woodcut No. 78) is a good specimen, published by Dr. Wiese. The elevation is a kind of pseudo-classic design with a central pedi-

78.—ELEVATION OF HÖHERE BÜRGERSCHULE, WIESBADEN.

ment, from which the addition of a belfry and pinnacles of gothic outline might have been omitted with advantage. The aula here also occupies its ordinary place in the principal front, and its internal position on the second floor is externally marked by five round-arched windows. The plan (woodcut No. 78) is simple, a long corridor, lighted chiefly at the ends, separating rooms placed on both sides. The entrance hall and grand staircase are extremely large and fine,—unusually so, even in Germany, for schools of this size. On the first floor (woodcut No. 79) is a large class-room devoted to instruction in singing, lecture-rooms and rooms for experiments in physical science, a natural history collection, and room for the rector.

I 2

It may be noted that the *pedell*, translated as "caretaker"

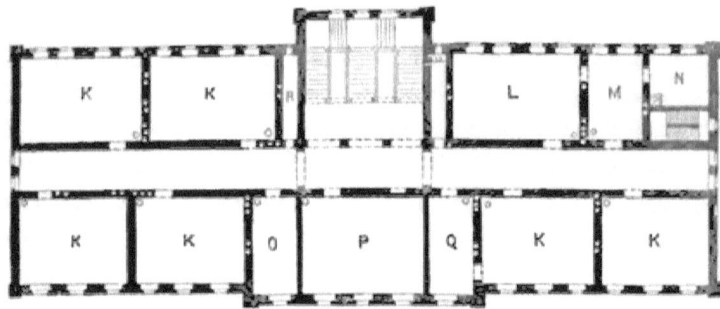

79.—PLAN OF FIRST FLOOR.

Reference.

K. K. Class Rooms.
L. Physical Lecture Rooms.
M. „ Apparatus.
N. Room for Experiments.

O. Rector's Room.
P. Song School.
Q. Natural History Collection.

really combines some of the functions of a proctor with those pertaining to the charge of the fabric. To him is committed the

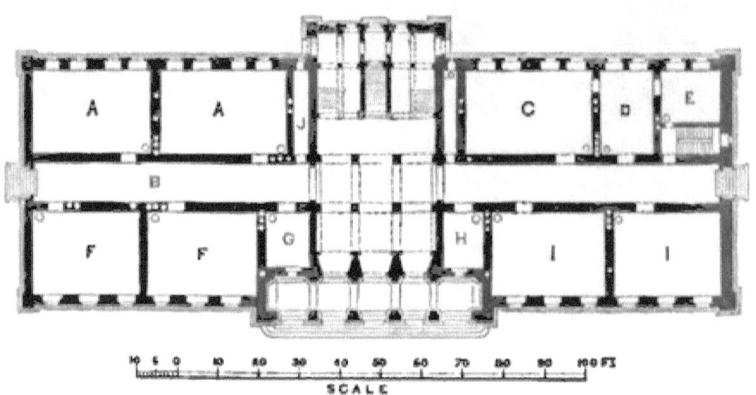

80.—HÖHERE BÜRGERSCHULE, WIESBADEN. GROUND PLAN.

Reference.

A. A. Class Rooms for Youngest Children.
B. Corridor of Preparatory School.
C. Lecture Room for Chemistry.
. Laboratory.
. Chemical Stores.

F. F. Class Rooms.
G. School Apparatus.
H. Porter (*Pedell*).
I. I. Reserve Class Rooms.
J. Fuel.

task of executing punishment on the refractory, and he is the gaoler who locks in the *carcer* the youths doomed to the reflection and meditation produced by solitary confinement.

Few provinces of Germany have so ancient a school history, or retain so many examples of old schools of well-tried reputation as Saxony. It is said that here the original foundation of schools extends as far back as the Carlovingian era, and the time of the Saxon emperors. The old foundations at Halle, commenced by Augustus Hermann Franké in 1696, and others, like Schulpforta, possess the greatest interest to the school student. Unfortunately, they must now be considered obsolete as specimens of scientific school planning, as well as standing somewhat apart from the scope of an enquiry leading chiefly among schools of humbler kind. By observing, rather, the improved provisions contained in recent specimens of national elementary and secondary schoolhouses, than the circumstances which formerly dictated the plans of the famous high schools, historical interest may be lost, but usefulness will be gained. The science of teaching or training, so far as it has attempted to control the forms of buildings and to engrave its history in the shape of brick or stone monuments distinctly expressing its opinions, is a thing of very recent date. The construction of schoolhouses specially and exactly fitted to particular methods of teaching, has never been undertaken in any country till modern times. Those which are newest ought, *primâ facie*, to be the best and most perfect in their arrangements, because of the knowledge and experience afforded by preceding examples. The old idea, now for ever exploded, was that any kind of building would do for a school, and that the shape of a barn was as good as any.

The early Saxon schools had, for some time before the Reformation, been in a state of decay, and the great movement against the excesses of the religious orders at the commencement of the sixteenth century, headed by Luther and Melancthon, exerted a beneficial influence on the school system of the time which was powerfully supported by the Saxon princes. Luther himself took considerable interest in the subject of the education of the rising youth, as proved by his letter to the Elector of Saxony in 1526, which foreshadows the policy of rate-supported schools, and the better utilization of endowments. He says, "Since we are all required, and especially the magistrates, above all other things to educate the youth who are born and are growing up among us, and to train them up in the fear of God and in the ways of virtue, it is needful that we have schools, and preachers, and pastors. If the parents will not reform, they must go their way to ruin; but if the young are neglected and left without education, it is the fault of the state; and the effect will be that the country will swarm with vile and lawless people, so that our safety, no less than the command of God, requireth us to foresee and ward off the evil. What is necessary to the well-being of a state should be supplied by those who enjoy the privilege of such state. Now nothing is more necessary than the training of those who are to come after us and bear rule. If the people are too poor to pay the expense, and are already burthened with taxes, then the monastic funds, which were originally given for such purposes, are to be employed in that way to relieve the people." Another point to which Luther attached great importance was the Christian teaching to be maintained in schools, which he urged in a remarkable address to the Common Councils of all the cities in Germany in 1524.

At this time an intense activity of thought was being created by several causes acting at once. The art of printing had been discovered in the middle of the preceding century. Books were beginning to be multiplied at prices enabling considerable numbers of the people (relatively) to avail themselves of study. The Bible, so long represented by a single copy chained to the lectern in the church, now became circulated freely.

The school regulation of 1528 known as "the Saxon school system"—the result of a visitation of the churches and schools of the land, in which more than thirty men had been employed a whole year—now became, in respect of education, the basis of administration from which sprang all future schools. Many of the older schools had been closed from time to time by state decrees, others had suffered in their buildings and ceased to exist, as the consequence of the ravages of the thirty years' war which commenced in 1618. Among those which have survived all change, is Schulpforta, which ranks, with Meissen and Grimma, as one of the most celebrated gymnasien in Saxony. Of this Mr. Matthew Arnold speaks as follows:* "No *Alumnat* in Prussia, or, indeed, in Germany, can compare with Schulpforta, which by its antiquity, its beauty, its wealth, its celebrity, is entitled to vie with the most renowned English schools. The Cistercian Abbey of S. Mary, Fforta, dates from 1137. It was secularised in 1540, and Duke Maurice of Saxony in 1543 established in its place, and endowed with its revenues, a Protestant school for 100 scholars. It stands near the Saal in the pleasant country of Prussian Saxony, and the venerable pile of buildings rising among its meadows, hills, and woods, is worthy of the motto borne of the arms of the old abbey: *Hier ist nichts anderes denn Gottes Haus, und hier die Pforte des Himmels.* (This is none other but the house of God, and this is the gate of Heaven. Gen. xxviii. 17.) It has a beautifully restored chapel, regular commemorative services, and a host of local usages. A Latin grace is sung in hall every day before dinner by the whole body of scholars. Every scholar has by ancient institution his *tutor*, every master his *famulus*. This is the German school where Latin verse has been most cultivated, and the *Musæ Portenses*, like those of Eton, have been published. The property is very large, and considerable Church patronage is attached to it. Up to 1815, when it passed into the possession of Prussia, the old Abbey estate had still its feudal privileges, and enjoyed full civil and criminal jurisdiction. The property is now entirely under the superintendence of the School-board of the province of Saxony, which appoints a procurator for it. The revenues of Pforta are from 8000*l*. to 9000*l*. a year ..... At Schulpforta

* "Schools Enquiry Commission. General Reports of Assistant Commissioners," vol. vi. Eyre and Spottiswoode.

they are very proud of their playing field, which is, indeed, with the wooded hill rising behind it, a pleasant place; but the games of English playing fields do not go on there; instead of goals or a cricket ground, one sees apparatus for gymnastics." The gymnasien of Saxony are partly boarding and partly day schools, some—like the three above mentioned—being reorganised, with the help of ecclesiastical funds, after the Reformation. These old boarding grammar schools are called, by Dr. Wimmer, the hearths of classical learning in Germany.

Coming to our own time, and to the people-schools, we find it generally admitted that the importance of an efficient and thoroughly national system of public instruction is as fully appreciated in Saxony as in any other province of Germany. The principle of compulsory primary instruction, first introduced to the country by the Elector John George in 1573, has prevailed ever since. Many of the existing buildings, hardly more than thirty to forty years old, are intended soon to be replaced by others planned on an improved system. Prussia has of late been making great progress in the improvement of her schoolhouses, and Saxony is not content to be left behind in the educational race. Having been, among German states, one of the earliest to convert the old ecclesiastical seminaries into schools of public character, and to provide for the training of teachers, she still maintains an honourable rivalry with other states. The present common schools are the result of the law of 1836. The various towns composing the kingdom are divided into districts, in each of which all the children of both sexes between the ages of six and fourteen must attend the school. No boy can, by law, be apprenticed to a trade before the proper age for leaving school. Every school district (*schul-bezirk*) must provide a schoolhouse and a dwelling for the teacher. One of the recent improvements in the *bezirk,* or district, system has been introduced to lessen the area of each, and thus obviate the necessity of compelling children to come long distances from their homes, which the large size of the old schools—covering a considerable area of population

—rendered unavoidable. A schoolhouse of this kind, visited in Chemnitz, was said to be capable of accommodating 2500 children at one time, while by arranging lessons and teachers at different times of the day, an area of population comprising 5000 children was actually embraced. The more distant of the children had thus to walk more than a mile from their homes. It is to remedy this evil and to supply a number of smaller buildings, conveniently placed in reference to the children's homes and containing each about 800 scholars, that the new buildings are intended. Through the kindness of Landbaumeister Canzler of Dresden, we are able to give specimen views and plans of school-

81.—BEZIRKSCHULE, DRESDEN.

houses, some being just completed, and some in progress of erection, but all representing the latest opinions on the subject in the Saxon capital.

The exterior (woodcut No. 81) of the Public School newly erected for the First District in Dresden, presents, with its German characteristics, some features borrowed from the banded brick and marble buildings of Italy. The *Bezirkschule* is not the lowest kind of Saxon elementary school, for here all pay some small fee—if only to the extent of a silver groschen in the week —and the architecture is rather higher in character. The gables with their flat pitch and ornamented surface, the square-turreted angles and suggestions of niches on the blank walls below, recall memories of many a church of southern Gothic style. The general group is straggling and lame from the dislocating effect of

82.—BEZIRKSCHULE, DRESDEN. PLAN OF SECOND FLOOR.

the two staircases, and the circumstance that the central block of the building has not sufficient height. A glance at the plans woodcuts Nos. 82, 83, and 84) shews the positions of the stair-

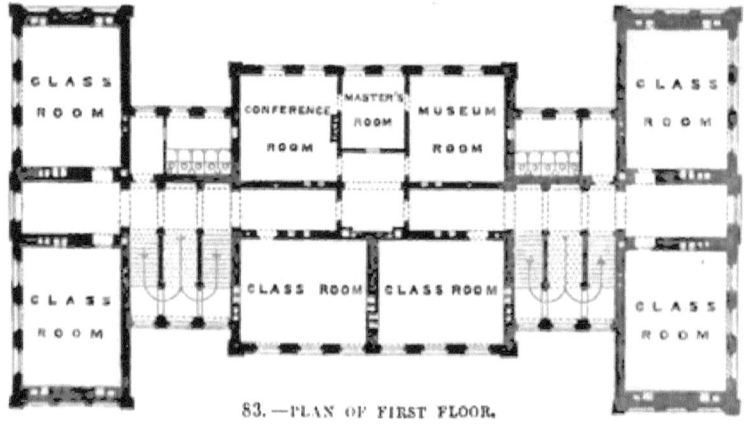

83.—PLAN OF FIRST FLOOR.

cases to have been dictated solely in reference to lighting the long corridor, which, otherwise, would have had no windows except at each extremity. But for this the staircases would have been better placed, both for compactness and use, at the outsides of the wings. The "grand" plan of staircase here followed is not quite satisfactory in its proportions, for if the middle flight be not too narrow, the two side-flights are clearly too wide. In dismissing the school, if both sides were filled with children descending, a block must ensue in the middle flight because of its inability to contain the two others. The presence within the building of the whole of the latrines for a large school is not a feature to be commended. In the present instance they are well contrived, but

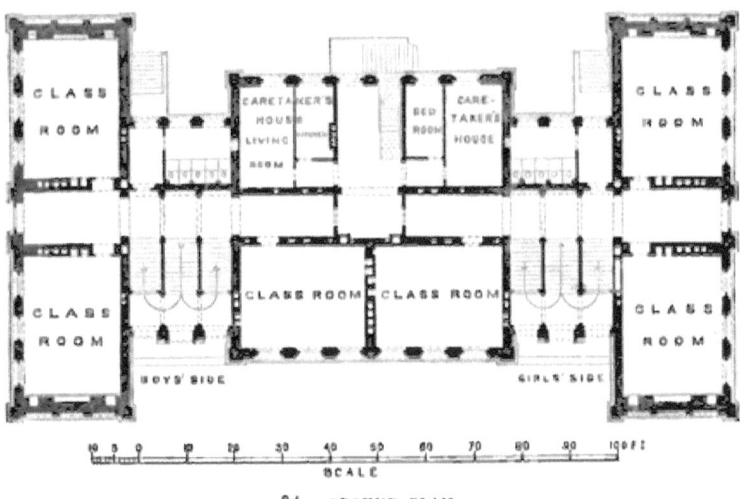

84.—GROUND PLAN.

they should have been placed *outside* in a separate building, as is now commonly advocated at Berlin. In every school visited in Germany the presence of this feature at the ends of the corridors, or elsewhere within the main walls, if existing, asserted itself in a manner which rendered enquiry superfluous. When we remember how carefully in that country the intense cold of winter is excluded by double windows, having every joint padded, and know that the process excludes also the fresh air required for the respiration of so many, the objection must be felt in a serious light. The air of the corridor first becomes impregnated with mephitic gases, and then that of the class-rooms occupied by the children. The building in question is arranged for six classes of boys, and six of girls, there is a reserve class-room for each sex, a conference room, a teacher's room, a museum or collection-room, a residence for the director, and another for the caretaker. The school seats are not shewn on the woodcuts, but these are of the long length kind, and the principle of lighting from the left side only of the children is here, as usual, rigidly maintained even to the detriment of the external appearance.

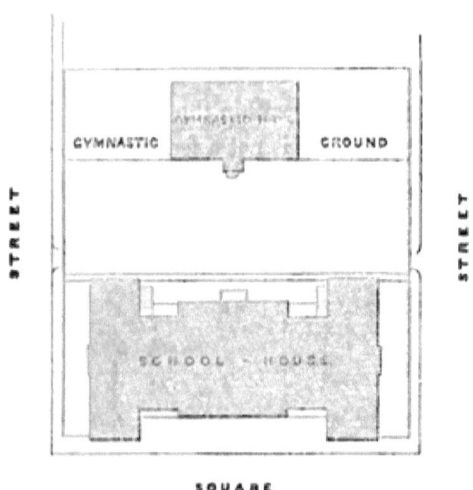

85.—BLOCK-PLAN OF BEZIRKSCHULE, DRESDEN.

There is no special hall for the examinations.

The block plan (woodcut No. 85) shews the general arrangement of the building on the land with the main frontage towards a large open space or square, the playground, and the turnhalle. Of the latter (woodcut No. 86), the elevation is also given with the object of shewing its character. It is no mere shed, but a building equal in permanent character to the schoolhouse itself.

86.—ELEVATION OF TURN-HALLE.

In the next example, the combination occurs of two kinds of schools in one building, viz., that of a parish with a district school, the two elementarschulen of Saxony. The principal front (woodcut No. 87) contains two inscriptions placed along the

87.—BEZIRKSCHULE AND GEMEINDESCHULE, DRESDEN.

parapet of the central block, one half marking the " VI. Bezirk-schule," the other " II. Gemeinde-schule," and a line struck down the middle would divide the building into two exactly equal and apparently similar establishments, only reversed in

88.—PLAN OF SECOND FLOOR.

position. The design attempts no particular style of architecture, but merely aims at avoiding a warehouse appearance by simple and good grouping of the several parts of the building. The roofs are always terminated by sloped ends, or "hips," instead of by gables. In the internal arrangements, the girls belonging to both schools occupy the central block (woodcuts Nos. 88, 89, 90). Adjoining this, and attached on each side, are two staircases, one for boys and the other for girls, having the courts and

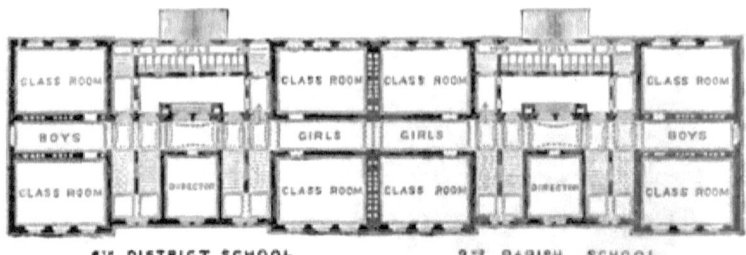

89.—PLAN OF FIRST FLOOR.

latrines in the rear. The two wings, forming the extremities of the building, contain the boys—one of the Bezirkschule, the other of the Gemeindeschule. The space occupied by the two entrances, the four staircases, and the latrines, is considerable. The Gemeindeschule differs from the other, where all pay, in

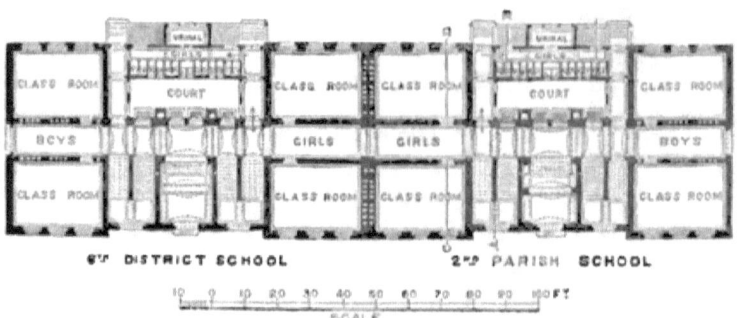

90.—SCHOOLS AT DRESDEN, GROUND PLAN.

having a lower scale of fees and in admitting a small percentage —the children of indigent parents—free. Two rooms for the director are shown on the plan of the first floor (woodcut No. 89) as being widely apart, but the schools are really worked together under one management. The arrangement of the latrines is better than in the preceding example and secures greater isolation from the school-house. This union of two kinds of elementary schools of slightly different class under the same roof is interesting, as marking an arrangement which has been found in Saxony desirable after long experience of public elementary schools and of the application of the principle of compulsory attendance.

The school system of Saxony is divided somewhat differently

91.—BÜRGERSCHULE, DRESDEN.

from that of Prussia. Bürgerschulen are designed to educate the children of parents in the middle ranks of society, and also those of the upper ranks where a public education is preferred. They are therefore separated into three kinds. Of the highest

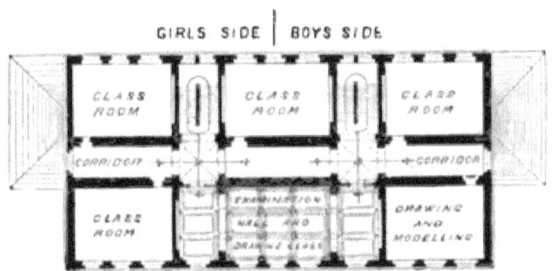

92.—PLAN OF SECOND FLOOR.

kind there are only 340, while of the second kind the number is 1300, and of the lowest kind 2000. The fifth Bürgerschule at Dresden presents in its architectural design (woodcut No. 91) nothing of higher character than is commonly found in a Prussian

Gemeindeschule, but forms a specimen of the class of school in question. The plan (woodcuts No. 92, 93, and 94) consists of six rooms on each floor separated either by the staircases or the corridor. In no case do two rooms immediately adjoin, and in

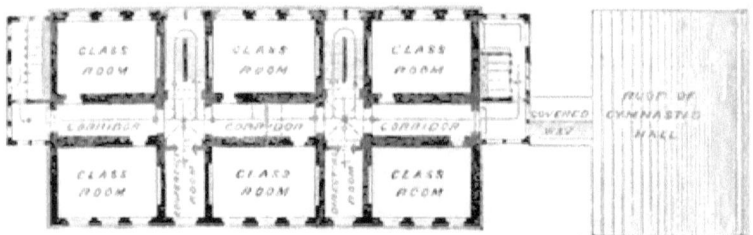

93.—PLAN OF FIRST FLOOR.

this respect the school presents a marked difference from those of Berlin, where the class-rooms lead out of one another. There are separate entrances and staircases for the sexes, and doors across the middle of the corridors on the two lower floors cut off one school from the other. On the first floor (woodcut No. 93)

94.—BÜRGERSCHULE, DRESDEN. GROUND PLAN.

a conference room and director's room are provided, but no rooms for special subjects except one on the second floor (woodcut No. 94) devoted to drawing and modelling. The aula is also on this floor and is used for a drawing class as well as for examinations. The latrines are placed outside the building in this case, but in immediate contiguity. The turn-halle has the rather unusual feature of a roofed connection with the schoolhouse, so that the children may reach it under cover.

The Royal Grammar School (*Königliches Gymnasium*) erected in 1871-2 at Chemnitz follows generally, in its internal plans, the arrangements common to the similar establishments in Prussia already described. With a splendid site in the suburbs, on the Kassberg Hill commanding the whole town, and surrounded by considerably more than an acre of garden and playgrounds, the building consists simply of a central block and two wings. Dr. Vögel, the rector,* tells us that, in accordance with the instructions of the Saxon ministry, the effort was not so much to create a monumental piece of architecture, as to contrive the most useful and perfect building for a gymnasium after studying the best recent examples, and to express this purpose worthily, yet simply, in the external appearance. The exterior (woodcut No. 45, page 69) is therefore plainer than some of the Prussian specimens, and the interior exhibits greater expenditure in decoration on those parts,—as the vestibule, staircase, and aula,—to which the public have access.

The second floor contains the aula, another large room or hall adjoining called the prayer-hall, capable of being used also as a combination-class for sixty, the library, the ante-room and librarian's room, also the Museum of Natural History. The prayer-hall is a most unusual feature. The "combination" class is chiefly used for singing-lessons.

The first floor accommodates two classes of forty pupils each, one of thirty, and one of twenty-four. It has also a combination-class-room for sixty, a room for physical science, a workroom, a laboratory for the Professor, a collection-room, a drawing-class-

---

* A "Director" may be a conductor or manager of anything—of a school as well as of anything else. The German School Director is simply the head-master. A "Rector" is so called because of his university diploma. The Rector Magnificus of a German—unlike a Scottish—University, can only be elected as the result of examination.

room, a room for plaster casts, and a singing-class-room. On this floor are, also, the *Carcer*, a small apparatus-room, and the study, reception-room and ante-room of the rector.

The ground floor provides four class-rooms each for forty scholars, two for thirty each, and one for twenty-four. Also the teachers'-room, the conference-room, and a small, or branch, library for the use of the pupils. Here also is the house for the caretaker or pedell, in Saxony called the Hausmann.

Without giving woodcuts of the plans of this excellent school, it will be seen from the foregoing that the ground and first floors are almost entirely occupied by class-rooms. The internal fittings, furniture, and school apparatus, here to be seen, are among the best and most carefully considered in Germany, and mark the steady progress taking place in educational appliances. In accordance with instructions from the Saxon Minister of Cultus and Public Instruction, the seats have been arranged on a principle since adopted by the School Board for London, viz., that of allowing each child to leave his place, without disturbing his neighbour. In the lower classes the seats are in pairs, and, in the higher, single, the backs and desks being fixed as in the Nicolai-schule at Leipzig. Thus, multiples of two give the class number in the one case, and, in the other, five scholars are placed in a line. In the vestibule is a clock provided with mechanism for sounding the bell of assembly. The basement has five heating apparatus, on what is known as the "Kelling" system. The cost of the whole building and fittings, exclusive of site, was 9450*l*.

The Turn-halle had not, in the spring of 1873, been erected. Its position is some distance behind the schoolhouse, and its internal size is intended to be 75ft. by 40ft. It is estimated to cost 1050*l*., and the windows are all to be placed 10ft. from the ground as an improvement on the ordinary plan. The principles, well known to be best in theory, are never all combined in the several parts of one schoolhouse, some examples containing the best features in one direction and others in another. It is well known that both in the class-room and turn-halle, the window-sills should

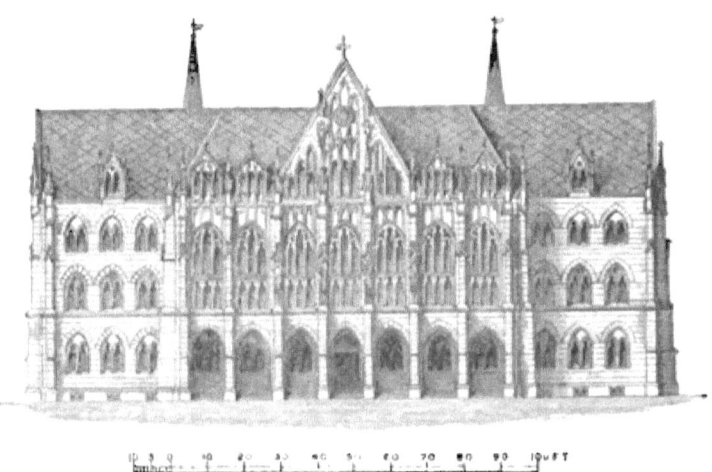

95.—THE KREUZSCHULE, DRESDEN.

be kept well up from the floor, but in practice they are often placed too low.

96.—TRANSVERSE SECTION.

The façade of the Kreuzschule at Dresden (woodcut No. 95)

is another specimen of the kind of gothic work produced by the German architect under favourable conditions. The lower story has a projecting arcade of seven pointed arches,—not treated as a *porte-cochère*, but only intended for pedestrians; which serves on the first floor (woodcut No. 98) to provide space for the large and handsome aula. The school is a gymnasium connected with the

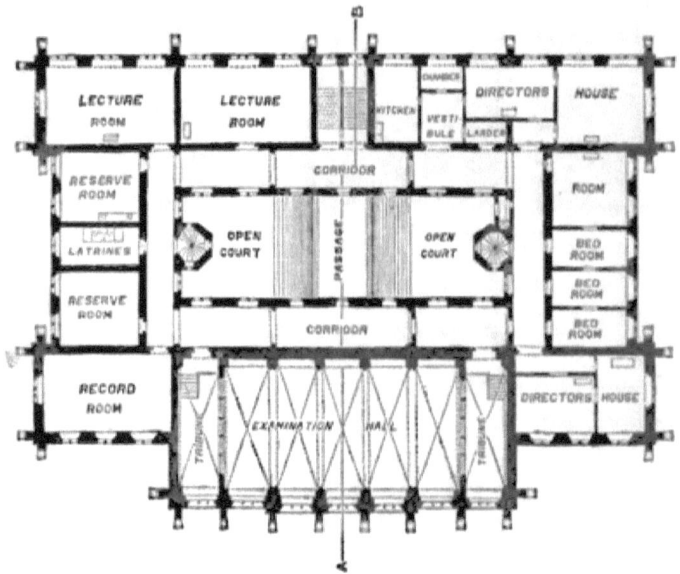

97.—PLAN OF SECOND FLOOR.

Kreuz-Kirche (Church of the Cross), and may possibly have more direct ecclesiastical objects than other *gymnasien*. The plan is sufficiently symmetrical, with a central court, crossed by the main staircase which reaches to the first floor in a straight line, and, for the upper stories, assumes smaller dimensions in a new position in the rear, as shown on the section (woodcut No. 96). All the floors are likewise reached by two small staircases placed in octagonal turrets on each side of the court. In its accommodation and objects it is similar to others already given. The various rooms are all set forth on the plans (wood-cuts Nos. 97, 98, and 99), and their uses can so easily be seen as to render description unnecessary. The latrines are here all placed unsatisfactorily inside the building, on each floor. The site is public, having a fine frontage to one of the principal streets of the capital.

The Gewerbeschulen, Trade and Practical Schools, deserve a study which the limits of this essay forbid. In England, trades

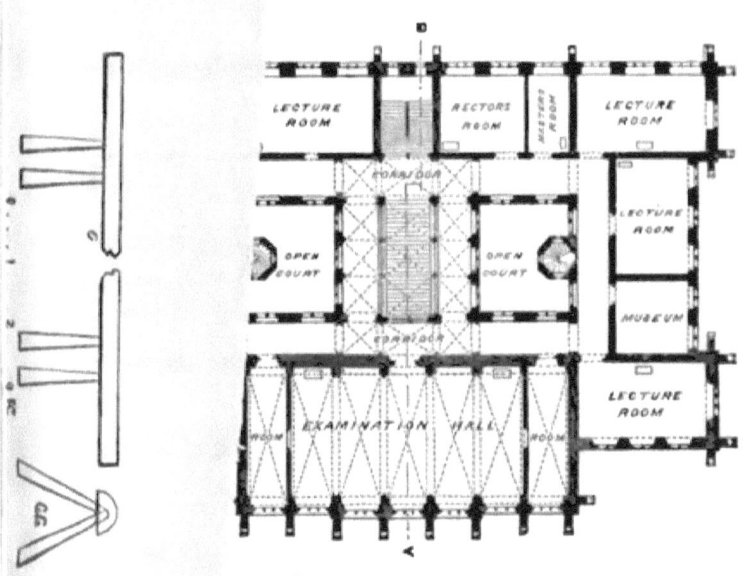

98.—PLAN OF FIRST FLOOR.

nt in the shops, by the sweat of the brow during
:iceship. And the success attendant on some
glish manufacture, may be fairly held as an argument against the introduction of similar schools in England. We have, for example, no schools like the Höhere Webschule at Chemnitz, a technical school for weaving, where the trade is taught theoretically and practically by skilled masters. In a large manufacturing town of the kind, it becomes of some consequence to secure the highest excellence in the various trades carried on. The school, accordingly, provides costly weaving machines—each one different—and endeavours to secure always a

specimen of the last improvement. Pattern drawing, designing, and the construction of weaving machines, are taught, as well as the chemistry of colours and materials used for weaving. Every effort is made to afford all the facilities for enabling the inhabitants of the town to compete successfully with the similar products of other parts of Europe. And, in visiting establish-

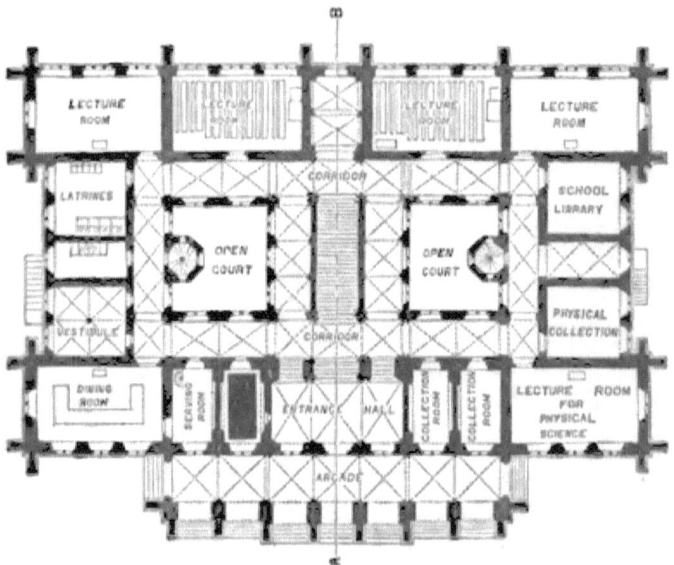

99.—KREUZSCHULE, DRESDEN. GROUND PLAN.

ments of the kind, it is difficult to shake off the impression that with such immense educational advantages, the foreigner must after all be a comparatively stupid fellow in not having driven the Englishman out of the field long ago. The only possible explanation is that in the numerous works always in progress in his own country, and in the large amount of manufactures carried on, he has schools second to none in value, which he can visit by the facilities afforded by shorter distances and swifter trains. This, be it noted, does not apply to the workman. The cost is beyond his means.

Polytechnic Institutes are as much beyond the scope of a work devoted to school-planning as Universities. Yet, in a material age, it is impossible to avoid some notice of these final schools of pure science and practical application, forming the apex of the course open to the professional student. Always carried on in large establishments, they contain ample libraries, plaster-cast models and drawing-copies, models of machines carefully executed to scale, large collections of botany, mineralogy, zoology, paleontology, &c., spacious lecture-halls, laboratories, analytical and technical, and other minor provisions. That at Zurich is especially excellent in the methods adopted for instruction, its remarkably complete and well-arranged laboratories, and other points. The Munich Polytechnic Institute bears out the above general description in its spacious building, its ample and well-lighted class-rooms and amphitheatres, its chemical laboratories, its drawing and model collection, its models of machines, constructions, bridges, and works of art. The Dresden establishment, besides being a technical school for training engineers for the public service and for general practice, partakes of the character of a normal school by undertaking the education of professors intended for teaching arts and sciences. This is considered an important advantage aimed to provide a more suitable education for professors than that afforded by the University. In a separate building the Modelling and Drawing school is placed, with a good collection of plaster models. The Polytechnic Institute at Berlin differs somewhat from others, and is called the Royal Trade Institute (*Königliches Gewerbe Institut*). It is divided into three sections: viz., one for mechanicians, one for chemists and metallurgists, and one for shipbuilders. Workshops and chemical laboratories are provided, with other appliances for experiments.

The example chosen for illustration is the Polytechnic School at Aix-la-Chapelle. Without the same advantages of commanding position enjoyed by its compeer of Zurich, its external architecture (woodcut No. 100), in the style of the Italian revival, is better. The basement is built of stone from the Siebengebirge mountains, which help to produce the picturesqueness for which

100.—POLYTECHNIC SCHOOL AT AIX-LA-CHAPELLE.

the scenery of the Rhine at Bonn is famous. The ground floor is of dressed sandstone from Trier. The upper stories of the building, consisting of columnar architecture, are of tuffstein, a common and excellent building-stone used in the Rhine provinces. On each corner of the projecting wings at both extremities are carved eagles. The four allegorical figures surmounting the central block are 9 feet high, and represent

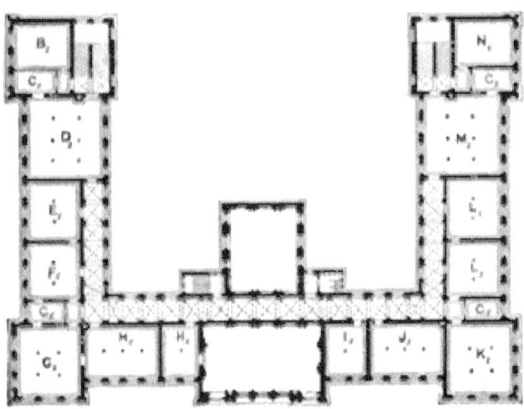

101.—PLAN OF SECOND FLOOR.

Reference.

B. 1. Instruments and Maps for Geodesy.
C. 1. Teachers' Rooms.
D. 1. Lecture Hall.
E. 1. Drawing Class Room for building Roads and Railways.
F. 1. Collection for Geodesy.
G. 1. Mineralogical Collection.
H. 1. Lecture Room and Collection for Engineer's School.
I. 1. Models for Building and Descriptive Geometry.
J. 1. Lecture Room for Building Science.
K. 1. Drawing Class for Descriptive Geometry.
L. 1. Drawing Classes for Building Construction and Hydraulic Engineering.
M. 1. Lecture Hall.
N. Drawing Class for Building Machines and Technology.

the town of Aix-la-Chapelle, the Rhine Provinces, the Province of Westphalia, and Borussia. The central group consists of a figure of Minerva as presiding goddess, with eagles. The interior is handsomely decorated in the more public portions. The aula, for example, is rich in design. The wall opposite the windows is ornamented with niches having, in the tympana of the semicircular arches, medallions of celebrated men (*Coryphées* in technical science), as Buch, von Humboldt, Klaproth, Mitscherlich,

Liebnitz, Gauss, Redtenbacher, Borsig, Hagen, von Decher, Beuth, Werner, von Liebig, Bunsen, Doré, Magnus, Karmarsch, Bessel, Schinkel, and Mellin. There are also life-size portraits of H.M. the Emperor and the Crown Prince.

The plans (woodcuts Nos. 101, 102, 103) shew the purposes to which the rooms on the three principal floors are devoted. In

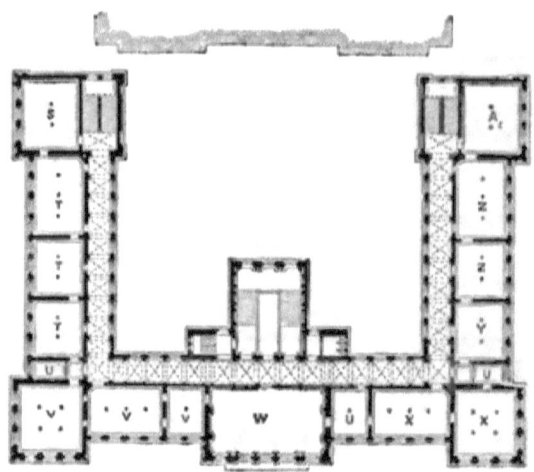

102.—POLYTECHNIC, AIX-LA-CHAPELLE. PLAN OF FIRST FLOOR.

Reference.

S.	Engineering Models.	U.	Reading Room.
T. T.	Drawing Class Rooms for Mechanical Engineering.	X. X.	Library.
		Y.	Building Models.
U. U.	Teachers' Rooms.	Z. Z.	Drawing Class Rooms for Bridges and higher Building Construction.
V. V.	Drawing Class Rooms and Collection for Practical Building.	A. 1.	Maps and Drawings for Practical Geometry.
W.	Aula.		

the basement, no plan of which is given, are provided rooms of useful kind to render the establishment complete. Here are the practical workshops for iron and wood, the smithies, forges, tool collections, shops for the wooden patterns from which forms are to be cast in iron, shops for modelling in clay and plaster, workshops for physical science and for articles requiring even temperature, models for mechanical engineering, and material stores, boiler and engine-room. In this case, the Caretaker's house is placed in the basement. The method of warming is remarkable, being effected by hot-water pipes carried from six different apparatus, each of which has three distinct systems capable of being shut off separately. The warming-power thus consists of a series of no less than eighteen separate systems each under control. The cost of producing heat for the whole of this large establishment is said to be only 1*l.* per day.

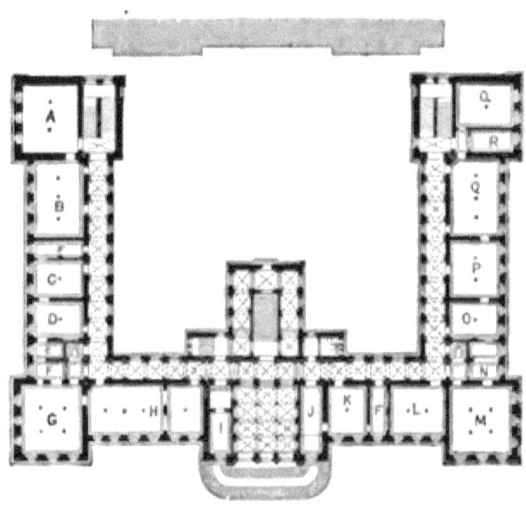

103.—GROUND FLOOR.

Reference.

A. Museum of Technical Appliances.
B. Technological Collection.
F. Ante-Room.
C. Professor's Laboratory.
D. Laboratory for Physical Science.
E. Camera Obscura.
F. Preparation Room.
G. Lecture Room for Physical Science.
H. Physical Collection.
I. Caretaker.
J. Payment Room (for Fees, &c.).
K. Director's Room.
L. Conference and Reading Room.
M. Lecture Room for Engineering.
N. Tool Room.
O. Collection Room.
P. Machinery Room.
Q. Q. Freehand Drawing Rooms.
R. Room for Plaster Casts.

In connection with, and immediately adjoining the Polytechnic School, but in a separate building, is the Chemistry School. Like the Zurich example, this is divided into two distinct parts or sections, one devoted to analytical, the other to technical chemistry, standing one to the east the other to the west. The uses of the various laboratories, lecture-rooms, &c., are all clearly set forth on the plans (woodcuts Nos. 104, 105, 106) and the

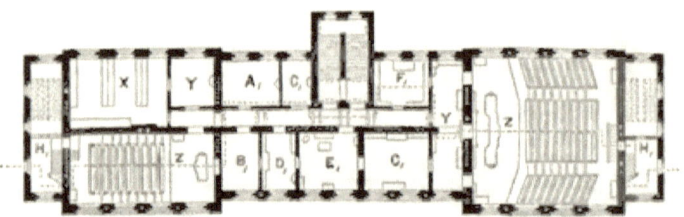

104.—PLAN OF FIRST FLOOR.
Reference.

X. Technological Collection.
Y. Y. Preparation Rooms.
Z. Z. Lecture Halls.
A. Gas Room.
B. Library
C. Scullery.
D. Private Laboratory } for Professor.
E. Reception Room
F. Collection of Preparations.
G. Apparatus Collection.

accompanying references. The room for sulphuretted hydrogen is carefully shut off from the rest of the building, so as to avoid any damaging influence upon other preparations or chemicals.

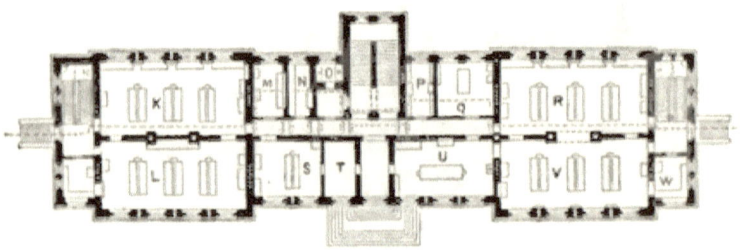

105.—CHEMISTRY SCHOOL, AIX-LA-CHAPELLE. GROUND FLOOR.
Reference.

K. Laboratory for Quantitative Technical Analysis.
L. Do. Technical Experiments in Quantitative Analysis.
M. Scale Room.
N. Chemical Preparations.
O. Water Closet.
P. Store Room.
Q. Scale Room.
R. Laboratory for Quantitative Mineralogical Analysis.
S. Private Laboratory.
T. Reception Room.
U. Operation Room.
V. Laboratory for Technical Experiments in Quantitative Mineralogical Analysis.
W. Condensation Room.

In calling attention to the methods adopted abroad for affording sound technical instruction, let it not be imagined that these can supersede the practical teaching of the workshop. The

English engineer, receiving into his establishment young men from the higher German schools, is apt to laugh at the apparent waste of time which has been incurred in acquiring knowledge of mere theory. To him practical experience is everything. The younger race of engineers (to take one profession as a sample)

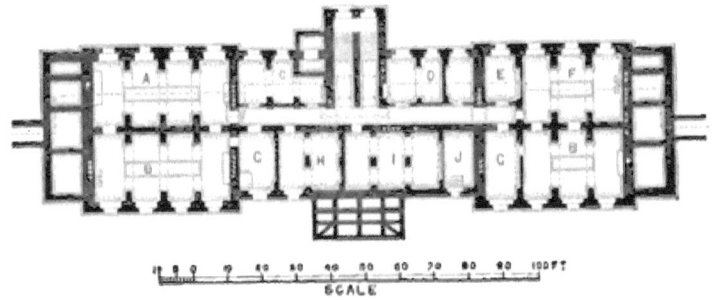

106.—BASEMENT FLOOR.
Reference.

A. Laboratory for Large Distillations.
B. B. Rooms for Experiments requiring Melting and High Firing.
C. Crystallization Room.
D. Store Rooms for Acid Preparations.
E. Gasometer for Oxygen.
F. Laboratory for Smaller Distillations.
G. G. Fuel.
H. Store Room.
I. Analyst for Law Cases.
J. Room for Sulphuretted Hydrogen, with External Door.

know that to rise to any position of eminence, *theory and practice must go together.*

Improvements in school hygiene appear to be contemplated in at least some parts of Germany. The ministry of Wurtemberg has recently published a decree on the special sanitary regulations to be observed in schools. Each school must have a playground and gymnastic apparatus, and separate rooms for the two sexes.

The length of the schoolroom must not exceed 12 metres, and the height must be at least 3·4 metres. Great care is to be observed that the paint used for walls and furniture does not contain any poisonous matter. Earthenware stoves are recommended in preference to iron. In every school there is to be a separate room for the teacher, and in larger schools a room for scientific collections. Every school must have a wash-bowl and towel, and a cloak-room. Rooms, staircases and entries must be swept daily, and scoured at least four times a year.

There must be good light and ventilation.

The temperature of the schoolroom is never to be less than about 62 degrees. If in summer the thermometer shews 77 degrees in the shade during the forenoon, there is to be no school in the afternoon. The scholars should be afforded an opportunity of changing their position by letting them sit and stand alternately; and the means of punishing is to be a thin switch, which must have the prescribed length of half a metre!

Some aspects of German education—whether directly traceable to the faults of a system or not—reveal great deficiencies which it is impossible to omit from this record. The exclusion of Infant Schools from the list of those forming the national system, and the too rigid uniformity prevailing in the latter, have already been mentioned. The now universal absence of the mixed system in all the great towns, and its careful omission in principle from all new schools except in rural districts where necessity compels, should be remembered, as contrasted with the great importance attached to its general use in America. It is to be feared that, in respect of this feature, the new system of elementary schools in London may have received too direct an influence from the German model. Some other features shew far greater defects. The condition and sphere of employment appropriated to women in Germany must be declared one of the greatest reproaches of the country. Their higher education is not what it ought to be. There is no lack of girls' schools of good quality, but the employment of men as the sole teachers instead of women, universal in Prussia (but not universal in Saxony and Southern Germany), is a serious evil. The spectacle of forty or fifty blooming young women taught as a class like so any children by a young man of thirty, suggests the thought that indolence is likely to take the place of discipline, and sentiment that of solid instruction. Teachers experienced in the schools are not slow to admit it. It would be easy to understand the reason, if the mixed system were in force, and the usual objection were urged that boys cannot be controlled except by men; but, with the invariable girls' and boys' departments separately arranged, it is inexplicable except on the suppositions either that a sufficient supply of women as teachers is not to be obtained, or, that a woman is regarded by the nation as an inferior being. In the lower ranks of life in Germany, women reach a hardship of servitude and depth of degradation very rare in England. No Englishman should forget that in the agricultural districts of Northumberland, for example, women are commonly found driving carts, working in the fields, and doing the ordinary duties of farm labourers, while instances have not been wanting of their working as "bank-men" at the mouth of a coal-pit. In Germany, the thorough character of the education, and the compulsory attendance of every child at school, do not prevent women from becoming the bricklayer's assistant on the construction of buildings, and carrying their loads like Irish labourers up the series of barred inclined planes which there do duty as ladders, from the bottom to the top of the scaffold. This feature is universal in Germany and Austria, and may be explained by the necessity of employing on military service the men who would otherwise act as labourers. The explanation forms no real excuse. The use of women as bricklayers' labourers, and their non-employment as teachers, are blots of the most serious kind on the fair face of German education as judged by its civilizing results. Separate schools for girls are on the increase, and the higher education of women receives more attention than formerly. The Victoria School for girls in Berlin, accommodates 950 pupils in eighteen classes, and has special rooms devoted to drawing, physics, &c. The presence of a fine aula may be taken, here also, to denote the importance of the principal examinations. Some have pronounced this to be the finest girls' school in the world. The Luisen Schule, just completed, is another Berlin specimen. The school was held in fairly good premises, which were not considered sufficiently good by the authorities; but which were certainly superior to many English schools receiving grants from the Privy Council. The new building is very complete, built of brick with the added red facing, and warmed and ventilated artificially. These two specimens are not alone, and show a real desire to afford facilities for the education of women. Another step is wanted, that of the introduction of teachers of the same sex. This is growing in some parts of the country, especially in Saxony. England has absolutely no school buildings for girls of the same high type as these two. Neither has she any girls' schools taught entirely by men.

While thus pointing out the defects visible to any ordinary observer, the general aspect of German education cannot fail to teach a useful lesson to those willing to learn. In the connecting of one school with another, link by link, so that the national system forms one complete chain having the common schools (free to the necessitous) at one end, and the Polytechnic or University at the other, the whole being guarded and ensured by the State itself and rendered accessible to the whole population, we have a principle applicable with advantage to our own land. In the care taken to produce buildings exactly fitted to methods of education, and in the definiteness and precise meaning observable in every part of a schoolhouse, the English architect may learn how, in his own sphere, he may greatly conduce to the development of national education by facilitating the work of tuition. In the universal treatment of education as a great question not to be dealt with in any trifling or unworthy manner, and in the dignity and importance always allowed to the buildings in which it is carried on, the school-boards and school-promoters of the United Kingdom may find new light.

The *Elementary* schools of Germany furnish information chiefly as to scale of cost and details of lesser importance. Their general plan and idea does not apply to similar schools here, conducted on another system of training. From those of higher class much more may be learned.

A consideration, even of the few specimens we have been able to give, must shew conclusively that, for *Secondary, Trade, and Technical* schools, Germany presents the finest models in the world. Those burdened with any lingering doubts on the subject may have them removed by visiting the country. The Teuton has been hard at the work of developing a complete educational system during long years while the Briton has been occupied in ruling the waves. Having been long used to good elementary education, he has gradually come to see the value of something more. The demand for higher education in Germany is now so great that the Secondary Schools cannot contain all who wish to enter them. It is thus that—there at least—the lower education produces a desire for the higher. The middle classes fully understand the value of knowledge and cultivation of mind—sometimes perhaps too much in its ultimate pecuniary aspect—yet often for its own sake and for the moral elevation it aims to produce. The noble is not necessarily the best informed, and there are not wanting those who class him among the ignorant as compared with the educated. German school buildings may not present, in all their details, absolute perfection. They are, without doubt, the foundation of good planning, and the source from which we may best study the results of definite intention and scientific meaning in its application to Secondary Schools in England.

In concluding an all too-imperfect notice of the school-buildings of Germany, the words of Mr. Matthew Arnold, than whom no Englishman writes with greater accuracy and force on educational subjects, apply with peculiar appropriateness. He says, "What I admire in Germany is, that while there too, industrialism—that great modern power—is making at Berlin, and Leipsic, and Elberfeld, the most successful and rapid progress, the idea of culture, culture of the only true sort, is there a living power also. Petty towns have a university whose teaching is famous through Europe, and the King of Prussia* and Count Bismarck

* Published in 1868, before the unification of the German Empire.

resist the loss of a great *savant* from Prussia as they would resist a political check. If true culture becomes at last a civilizing power in the world, and is not overlaid by fanaticism, by industrialism, or by frivolous pleasure-seeking, it will be to the faith and zeal of this homely and much-ridiculed German people that the great result will be mainly owing."

107.—GYMNASIUM AT COTTBUS.

CHAPTER VII.

AUSTRIA.

Vienna Primary Schools—Theory of Austrian Education—The Handel's Academy at Vienna—The Imperial Gymnasium—Stadische Schools—Polytechnic Schools.

LIKE that of Germany, the capital of Austria has spread far beyond her ancient boundaries. The line of fortified wall which formerly surrounded old Vienna and separated the city from the miles of suburb beyond, is only traceable by the splendid boulevard now occupying its site and called the Ringstrasse. A double line of tramways,—here exactly suitable, if suitable in any public street,—is laid along the whole line. One element of suitability is, the provision of a width sufficient to allow two carriages to pass each other without being forced on to the iron ways to the detriment of springs and the discomfort of occupants: another is that the Ringstrasse forms the leading artery of communication, spacious, airy, modern, and devoted to pleasure, as contrasted with the streets of the old city, narrow, tortuous, picturesque, and devoted to shops and commerce. The city is beautifully placed on the little river Wien, from which it derives its name, within a short distance of the mighty and swift-rolling Danube, and with a background of purple mountains simply glorious to behold. Contrasted with its northern German rival, located in a howling wilderness of sand, flat and bleak, the site of Vienna seems perfect. Population has considerably increased of late years, but not at the same enormous rate which has characterised the increase of Berlin. New schools have been built, and others are in contemplation, but the circumstances of the case have not required so large a number as were found necessary in the northern city. The municipality voted last year (1873) a sum of 7,000,000 florins to be applied to primary schools of various kinds, and appear to do their work most efficiently in furnishing means for providing buildings and teachers. Some of their more recent specimens are satisfactory enough. In point of plan they present no new features, or indeed anything different in principle from those already given so profusely under the head of "Germany." Elementary schools are here divided into three, and Burgher Schools into four kinds, marking a local difference in system of teaching, or of arrangement. One of the former, recently built from the designs of Herr Hansen, on a site not far from the Polytechnic, forms a good example of the central court plan. Although erected at the expense of the Protestant Commune, it is attended by Jews, Greeks, Roman Catholics and Protestants. The court itself is 30 feet square, and extends upwards through the three stories of the building, terminating in a glass roof. Surrounding this on all the four sides are corridors 9 feet wide on every floor, each corridor leading to the class-rooms on one side, and having an open arcade of three arches on the other. The ground floor of the court is used for play alternately by boys and girls. The building accommodates 1,100 children, and the large room provided for instruction in drawing is used, like the covered playground, alternately. Latrines, and also drinking fountains, are provided on each floor. These Communal Schools, of which there are several new specimens, are among the best in Vienna, if we except a few special examples.

The theory of Austrian education has been ingeniously set forth by a diagram (woodcut No. 108), which originally accompanied the Report of Baron Helfert in 1862, and which explains it better than many pages of type and ink. In the form of a genealogical tree, the course of the student from the Primary Schools is traced up the two sides, and shown as terminating with the Polytechnic in the one case and with the University in the other. One line indicates the progressive education provided for

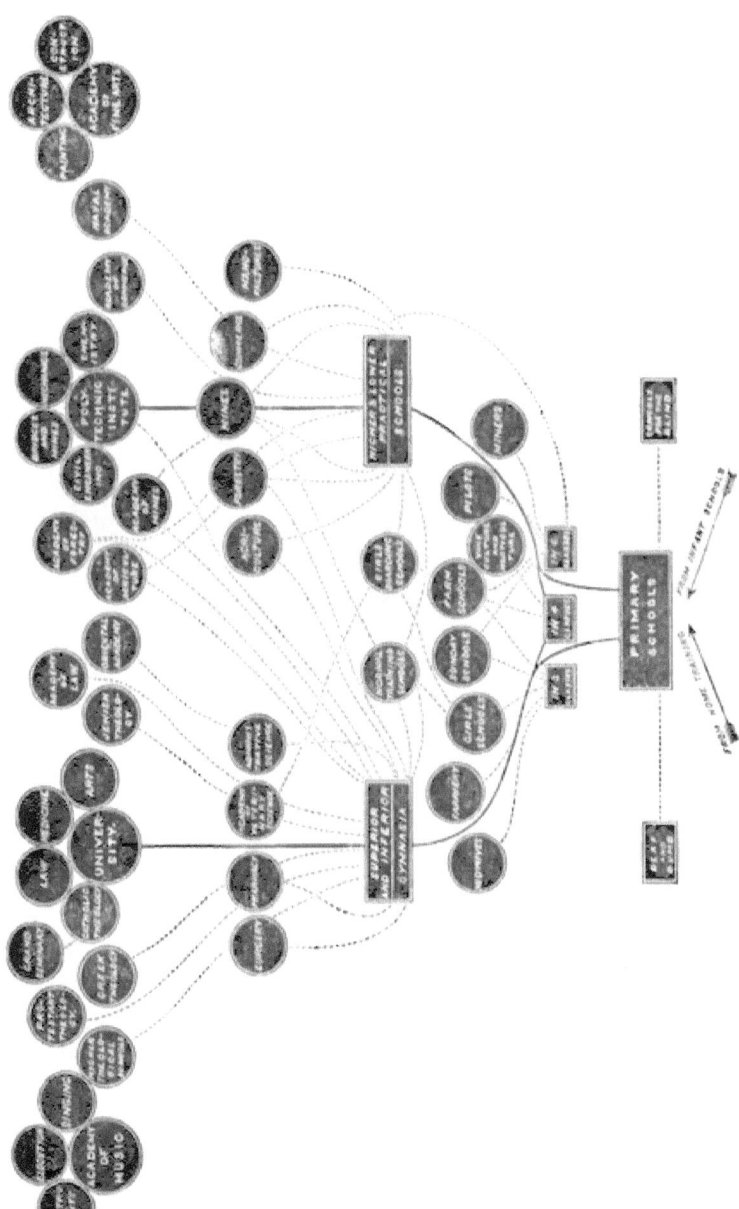

108.—THE THEORY OF AUSTRIAN EDUCATION.

those intending to pursue any of the higher professions, and is a feature which deserves to be carefully pondered in England. The other marks the course of the classical education which, among us, would lead commonly to the Church or the Bar, or would be sought by the wealthier classes who desire higher education.

This theory, as complete as that of the German Empire, has somehow not caused the people of Austria to prize education so highly as might have been expected. It has not taken deep hold of the popular mind and come to be regarded as the citizen's birthright in the same way that we find it to have done in Prussia and Saxony. Nor do we find the people taking the same pride in their *public* schools. Can it be that direct compulsion has not existed for a sufficiently long period? It is said that, formerly, compulsion was only applied *indirectly*, or, by laws which rendered impossible without a certificate of scholastic attainments many of the common steps in life such as apprenticeship to a trade, employment as a workman, occupation of any kind in the service of the state, and even marriage. In any event, direct compulsion is now applied with sufficient vigour, and the street urchin—the pest of every town in England—is no more seen in Vienna than in Berlin or Dresden. Vigilance committees look after absentees, who, aloof without proper reason, are liable to be fined, for the first offence ten florins, and for succeeding offences higher sums up to thirty florins.

Great care appears to be taken in directing the studies so that the pupil is fitted in the best manner for any future career. As the examinations of the elementary schools are successively passed, the child is moved into the secondary school of the kind suitable for him. Again, as he is passed from the care of one teacher to another, the transfer is effected with due reference to the kind of pupil and the kind of teacher. Children under twelve are not allowed to be employed in the manufactories as "half-timers" except by permission of the authorities and under certain regulations. The number under this head is therefore small in the schools. Under the age of ten, half-time labour is not permitted. Between twelve and sixteen it is allowed, if school also be attended regularly.

The aula is generally lacking in the lower schools just as we have found to be the case in Germany. Even in such an example as the large Communal School in the Eschenbach-Gasse in Vienna it does not exist. One feature is to be observed as marking a contrast to the northern schools. Staircases are not

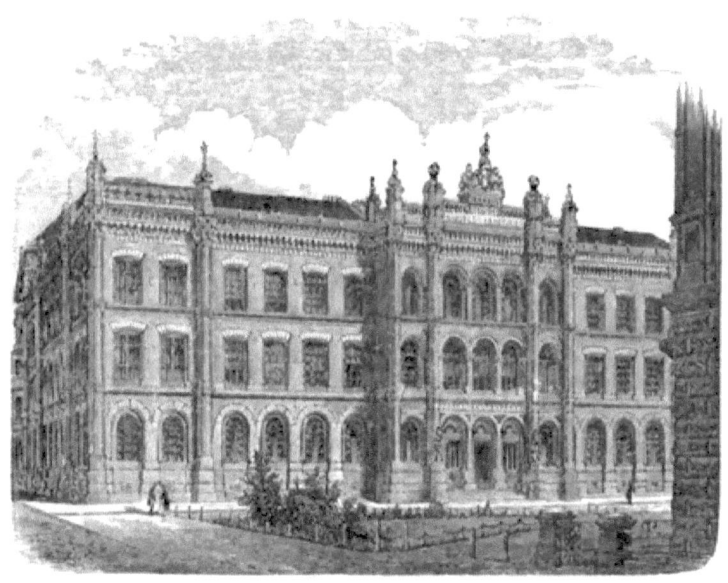

109.—THE HANDEL'S ACADEMY, VIENNA.

made of the same prodigious and useless size. An entrance lobby 6ft. 6in. wide, and a staircase 5ft. wide are the usual, and very liberal, allowance.

Although the primary school-buildings of Austria may furnish us with no new plans or principles of planning beyond those already noticed in Germany, some of the special schools of the capital will bear comparison with any in Europe, if richness of architectural embellishment, completeness of arrangements on plan, and perfection of internal appointments have any importance. It is, therefore, rather to some of the exceptional than to the ordinary schools, that we would draw attention. Among those which must be regarded as among the most singular is the Handel's Academy, of which we give a view (woodcut No. 109). The French Imperial Commission give the following account of this establishment:—" Under the title of Academy of Commerce, there was founded at Vienna, in 1857, a very remarkable establishment for the instruction of young men intending to follow commercial pursuits. A capital of 400,000 florins was subscribed and suitable premises built for the purposes. The school is provided with technological collections, a museum of natural productions, and complete chemical laboratories. A committee composed of nine members presides over the general management. The instruction is given in two divisions, one of them preparatory—requiring two years' study, the other technical—occupying the same length of time.

"The number of hours per week devoted to the different branches of instruction is shewn in the following table:—

PREPARATORY DIVISION.				TECHNICAL DIVISION.			
Subjects taught.	Number of hours.		Totals.	Subjects taught.	Number of hours.		Totals.
	1st year.	2nd year.			1st year.	2nd year.	
Religion	2	2	4	Commercial Calculations	3	3	6
German	4	3	7	Book-keeping	2	–	2
Arithmetic	5	4	9	Commercial Correspondence	3	–	3
Geography	4	3	7	Political Economy	3	3	6
History	3	3	6	Commercial Law and Exchanges	–	–	–
Natural History	4	2	6	Geography, Commercial and Statistical.	2	2	4
Caligraphy	2	4	6	Commercial History	3	2	5
Book-keeping	–	2	2	Chemistry	3	2	5
Physics	–	2	2	Physics	2	–	2
				Study of Merchandise and Technology	3	4	7
				Austrian Commerce and Manufactures	–	–	–
				Model Counting-house	–	8	8
TOTALS	24	25	—	TOTALS	24	24	—

"Besides this compulsory curriculum there are French, English, and Italian classes, one or other of which every pupil must attend, or two, or all, if he pleases. There are excellent laboratories for those pupils who wish to learn how to analyse different kinds of merchandise. This study is altogether optional. In winter qualitative analysis is taught, and quantitative in summer. The school fee is 157 florins 50 kr. a year for all the courses.

"At the close of the courses there are examinations for those who please to present themselves, and certificates of capacity are given to all who pass satisfactorily. Among the optional branches of instruction are stenography, to which some importance is attached, and drawing, which is cultivated both artistically and commercially.

"Besides the regular classes during the day, there are evening classes for persons already engaged in business. These are held from 7 to 9 o'clock from October till Easter, and are attended by about 250 persons, who pay four florins for each course, with the exception of the living languages, which are only two florins, and stenography, fixed at one florin. The subjects taught in these classes are commercial arithmetic, book-keeping, commercial correspondence, the rules of commerce, and exchange, &c., the living languages and stenography. The majority of the persons attending the evening classes present themselves for examination to obtain certificates. In this department discipline is maintained by the professors under rather severe regulations."

The Imperial Gymnasium at Vienna, opened in 1866, presents a remarkable example (woodcut No. 110) of a higher Roman Catholic School. Like the Communal School near the Polytechnic, its plan is a quadrangle surrounded by corridors leading to the several class-rooms (woodcuts Nos. 111, 112, and 113). In this case the central court is not roofed in. The Aula or examination Hall—here called the *Prüfungssaal*—(woodcut No. 114) is also used as a Chapel, and has at one side a groined polygonal niche wherein stands an altar separated from the hall by a

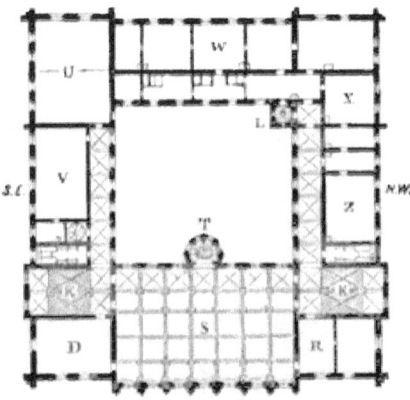

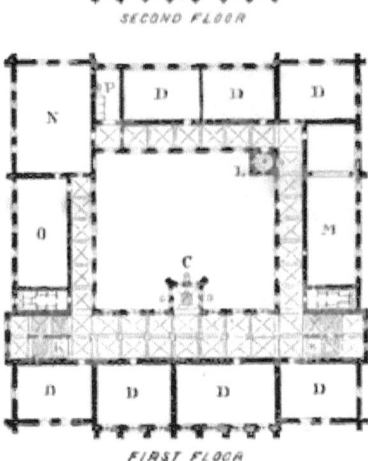

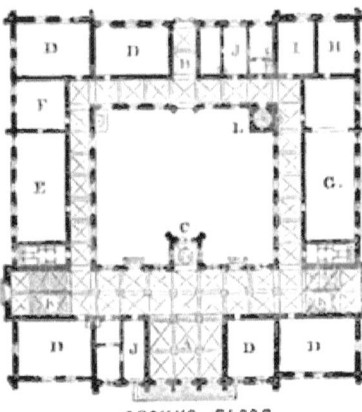

Reference.

A. Hall.
B. Carriage Entrance.
C. Well.
D. Class-rooms.
E. Library.
F. Reading-room.
G. Gymnastic-room.
H. Gymnastic Apparatus.
I. School Apparatus.
J. Caretaker.
K. Principal Staircase.
L. Director's Staircase.
M. Drawing-class.
N. Physical-class.
O. ,, Apparatus.
P. Laboratory.
R. Sacristy and Parlour.
S. Examination Hall.
T. Altar.
U. Lecture-room for Natural History.
V. Natural History Collection.
W. Director's House.
X. Director's Office.
Z. Conference-room.

111, 112, 113.—THE IMPERIAL GYMNASIUM, VIENNA, PLANS OF FLOORS.

curtain, except when in use. Towards the open court, this projection forms a handsome turret. The hall itself, with its open-timbered roof of hammer-beam construction, has rather the air, and is certainly modelled from an English college type. The ceilings of the principal entrance, staircase and corridors are groined in stone, and polished shafts of granite are freely used. Coloured decoration—far inferior to that in the principal salon of the Grand Hotel, and indeed abominable—is boldly used throughout, especially in the roofs and ceilings of the hall and corridors. The unusual feature of a class-room capable of holding 100 pupils here occurs, although the other rooms follow the German rule, and are made for 40 to 60. This special room is used for collective lessons which never exceed one hour each. In all the class-rooms the windows have double casements. In the corridors they are single. The class-room for physical science is fitted like a lecture-room, with circular seats raised in tiers. Besides open playgrounds, there is attached to this fine school-house a completely fitted turn-halle. The fees to students are from twelve to fifteen florins. The total cost was 500,000 florins, of which sum 25,000 was devoted to furniture and fittings.

The Vienna Stadische Schools consist of three divisions, a Burgher School for 500 boys and girls, having eight class-rooms for each sex; a Grammar, or Classical, School for 100, and a Training School for 60 teachers of both sexes; the whole being under one roof. The girls' Burgher School has a special room for teaching sewing, fitted with flat tables 3 feet wide, round which the girls sit. They are substantially built, and are in every way well worthy of a visit. The necessity for trained women-teachers is, at last, being felt here as in Germany.

There are six Polytechnic Schools in Austria, viz., at Vienna, Prague, Gratz, Ofen, Brünn, and Lemberg, among which that of Vienna is the best and most complete, and, in method of organization, somewhat resembles that at Carlsruhe. As a building it has no architectural pretensions, and is neither so recent nor, although in one of the new quarters of the town, placed in so commanding a situation as the Polytechnic at Zurich. It accommodates 1,000 pupils, and has large drawing-schools of engineering and architecture, laboratories for chemistry, large rooms crowded with models, chemicals, machines, &c., and generally aims at the same kind of provision as that already illustrated by the woodcuts relating to the Polytechnic School at Aix-la-Chapelle (Nos. 100 to 106). The School is divided into five Sections, as follows : viz.

(*a*) General. In which mathematical instruction is given preparatory to the following schools. (Two years.)

(*b*) Engineering. Wherein the making of roads and bridges and hydraulic works is taught. (Three years.)

(*c*) Architecture. Which comprises instruction in building materials and construction, drawing, and architecture as a fine art. (Three years.)

(*d*) Machinery. (Two years.)

(*e*) Chemistry. (Three years.)

The preliminary professional education is usually obtained in the Realschulen (here divided into *Unter* and *Ober*), and pupils then pass into the Polytechnic, where, after the probation of the General School, they continue their studies in one of the four special sections. The fee is fifty florins—only 5*l.*—per annum, and even this small sum is sometimes remitted in cases of poverty or of extraordinary success in study. Before being admitted, an examination must be undergone. Periodical examinations are made to test progress. And, before commencing his professional career, the pupil must shew, by the ordeal of a strict final examination, that he is properly qualified to practise. In addition to the regular pupils, there are others who only attend certain courses of lectures.

CHAPTER VIII.

THEORY OF ENGLISH ELEMENTARY SCHOOLS.

Necessity of clear preliminary arrangements—Controlling effect of the Code—Pædagogy—Size of class-rooms—Size of school-room—Division of Departments—Points to be settled before building—Compactness of Plan—Economy—The Site.

HAVING, in the foregoing pages, drawn attention to some of the leading characteristics of schools in other countries, and discussed the plans, general arrangements, method of lighting, and other features, we are the better enabled to approach the consideration of the important question, "How should we build schools in England?" Before beginning the erection of any school, or any group of schools, something more is necessary than merely to count the cost. The aim and object of the enterprise should be clearly defined, and the theory of the school determined completely. Above all, the system of teaching should be settled, so that every facility may be provided in the building for its successful results. Our inquiry leads us principally to the subject of *Elementary* Schools, because on this the public mind is much exercised, with a desire for genuine progress and real improvement.

If the Public Elementary School—the school for the rudimentary education of the poor—has, in the nature of things, no remote history as an English institution, neither has the scientific, constructional, or artistic aspect of its buildings any annals. Unlike the Secondary School or College, it has no list of famous establishments in which the numerous little points of arrangement and building contrivances, perhaps adopted one by one in the course of years, have, by long experience in each case of their peculiar fitness, imperceptibly settled down into a system and combined to establish the main principles on which future schools should be built. There are no English Elementary schools of the precise kind so suddenly required in large numbers by our School Boards. A careful consideration of the plans of those erected in other countries, while affording hints on isolated features, do not furnish in point of general scope and idea anything of the kind wanted in England. They are, one and all, un-English in spirit, and based on systems of training not in favour among us. When it has thus been clearly ascertained that, for them, foreign examples are no models, and that our own are beneath the standard aimed at, the course is clear enough. We must think for ourselves in the matter, and, so to speak, build on our own foundations.

For a group of new schools, then, the first necessity is to decide the number of departments, the number of children to each department, and the relation of these numbers to each other. Then the number of class-rooms for each should be considered. A distinct aim and intention as to the meaning and use of every part of the building and its furniture carried down to the most minute items should be maintained. And, before proceeding far, we shall find it necessary to acquire precise information as to the meaning of the term "Public Elementary School."

The Elementary schools in England which are regarded as "public," are so regarded because placed under Government inspection, and assisted by grants according to the success of the teaching as tested by examinations or number of attendances. No individual examination takes place before the age of seven, and the results of Infant training are measured by the number of attendances. This inspection pre-supposes that certain preliminary conditions, both as to the building and the method of instruction, have been complied with. It is with the former that we are chiefly concerned; but some notice in outline of the latter is necessary for elucidation.

The Revised Code of the Education Department, which first came into force on the 1st of July, 1862, is now published afresh every year, with any necessary emendations, under the name of the "New Code;" and it is this document which regulates the teaching, and, therefore, in large measure, the architectural plan of our Elementary schools. From its pages, also, can be learnt what is meant by a "pupil-teacher," and under what regulations the learning and teaching of this assistant is to be carried on.

The difference between the German and the English methods of conducting Elementary schools lies deeper than the apparent mere variation in the mode of teaching. It arises from fundamentally different opinions held on the science of pædagogy. In Germany, the principle of educating teachers entirely in the training college is adopted both for primary and secondary schools. In England it is held that what is desirable for higher, is not so important for elementary, schools. The possession of knowledge does not necessarily carry with it the power of imparting it to others. And it is believed that, for teaching in the earlier stages, much greater facility is necessary than can be acquired in the ordinary course of the training college. Such facility requires long practice, only obtainable by the means of a regular apprenticeship as to a trade. Hence the pupil-teacher system, which allows boys and girls who have entered upon their fourteenth year to be apprenticed for a term—the full course being five years—as laid down by the Code. During this term they are engaged in teaching, in themselves receiving further instruction, and in qualifying themselves for the successive examinations through which they obtain certificates which enable them to fill, one after another, the various positions open to their profession in schools of the kind.

The Code divides schools above the Infant stage (article 28) into six grades or "standards of examination," probably commencing at the age $6\frac{1}{2}$ to 7, and corresponding to six successive years. It also contemplates (article 32) that the number of children to be taught by a certificated teacher, assisted by one pupil-teacher, shall be 60. That, for every additional 40 children there shall be an additional pupil-teacher. And that, for every additional 80 children, the increase of teaching power shall be either one assistant (certificated) teacher, or two pupil-teachers.

From this we gather that for a Graded school the largest allowable double class—that is, two classes taught by a teacher and pupil-teacher working together—is 80. This maximum number of the Education Department need not necessarily become the minimum of school managers. Because this is the number above which nothing will be counted in the grants made as payments on the results of teaching, it does not follow that it is the number which would ensure the most efficient kind of teaching. It is simply recorded as an extreme limit; and, in planning a new school, 60, 70, or 80 may be taken for the numbers of a class, just as convenience of plan may dictate, 60 being certainly the most reasonable number for the senior classes.

In an English elementary school the principal teacher of any department is expected, not only to be responsible for the management of the whole department, but to be actually engaged in the work of teaching, and not, as under some other systems, to be merely a general superintendent of the work of others. Assistants or pupil-teachers (in number according to the size of the school) are regarded only as aids to, and are not appointed to supersede, the head teacher.

It will thus be seen that a school planned to consist entirely of class-rooms, separate from each other and approached from a general corridor, would not be in strict unison with the intention of the framers of the Code, because not easy of supervision on the part of the head teacher. This intention is further made clear in the Code (article 17c), where it is required that the principal school-room shall be sufficiently large to contain at least 80 cubic feet of space per child, calculated on the whole number in the school. And its reason is seen by a visit to any of the Elementary schools carried on under Government inspection, where the practice of assembling the whole of the children at least once in the day—viz., in opening the school in the morning and in many cases again in the afternoon, is maintained, and where a general room is thus indispensable. The general tendency of opinion, however, is decidedly in favour of a much larger number of class-rooms than have hitherto been usually provided; indeed, of as many as possible, subject to the maintenance of the above limitation, and to the difference of organization between Primary and Secondary Schools.

Experience has shewn that the separation or isolation of classes in separate rooms has an important bearing on results, and the lessons contemplated under the six standards should, *as far as practicable*, be taught in separate class-rooms. But, as each school is under the general supervision of one master or mistress, actually and personally engaged in teaching, this principle must be subordinate to the necessity for such supervision combined with teaching.

The numbers to be allotted to each department of the school will vary in different localities; and even the number of departments is not always the same, some school promoters preferring the simple separation of boys, girls, and infants, and others advocating the division of the Graded schools into senior and junior, whether "mixed" or with the sexes separate.

The following extract from the Government statistics of the whole of England is interesting and useful, as shewing the general relation of attendance to age in each thousand of population before the passing of the Education Act:—

Between 3 and 4 years of age 111
[From 3½ to 4—say 55.]
,, 4 ,, 5 years of age 110
,, 5 ,, 6 ,, ... 105
,, 6 ,, 7 ,, ... 103
[From 6 to 6½, or 6½ to 7—say 52.]

Between 7 and 8 years of age . . . 100
" 8 " 9 " 98
" 9 " 10 " 96
" 10 " 11 " 94
" 11 " 12 " 93
" 12 " 13 " 90

Similar statistics, when taken in particular localities, will be found to differ widely according to the state of population. Yet, taking it for granted that no child under three years of age will be admitted to an Infant school, the numbers may, for the sake of convenience, be placed as follows, viz. :—

Infants' Department,		six-fifteenths.
Junior	do.	five-fifteenths.
Senior	do.	four-fifteenths.

If, however, the Graded schools be divided simply into two departments, one for boys and the other for girls, then the readiest plan will be to consider them as of equal numbers. Unless a one-storied building throughout be contemplated, this plan is almost imperative, because otherwise one school will not, in the building, fit naturally over the other, and unnecessary difficulties for the architect and increased expense in the fabric will follow.

In London, where a public elementary school of one story throughout is a very rare exception, and where the simple division into boys, girls, and infants is preferred, the numbers usually are as follows, viz. :—

	1,040 school.	780 school.
Infants' Department ..	400	300
Boys' do. ..	320	240
Girls' do. ..	320	240

In view of the conflicting requirements of having as many separate class-rooms as possible, and yet providing one room sufficiently large to accommodate the entire school-department at one time, while at the same time arranging our plan with due regard to economy of superficial area, perhaps the simplest rule is to provide as many class-rooms as the school-room itself will usually accommodate classes. The sizes of the desks being usually arranged for writing with comfort, the children can easily sit closer together when the general assemblage takes place.

Having determined the number of departments and of class-rooms, it remains to place these on the site in the form of a building to the best advantage. Shall the building be of one, two, or three stories? Shall the Infant school (the plan of which differs in principle) be placed under the Graded schools, or apart by itself? Where shall the entrances be? Where the staircases? In what manner shall we secure side-lighting without destroying the plan of an Elementary school? Shall we warm and ventilate by open fires or by some other method? Where shall the latrines be, so as to be near, and yet not too near the main building? Is a master's house to be built on the site? These and others are among the first questions to be asked and answered at the outset. The answers to many will appear in the course of our discussion. Others can only be decided in each case according to the peculiarities of site and the different controlling causes.

Compactness of internal arrangement is one of the first essentials of school planning, because it bears directly on the question of cost. There is always one method of obtaining a desired result better than any other, and it can only be arrived at by the careful excision of everything superfluous. In the plan of a house, the first item of luxury or waste which creeps in is usually the corridor, wherein we neither eat, sleep, nor live. Economical reasons exclude it in the cottage. As the number of rooms increases, some means of reaching each without passing through any other is desired, and hence the use of the corridor. In the dwelling-house, an economical plan reduces the amount as much as possible, till unconsciously every one admits the principle that the style and importance of a house is much affected by the indication of luxury and wealth expressed in the greater or less size, importance, and stateliness of the hall, corridor, and staircase. The same principle holds good in the schoolhouse. Unless some amount of style and show is sought, no corridor should be provided unless necessary for the convenient and economical working of the school day by day. In a school conducted on the separate class principle, some communication becomes necessary to a certain extent, as in a dwelling-house. Where the system of teaching used in English Elementary schools is adopted, the corridor considered only as a passage should be eliminated as far as possible, because it increases the expense and renders the *through* ventilation and sometimes the proper lighting of the school-room more difficult, if not impossible. In a house, the greater number of corridors and halls will involve the larger staff of servants. In a schoolhouse, want of forethought and absence of compactness in planning, will surely entail a larger staff of permanent teachers, or, what is nearly as bad, continual loss of time in daily work.

In the whole of the preliminary arrangements, the annual working cost of the establishment must be economically considered. To this end, the building should be planned with facilities for economical teaching, and the yearly expense of repair, maintenance, warming, and ventilation should be fully studied.

The site itself must be freehold, and must contain (Code, article 29) not less than 1,200 square yards of land (unless the price be prohibitory), and must be quiet, healthy, and conveniently near the children's homes. Regarded as public buildings, the schoolhouses should, if possible, stand detached from adjoining structures, and the boundary-walls should never, when avoidable, be party-walls.

The general planning of the buildings will be very much controlled, not only by the frontage or frontages and the surrounding buildings (if any), but also by the aspects of the site in relation to sun and air. Much difference of opinion has prevailed on this latter point. It is well known that the rays of the sun have a beneficial influence on the air of a room, tending to promote ventilation, and that they are to a young child very much what they are to a flower. Acting on this known fact, the builders of some schools have sought to secure as much sun as possible, and produced results of light and glare painful in hot summer weather, either to pupils or teacher, or both. On the other hand, advocates for the entire exclusion of the sun's rays have not been wanting (especially among the members of provincial school-boards), and these have urged that a school-room should have no windows on the south or west! In this, as in so many other disputed questions, the real truth lies somewhere between the far-distant extremes. The *main* lighting of the school-room should never be from the south or south-west, although some sunny windows should always be provided. The coolest, steadiest, and best light is that from the north, and the principal aspect of the boys' and girls' school-rooms should first be selected as near that quarter as may be practicable from the nature of the site. In this sunless climate of ours it is difficult to make a school-room too sunny; yet this may be done if the sun be admitted at the wrong places, as, for instance, right in the eyes either of teacher or children, and without the most absolute power of control.

It may sometimes happen that the plan of the buildings is so determined by the conditions of the site, as to leave only sunless playgrounds. Under ordinary circumstances, the aspect of these latter is quite as important as that of the school-rooms. Sun is here a necessary of life. Without it, the playgrounds will be mere draughty yards, conducive to colds, which are the seeds of so many disorders.

CHAPTER IX.

SCHOOL SEATS AND THEIR LIGHTING.

School Desks control Dimensions of School-room and Class-rooms—Difficulty in massing a Class conveniently and compactly—The Dual Desk with Lifting Flap—Extra width required to Rooms—Eye Diseases occurring during School Life—Proportion of School Desks—Summary.

NEXT in importance to the method of teaching, considered in relation to its bearing on the general plan of a group of schools, are the elements which control the shape and size of the school-rooms and class-rooms composing the building; and, chief among these, with a powerful influence on the final result, we find that of the school-desk.

The grouping of each single class in the best manner for effective teaching is of such extreme importance to the architect, that it may be said to lie at the very threshold of the subject, and ought largely to govern his plans. It is yet a feature seldom thought of at the outset, and not until the building is finished and ready to receive the furniture, is it found how much more suitably the schoolhouse might have been planned had the desk question been first decided. Too frequently the complaint is similar to those so commonly heard against the houses run up by speculative builders, in which the bedroom has no proper place for the bed, and most of the doors are either in the most unsuitable positions, or are made to open the wrong way.

There are two parties to be considered—the teacher and the children composing the class. The former should be so placed that his angle of vision in teaching shall not much exceed 45°;

that he shall be able at any moment to inspect the work of any child without disturbing the rest; and that his voice shall reach every child day by day and during the whole day, without unnecessary effort and fatigue to himself. If there be unnecessary strain upon the teacher; if he is giving a lesson under a constant sense, however unconscious, of hardship or discomfort—we may be certain that his influence is impaired and his teaching lessened in value. Everything should therefore be done, within reasonable limits, to contribute to the easy performance of his work. The children, for their part, should have benches and desks so contrived as to insure comfort—not for sitting at *or* for standing in—but for both. And herein lies a difficulty nearly as great as that already referred to and affecting the number of class-rooms. If the bench and desk be made comfortable for sitting at and be immovable, then the child cannot conveniently stand up in it. If made with sufficient space for standing in (the usual practice hitherto), then it is wretchedly uncomfortable for sitting at. And if the long length of desk, say for four, five, or six children, be adhered to, the necessary access and egress for both teacher and children can only be provided by longitudinal gangways behind each row. This system of furnishing a school with desks in long lengths has, further, the great objection that, for seating a proper number of children, it entails so great a length as to limit the class to three, or at most to four rows in depth. In such case, the teacher's angle of vision is increased far beyond that of 45°, and the shape becomes highly inconvenient. To the great difficulty connected with the shape of the desk itself, and the consequent provision of longitudinal gangways, must be attributed the advocacy of the Education Department of the depth of three rows. For, by the new kind of desk now introduced (woodcuts Nos. 115, 116, and 117), five rows now occupy scarcely more than the three rows of the old method, and the class is thus shortened and brought more compactly under the eye of the teacher.

The whole question has been for years confused by the incessant efforts, both of school-promoters and desk-manufacturers, to render what ought to be one of the simplest articles of school furniture a species of harlequin, capable of assuming a new character at a moment's notice. When too much is attempted, the result is never satisfactory. What is intended to suit everything, generally succeeds well in nothing. By keeping the object steadily in view, and endeavouring to make the school-desk really fit for school purposes, while avowedly ignoring all other purposes, the recent improvements have alone been effected. The developments of school-planning have been produced in some measure by the desire to isolate each class by means of separate rooms, so far as compatible with an occasional general assemblage. And, in like manner, the improvements in school-desks have arisen from a desire to isolate each scholar so far as compatible with a convenient form of class. Both have for their object the increased concentration of the child on his lesson, without unnecessary sources of distraction. And in both it has been sought to maintain the valuable influence known as the "sympathy of numbers."

The Americans appear to have preferred hitherto a separate seat and desk for each child, for primary schools, as in woodcut No. 8, page 29, and the dual arrangement with fixed flap for others, though sometimes the seat only is separate, forming a fixed chair, while the desk is continuous. Now, however, many of their writers urge the expense and waste of space entailed. In Sweden, a separate desk is found, in which the seat rises simultaneously with the child by means of a lever action produced by an iron weight. This contrivance, excellent for theatres, lecture-halls, or other places designed for adults, is unsuitable for children. In Germany and Switzerland, a length for four children is the most usual arrangement. But in Holland we find the grouping *in pairs predominant;* and it is the adoption of this method which must mark the first great improvement in our English school-desks; for, however much we may, in theory, approve the single principle, it is clear that for young children nothing is gained by having a gangway on both sides, while the great cost of providing a seat and desk for each child, in large numbers, is practically prohibitory. It may be that for higher schools, where a still greater degree of concentration on work is required, the single system has advantages; but, for Elementary schools, seating in pairs is to be preferred. The Dutch desks are placed six, or even seven deep, and are fixed on frames. Careful examination, however, and experiment in actual schools of our own, has shown that with any number in depth greater than five, the effort for the teacher is too great to be continuously exerted, and in practice, the back rows soon cease to be used. Again, the Dutch desk is so constructed that its front forms the back for the two children sitting immediately in advance. This has been found objectionable from the unsuitability of the height of the back, and from the greater liability to be shaken by the children in front. The desks given in our illustration stand entirely free from each other, and are carefully proportioned in all their parts to the anatomy of the child. Three sizes, Nos. 1, 2, and 3 (woodcuts Nos. 115, 116, and 117), having sloped tops and lifting flaps, are devoted to the Graded schools, and in these a slightly higher backrail is given for the girls than for the boys. Two other sizes, Nos. 4 and 5, page 190, with flat tops, are intended for use in Infant schools, and in these no difference is thought necessary in the backrail, nor is the top made to lift. In point of construction, rigidity has been obtained by the use of "lugs" cast on the iron standards, where shewn, to enable the whole to be screwed firmly together. No fixing to the floor is necessary, the weight alone being sufficient to secure immovability. In reference to the lifting-flap shown for Nos. 1, 2, and 3, it may be remarked that, without it, a child cannot possibly stand upright in his place, the edge of the desk-top being perpendicularly above the front of the seat. According to the Dutch method, when children are required for drill or other purpose to rise, they must first move out into the gangways—one to the right and the other to the left in each pair—and stand behind each other in Indian file. This simpler method is perhaps the best suited to Infant schools. But with the flap comes another improvement, highly appreciated by experienced teachers in its application to

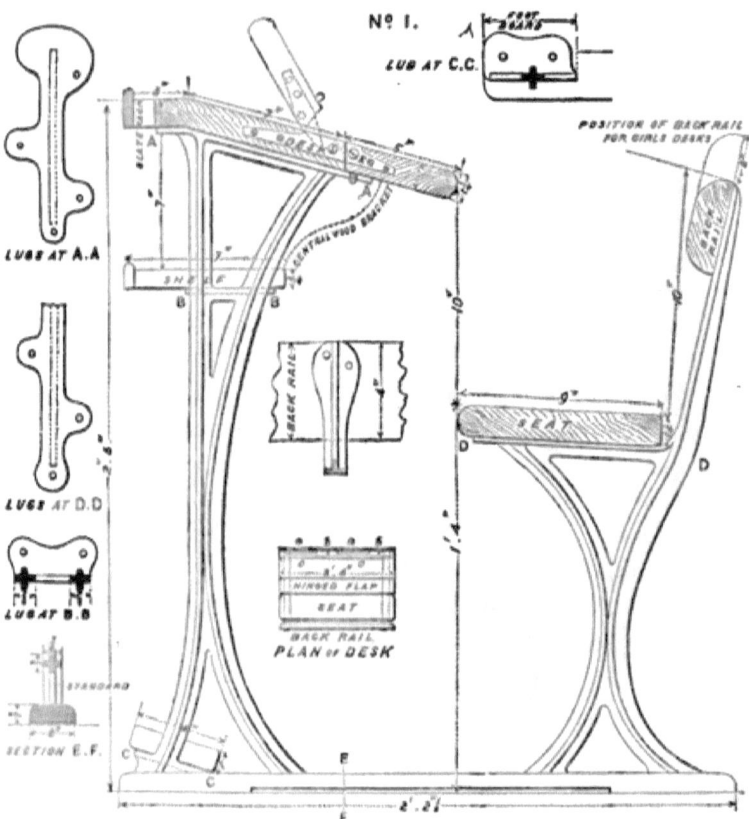

115.—DESK FOR GRADED SCHOOLS.

Graded schools. At a word, the flap is lifted, the children rise, and can then go easily through any manual exercises without leaving their places. This improvement solves the problem of making school-desks and benches suitable both for sitting and standing. Noise, the great enemy of such an arrangement when attempted with long-length desks, is greatly reduced when the

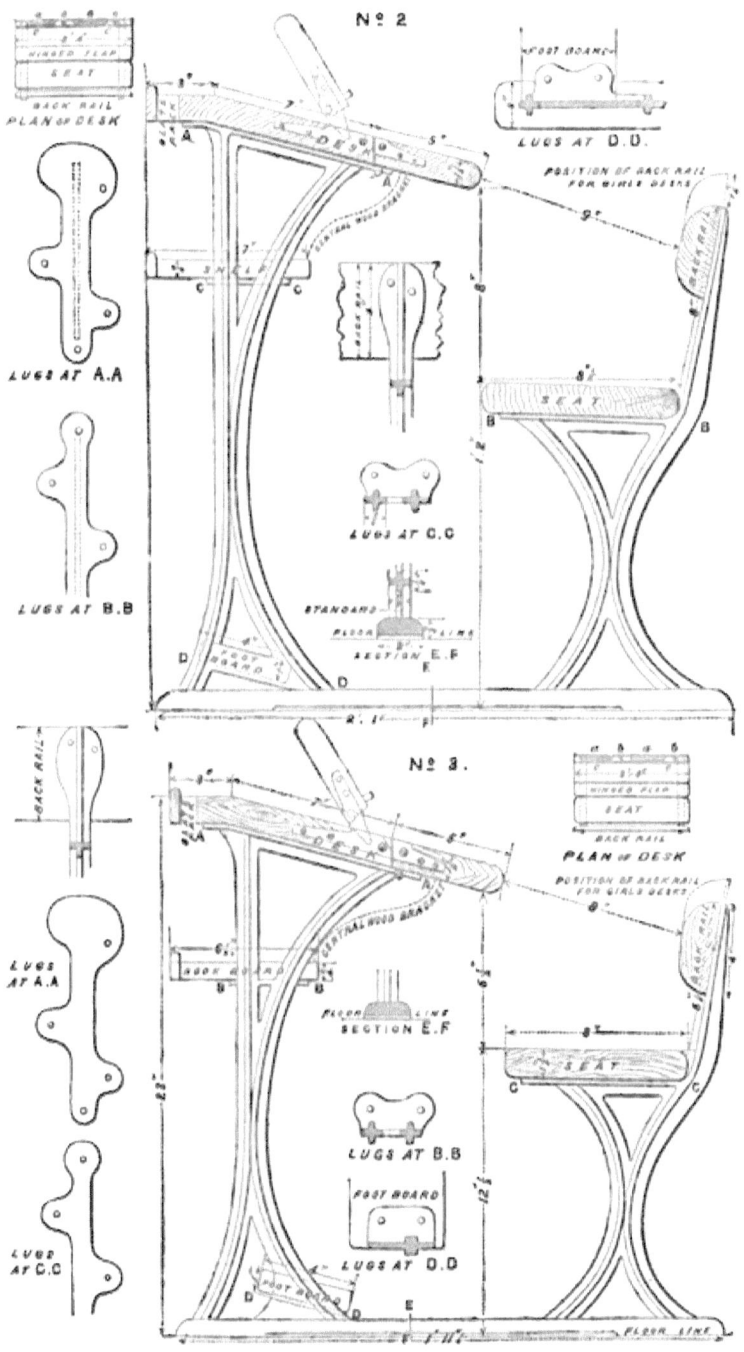

116, 117.—DESKS FOR GRADED SCHOOLS.

flap is applied to short ones, and it is still further reduced, indeed rendered imperceptible, by the use of end hinges, which themselves stop the flap when turned over, prevent it from falling

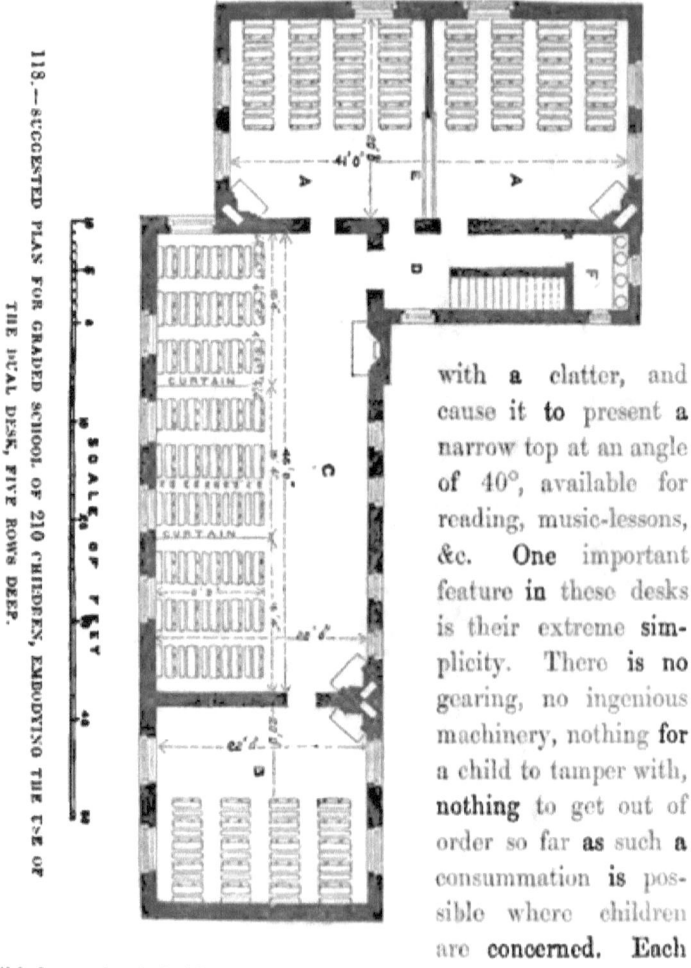

118.—SUGGESTED PLAN FOR GRADED SCHOOL OF 210 CHILDREN, EMBODYING THE USE OF THE DUAL DESK, FIVE ROWS DEEP.

with a clatter, and cause it to present a narrow top at an angle of 40°, available for reading, music-lessons, &c. One important feature in these desks is their extreme simplicity. There is no gearing, no ingenious machinery, nothing for a child to tamper with, nothing to get out of order so far as such a consummation is possible where children are concerned. Each child has a bookshelf, a slaterack, a groove for pens and pencils, and last, though not least, an inkwell at his *right* hand. The general plan given shews the grouping of the desks both in school-room and class-room. A, A, is a double class-room for eighty.

B, a single class-room for forty. C, the school-room, containing three classes of thirty each. D, the landing. E, sliding partitions. F, a lavatory. In the school-room a gangway of twice the usual width is provided between each class, and, as far as possible, the children are placed with their side to the light. Curtains may be used between the classes by those who like them. The three back rows may, if preferred, be raised one above another, by fixing the iron standards of each desk to wooden bearers, the level floor being still maintained throughout. As a rule, it will be found sufficient to use a larger size of desk for the *two* back rows, and to place the biggest children there.

Some time ago the School Board for London adopted the principle of arrangement in pairs, and the drawings we give represent the desks used in their new schools. In all the instances where they have been hitherto introduced, an experience of three or four weeks has resulted in the entire approval of both managers and teachers, although, in most cases, the slight change required in the drill has led to some hesitation at first.

The room in which these desks are to be used, five deep, must be itself deeper than has previously been common. If a classroom, it should be from 18ft. to 20ft., the latter being preferable. If a school-room (always used more or less as a passage-room), the size should never be less than 20ft. for the smallest rooms, 21ft. or 22ft. being preferable. For large school-rooms 22ft. must be regarded as the minimum. The School Boards for London and for Liverpool, and possibly others, have received the consent of the Education Department to build school-rooms 22ft. wide, in view of the new kind of desk to be used; and, in giving such consent, the Department is merely approving the natural development of its own principle—that of arranging desks in the best manner that can be devised *along one wall*, with sufficient space in front for drawing out the classes.

Some may think that so apparently trivial a question as that of school-desks could not need so much discussion. Medical authorities think otherwise, and lay the greatest stress on the proper shape and proportion to be used in every part, as well as on the admission of suitable light in a suitable manner to the children seated at the desks.

According to Dr. Leibreich, the ophthalmic surgeon of St. Thomas's Hospital, the change in the functions of the visual organ developed during school-life are threefold, viz. :

1. Decrease in range, or short-sightedness (*Myopia*).
2. Decrease in acuteness (*Amblyopia*).
3. Decrease of endurance (*Asthenopia*), and are owing chiefly to two causes,—improper method of lighting and improper shape of the school-desk. Confining ourselves to the first, great importance should be attached to the opinion of Dr. Leibreich as an eminent oculist. Is *myopia* developed chiefly during school-life? If it is, and to a greater extent in schools injudiciously lighted, the question assumes a practical aspect. It is easy to understand that eye-disease may be on the increase, and that careless or ignorant arrangements may tend to aggravate it, but other causes must surely be at work besides bad lighting and unsuitable desks. Education of almost any kind must to a certain extent affect sight. Civilized man never has the perfect vision of the savage. Constant poring over books and white paper is known to be injurious. When continued through the many years required for completing a higher education, the results may easily become marked unless the education be conducted with the greatest care and discrimination in suitable premises.

Diminution of, and injury to, sight cannot be entirely attributed with fairness to improper positions of windows and bad shapes of desks. In no country in the world is there a more complete system of national education, or one which has existed longer in its popular form, than in Germany. And in no country has closer attention been paid to the judicious lighting of school-buildings and to the proper shape of school-desks. The light is invariably admitted *from the left side only* of the children. The desks are the result of long study on the part of their anatomical authorities. Yet in no country is *myopia* so common. The short-sighted, spectacle-wearing German is a well-known type on the stage of any London theatre, just as he is in the streets of Berlin itself. Increase of knowledge must have some attendant drawbacks, however comparatively slight, and impaired eyesight and crooked shoulders may be among them if due care be wanting. This is one reason why instruction in the hall for gymnastics has become an inseparable part of the regular school course in Germany.

The second change in quality of eyesight (*amblyopia*) occurring during the school age may well be left to take care of itself. Decrease in the acuteness of vision must always rank in the same category with decrease in youthfulness, as one of the ills to which flesh is heir.

The third abnormal state of the eye arises—we are told—principally from two causes; one, a congenital condition which can be corrected by the use of convex glasses; another, a disturbance in the harmonious action of the muscles of the eye, often caused by unsuitable arrangements for work. "Insufficient or ill-arranged light obliges us to lessen the distance between the eye and the book while reading or writing. We must do the same if the desks or seats are not in the right position, or of the right shape and size. When the eye looks at a very near object, the accommodating apparatus and the muscles which turn the eye so that the axes converge towards the same object, are brought into a condition of greater tension, and this is to be considered as the principal cause of short-sightedness and its increase. If the muscles of the eye are not strong enough to resist such tension for any length of time, one of the eyes is left to itself; and, whilst one eye is being directed on the object, the other deviates outwardly, receives false images, and its vision becomes indistinct—*amblyopic*. Or perhaps the muscles resist these difficulties for a time, become weary, and thus is produced the diminution of endurance. How can these evils be prevented? The light must be sufficiently strong, and fall on the table from the left-hand side, and, as far as possible, from above. The children ought to sit straight, and not have the book nearer to the eye than ten inches at the least. Besides this, the book ought to be raised 20° for writing and about 40° for reading."

The question of lighting has been much discussed in Germany for some years. The recent researches of Dr. Cohn give us the fact that of 410 students whom he examined only one-third were found to possess good sight. Nearly two-thirds were short-sighted. Among 244 cases of *myopia*, only 59 were hereditary. He visited many schools, and found generally a large percentage of short-sighted persons. He considered the reason to be the defective lighting of the schools, because the relative number of the persons whose sight was injuriously affected was found to be smaller in the better-lighted buildings. It is therefore argued that a class-room is only well lighted when it has 30 square inches of glass to every square foot of floor-space. Taken in conjunction with other considerations, this would shew that each scholar should have the advantage of about 300 square inches of window glass. The calculation is very rough, and cannot be accepted as a rule, for much depends on the position of the glass. It serves, however, to shew the kind of attention now paid to this branch of school planning.

The taxes on knowledge, payable by children in the shape of weariness and fatigue, are sufficiently heavy to justify all the pains which are being, or can be, taken to alleviate them. It is yet difficult to believe that, although 20 per cent. of all school-boys and 40 per cent. of all school-girls in Switzerland may have one shoulder higher than the other, the cause is to be found in the improper shapes and positions of seats and desks in days gone by. In England we have, in the past, always neglected the question of lighting our school-rooms scientifically. Provided the quantity were sufficient, little care was used as to its source or direction.

To summarize the results arrived at on the subject of school-desks and their lighting, we may point out—

(*a*) That a desk for two, 3 ft. 4 in. long, with intervening gangways 1 ft. 4 in. wide, has proved to be the best for Graded schools, and that five rows have been found practically sufficient in the direction of depth or distance from the teacher.

(b) That the full-size section should be carefully studied in every part, and adapted to the anatomy of the human frame in its varying sizes.

(c) That lighting from the side—especially the left side—is of such great importance as properly to have a material influence over our plans.

The first and last, tending to determine specific sizes of rooms and to affect the general principles to be followed, have an important bearing on the arrangements of plans hereafter given, and cannot therefore be too clearly remembered.

CHAPTER X.

INFANT SCHOOLS.

Age of Commencement—School for 120—School for 170—School for 300—Galleries, their Proportion, Arrangement, and Lighting—French Infant Gallery—Desks for Infants—The Playground or Uncovered School-room—Five kinds of Apparatus for Infant Playground.

THE Infant department of a group of English Public Elementary Schools has no real counterpart in other countries. It possesses points of similarity to institutions like the Kindergarten Schools of Germany, to the French Halls of Asylum, and to the Alphabet Schools of the United States, and yet differs from all. Other nations extend a kind of half-recognition to this earliest species of training, but only in England is it carried on in the same building as one department of a public school, and regarded as a part—a very important part—of the National educational system. The German view apparently is, that teaching should not begin at too early an age, so that the development of the body may have a fair start over that of the mind. A little later we find their course of study to be sufficiently severe.

With us, under the Education Act, children cannot be *compelled* to attend school before the age of five years, and they are expected to pass into the Graded school at the age of six and a half to seven. But, in fact, large numbers do attend before that age. So glad are the labouring poor to provide for the care of their young children during the day at a small weekly fee of a penny or twopence, thus enabling the mother to earn wages, that, in many of the London schools, the babies' room is always crammed and numbers are refused admittance.

It is doubtful whether any limit of age can be fixed under which a child shall not be admitted to a school, for experience has shown the test of mere age to be uncertain. Perhaps that of ability to walk and talk is the most simple and practicable, and the most easy of application by the Infant Mistress.

Considerable difficulty therefore arises in fixing precisely the proportion of children under four (who may be classed as *babies*) to the total number of the Infant school. An estimate of one-fourth should be sufficient, and one-third ought certainly never to be exceeded. Although the Act has been so framed as not to exclude children, however young, from the advantages of training, to exceed this proportion would be to contravene its spirit and to organise a nursery, or *Crèche*, rather than a school. It is open to question whether Infant schools should not be more numerous than Graded schools, so that children of the smallest size may not have long distances to walk from their homes. If two were built a little distance apart, and on different sides of the latter, the elder children from opposite parts of a district could with greater convenience bring the younger.

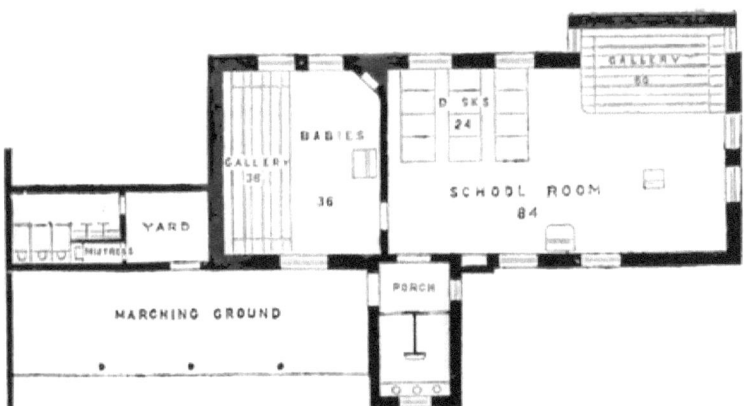

119.—SUGGESTED INFANT SCHOOL FOR 120.

The minimum size of school may be considered to be that for 120 (woodcut No. 119), consisting of a school-room, and one class-room for the youngest infants. No Infant school, however small, can be regarded as complete which does not at least provide a separate room for "babies" apart from the general room. The instruction given herein is little more than amusement under direction, but the marching and other exercises create noise, and complete separation by a solid wall is therefore indispensable. The babies' room should also have direct access to the covered playground and latrines without the necessity of passing through, and disturbing the work of, the school-room. A pane of clear glass should be provided in the wall of division to enable the mistress to see how these youngest children are being interested. The pupil-teacher intended for the management of an entirely untrained class should never be a beginner.

The plan which provides only one class-room, though sufficient for small schools, becomes incomplete when the numbers are considerably increased. When 170 has been reached—that is to say, when another class-room can be usefully employed without rendering the general room inconveniently small by too great

120.—SUGGESTED INFANT SCHOOL FOR 170.

curtailment—the senior infants (a small class from twenty to thirty) should, in turn, have a separate room (woodcut No. 120); and, in proportion to the increase of numbers, the class-rooms should themselves be increased in number, till, in a school of maximum size, the senior division would demand two rooms capable of being thrown into one at pleasure, by sliding partitions. In this case, two entrances or exits, also, are desirable to prevent the children from one end having to disturb the whole school before getting out. In the plan for 170 children, a window, not shewn on the woodcut, would be placed in the wall between the schoolroom and babies' room for the purpose of supervision.

The greatest number of infants which can be managed with comfort by the average mistress appears to be 250, though some can control 300. We, therefore, give a third plan (woodcut No. 121) of this largest number, one end being treated as for the Junior and the other for the Senior section, and with duplicate provision of entrance, cloak-room, and lavatory. The large galleries are also in duplicate, placed side by side with top-light and an intervening sliding partition, so that on occasion a collective lesson can be given to a very large number. No larger Infant School than that for 300 should ever be permitted. When the numbers rise above this, a second department should at once be organized.

The noise created by gallery lessons, drill, and above all by marching, is so considerable, and tends so much to disturb other classes engaged in lessons, that no sliding or revolving partitions or shutters should be used in Infant schools, except between two class-rooms of similar use, or between two galleries intended to be occasionally thrown together.

In point of position, an Infant school, if single, should always be on the ground-floor. Where, as in the case of two schools desired to be placed on a very limited site, it is impossible to avoid locating one on an upper story, the approach to the latter may be by an inclined plane if a very slight gradient can be obtained, or by a specially easy staircase, or by the union of the two. The "babies" properly belonging to both schools may then be all placed on the ground-floor, and the elder children upstairs.

The general shape and proportion desirable for the buildings

[CHAP. X.

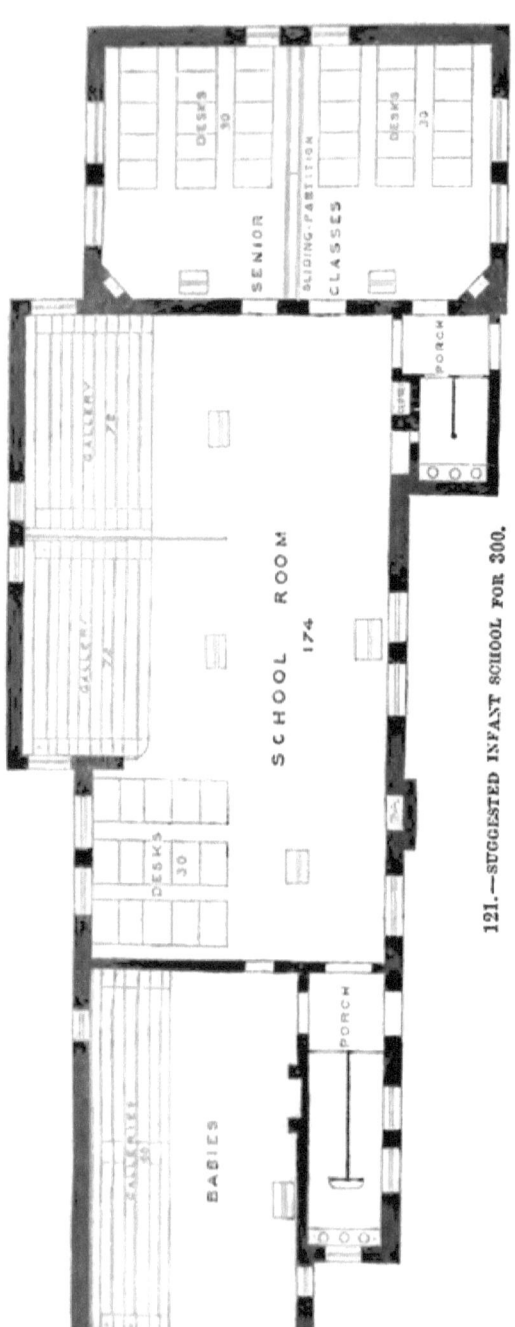

121.—SUGGESTED INFANT SCHOOL FOR 300.

may be judged of by the three illustrations given (Nos. 119, 120, and 121). The rules which govern the width of schools for children beyond the age of seven do not apply to plans for those of smaller size, to whom a wholly different method of teaching is applied. The superficial area, measured over school-rooms and class-rooms, should never be less than eight feet to each infant. Below this standard as to space no government grant on results will be paid, and the school will also be condemned as overcrowded.

The fittings will always comprise two galleries of unequal size, the larger being never so large as to accommodate more than 72 children, allowing 14 inches to each. On no account should they be without proper support to the children, in the form of backs of a low height, so as not to interfere with the free action of the elbows. The gangways should be arranged for facility of access and egress at each side, a central gangway, producing a hiatus in the direct line of the teacher's eye, not being desirable. The woodcuts Nos. 122, 123, and 124 shew an Infant gallery of maximum size, six rows deep, the seats being carefully graduated in height from $7\frac{1}{2}$ to $9\frac{1}{2}$ inches, according to the varying sizes of the children, and in each case fitted with a low back 8 inches high. The gangways are 18 inches wide, placed at each end of the gallery, and the steps are arranged so as to be out of the way, and not too high for the little legs intended to use them. The width from back to back has been fixed at 1 ft. 11 in. to afford room for freely passing in and out. Next the wall, at the rear, is panelled framing or match-boarding. This design of gallery is being universally used in the Board schools of London. Some may prefer a depth of five rows to that of six. In that event the back row of seats would be omitted, and the accommodation thus limited to 60 children. The second, or smaller, gallery may be made to run on wheels, so as to be placed near the first on the occasion of a collective lesson, if the shape of the room will admit. The gallery for the babies' room will be more suitable if made only four rows—certainly not more than five—deep, according to the shape and size of the room, and of rather greater length. The detailed sizes should be taken from the lower portion of the illustrations in view of the use being for the smallest of the children.

Galleries must always be amply lighted from the side or top, never on any pretext from the back or front. Unlike the Graded schools, where light for the writing-lesson is the chief consideration, the windows should be on the *right* of the children, because then the teacher has not to interpose herself between the light

186 SCHOOL ARCHITECTURE. [CHAP. X.

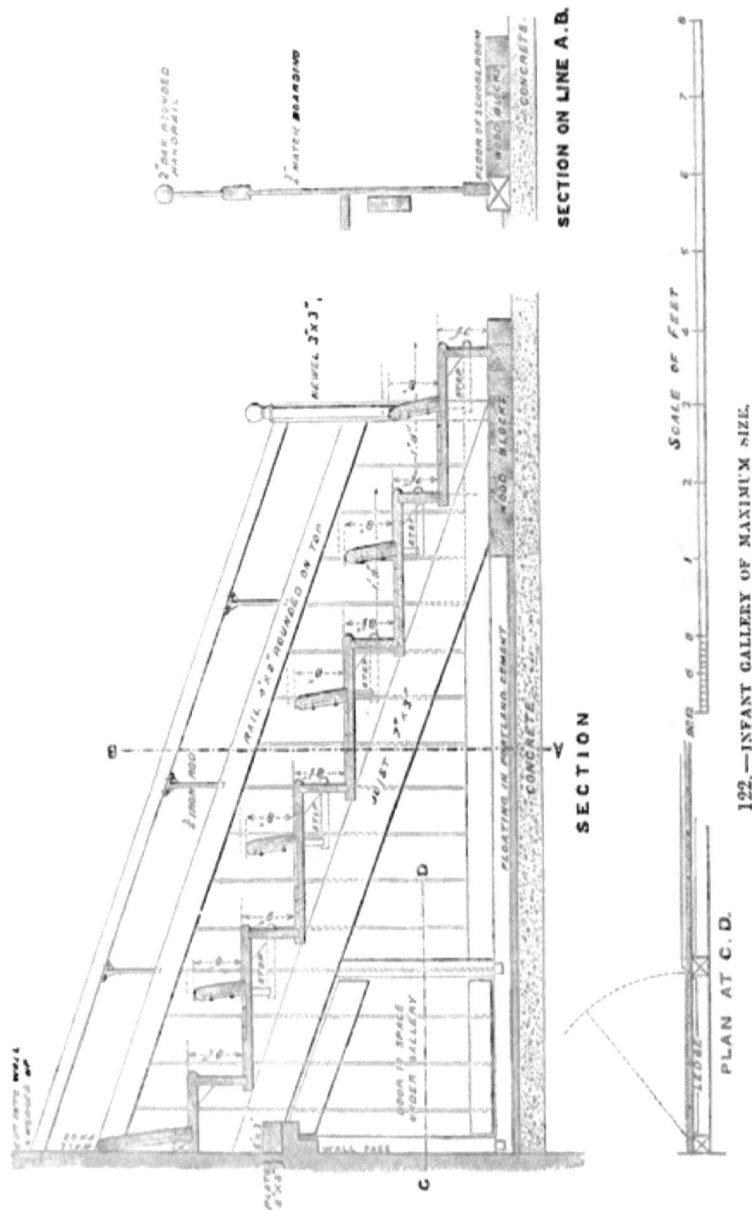

122.—INFANT GALLERY OF MAXIMUM SIZE.

INFANT SCHOOLS.

16"
WITH WOOD   L IRON

CENTRE LINE OF GALLERY

CENTRE LINE OF GALLERY

MUM SIZE.

and the black-board in giving a lesson. The use of a gallery depends on the mistress being able to see the expression of face of each child, and each child that of the mistress, so that without side, or top lighting it is simply useless. Galleries should not be placed in the school-room so as to face each other, unless the

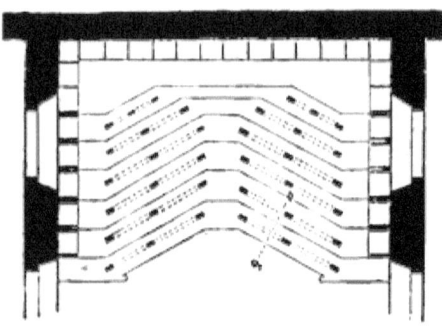

125.—PLAN OF FRENCH GALLERY.

distance apart is too great for mutual distraction by the respective groups of children. With the effort to place always one side to the window naturally comes this danger, and all the greater care is therefore required to avoid it. The children in one class should never be able to attract the attention of those in another, and the mistress should be so placed as to be able to see the whole at a glance. In

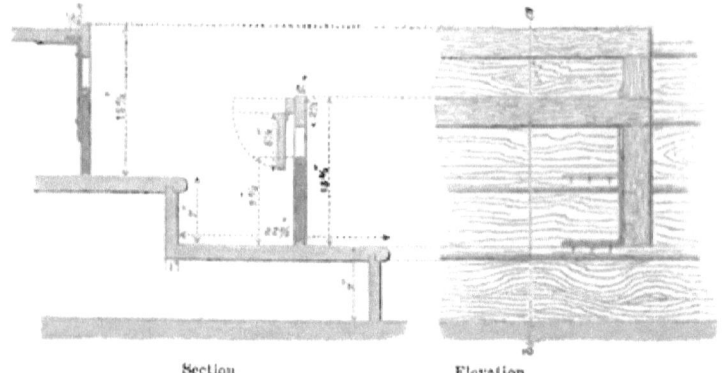

126, 127.—DETAILS OF FRENCH GALLERY.

point of economy of space, it is wasteful to place a gallery across the end of the room, because so much is then taken up with the gallery lesson.

Among his studies, M. Uchard gives an interesting specimen of a gallery (woodcuts Nos. 125, 126, and 127) from the *Salle d'Asile* in the Rue des Ursulines, Paris. The general shape is ingenious and pretty, being polygonal, so that the faces of the children may be turned towards the teacher, but it depends for beauty partly on the adoption of a central gangway to separate the boys from the girls, neither the gangway nor the separation being according to the views of English educationists, and the shape being naturally more expensive than a gallery of the straight form adopted among ourselves. Against the walls, both at sides and back, are ranged stalls or arm-chairs, which turn over the steps of the gallery, and which are provided for the use of the pensioners who assist with the lessons of a practical course. The backs of the other seats are perfectly upright, and have hinged flaps convertible into flat tables for the use of the children sitting behind, and available for the toys, little houses, and wooden cubes used in Kindergarten lessons according to Froebel's method. Usually these tables occupy a vertical position, but are fixed at will horizontally by means of a square rule turning with a segmental joint. When replaced in their upright position, the square rule presses against and assists the back.

The more advanced infants should have a small group of benches and desks for writing, which, in the case of schools of small size, would be placed in one corner of the room, and in larger schools, in the class-room provided for seniors.

In consideration of the smallness of the children in the writing classes, the lifting flap, involving more drill and trouble, is not recommended for the desks of Infant schools. The sloped top is also less convenient for the use of wooden bricks, Kindergarten exercises, &c. Two sizes of desks are sometimes necessary, and are shewn on the woodcuts Nos. 128 and 129. They are simpler than those designed for Graded schools, and have fixed flat tops. Desks form so small a proportion of the fittings to an Infant school that they cannot be said to govern the general plan of the building seriously, yet as an item of detail they deserve the same

190 SCHOOL ARCHITECTURE. [CHAP. X.

amount of consideration accorded to any other. Where one size only is used, the larger (No. 4) is to be preferred.

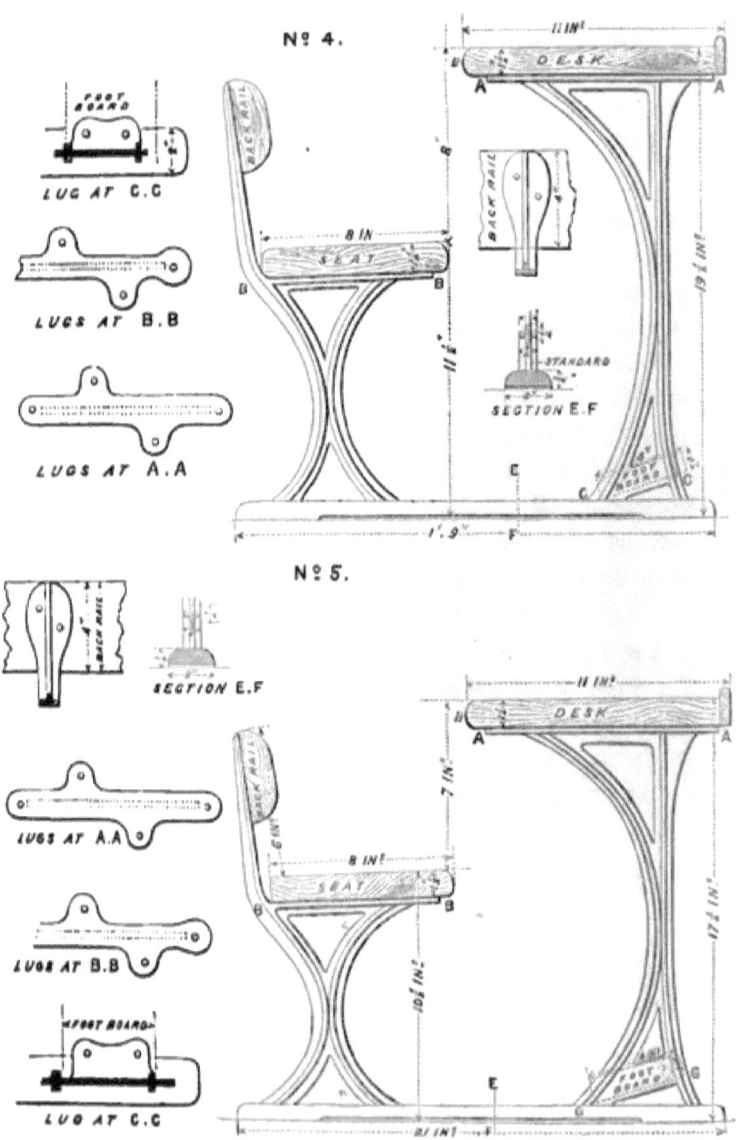

128, 129.—DESKS FOR INFANT SCHOOLS.

CHAP. X.] INFANT SCHOOLS. 191

Many examples of Infant schools, actually erected, will be found in Chapter XVI. on the Board schools of London.

An exercising or marching ground, and a playground, are both necessary adjuncts to an Infant school. The former should be covered, and, if not itself in communication with the school-door, should be connected therewith by means of a covered way. The children attending school should be able to have marching exercise all the year round. It is important that they should breathe fresh air frequently, and that the feet should be kept dry.

130.—UNCOVERED SCHOOL-ROOM.

The woodcut (No. 130) shews the kind of discipline during play, the maintenance of which Mr. Stow advocated under the name of "the uncovered school-room," and its neglect yet remains as one of the practical difficulties of school management to be grappled with and overcome. Neither managers nor teachers like the trouble of exercising supervision over the pupils during the hours of play, although it is admitted to be of the highest importance to thorough training. The consequence is, that a spirit of lawlessness often reigns supreme in the playground, slates, windows, and other property become continually damaged, and the more timid children are kept in a state of terror.

The playground should always be supplied with proper appliances for play. In the case of infants, these should no more be omitted than the furniture of the school-room, yet it is very rare to find a really well-appointed ground.

We give, without intending to quote the building as a specimen (woodcut No. 131), the plan of a playground fitted as recommended by the Home and Colonial Society, whose long labours in

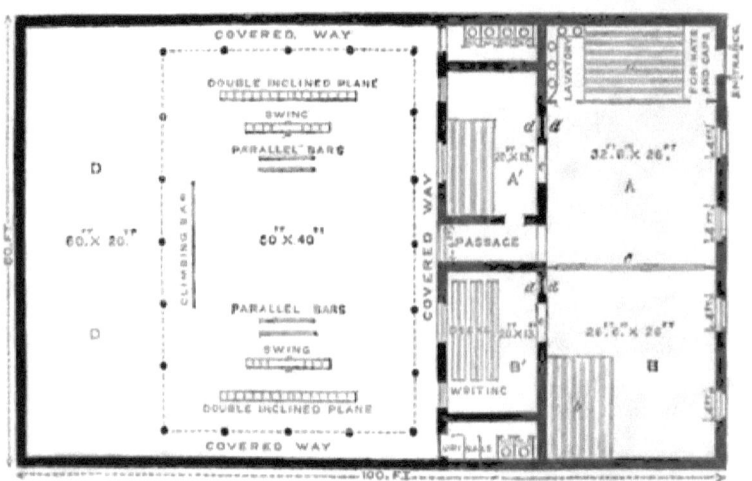

131.—PLAN OF PLAYGROUND SUGGESTED BY HOME AND COLONIAL SOCIETY.

the field of Infant education entitle their opinion to some weight. There are five different kinds of apparatus recommended, the fifth not being shown on the plan. When used, it would naturally be placed in a central position. The other four are all shown, symmetrically arranged. The whole comprise in detail—

1. (Woodcut No. 132.) A Double Inclined Plane for the use more particularly of the youngest children. In the construction, each length of plank should be about 12 feet long, and should be

raised about 2 feet 6 inches from the ground. On the upper sides of the planks should be nailed small cross-pieces, at intervals of 8 or 9 inches, to prevent slipping.

2. (Woodcut No. 133.) Wooden Swings, or see-saws, also for the use of the younger children. These consist of plank, 2½ inches thick, about 12 feet long, fitted with three handles or divisions at

132.—DOUBLE INCLINED PLANE.

each end and made to move on a fixed pivot of thicker plank standing about 18 inches out of the ground. Between the handles, the plank should be rounded so as to make an easy seat.

133.—WOODEN SWING.

3. (Woodcut No. 134.) Parallel Bars. The bars themselves should be from 6 to 8 feet long with rounded corners, and project-

134.—PARALLEL BARS.

ing about 4 inches beyond the posts. There should be two sets provided, one 2 feet 9 inches high and the other 4 feet 9 inches high. The posts should be 18 inches apart.

4. (Woodcut No. 135.) The Horizontal Bar about three inches in diameter and 6 feet long, made smooth and round, of hard

135.—HORIZONTAL BARS.

136.—THE CLIMBING STAND.

wood. There should be three of these, one 6 feet from the ground another 5 feet, and a third 4 feet.

5. (Woodcut No. 136.) The Climbing Stand, consisting of a simple frame, supporting ropes for climbing, the latter being always attached at the bottom so as to hang loosely.

Additions, such as Giant Steps, Rope or Circular Swings, &c., may be made to the list, in cases where the recreation ground is constantly under the supervision of the mistress. Experience, has shewn that, otherwise, they are not free from liability to accidents. And, in ordinary cases it is best to limit the provision to the five articles recommended. They may be used for young children quite safely, and with much advantage to their health and physical development.

CHAPTER XI.

ELEMENTARY GRADED SCHOOLS.

Division of Schools into Infant and Graded Departments—Examinations—Number of Class-rooms in relation to School-room—Their Size—The School-room—Position of Class-rooms in reference thereto—Study of Internal Details.

PUBLIC Elementary Schools are not, in so many words, divided either by the Education Act or the Code into "Infant" and "Graded" departments, but such a division is, in practice, found to work satisfactorily. When education has fairly begun, all schools attended by a regular succession of children, and with work duly arranged, must be in some sense "graded" schools. On leaving the Infant school, therefore, where results are chiefly tested by number of attendances, children pass into the Graded school, where they undergo examinations from time to time. There are six standards of examination, the essential subjects in which are set forth in article 28 of the Code, viz. :—

	STANDARD I.	STANDARD II.	STANDARD III.	STANDARD IV.	STANDARD V.	STANDARD VI.
READING.	A short paragraph from a book used in the school, not confined to words of one syllable.	A short paragraph from an elementary reading book.	A short paragraph from a more advanced reading book.	A few lines of poetry selected by the inspector.	A short ordinary paragraph in a newspaper, or other modern narrative.	To read with fluency and expression.
WRITING.	Copy in manuscript character a line of print, and write from dictation a few common words.	A sentence from the same book, slowly read once, and then dictated in single words.	A sentence slowly dictated once by a few words at a time, from the same book.	A sentence slowly dictated once, by a few words at a time, from a reading-book.	A short paragraph in a newspaper, or 10 lines of verse, slowly dictated once by a few words at a time.	A short theme or letter, or an easy paraphrase.
ARITHMETIC.	Simple addition and subtraction of numbers of not more than four figures, and the multiplication table, to 6 times 12.	Subtraction, multiplication, and short division.	Long division and compound rules (money).	Compound rules (common weights and measures).	Practice and bills of parcels.	Proportion and fractions (vulgar and decimal).

The children are "*graded*" or classed accordingly. Singing must be taught, and in Girls' schools, plain needlework and cutting-out must form part of the ordinary course of instruction. In addition, there are other specific subjects in respect of which extra grants may be earned for the school by children presented in Standards four, five, and six. These may be geography, history, grammar, algebra, geometry, natural philosophy, physical geography, the natural sciences, political economy, languages, or *any* definite subject of instruction extending over the classes named, and taught according to a well-developed and complete scheme, of which H.M. Inspector can report to the Education Department that it is well adapted to the capacity of children, and sufficiently distinct from the ordinary, or "essential" subjects, to justify its separate description as a "specific" subject.

Drawing may also be taught, and special grants obtained from the Science and Art Department, for proficiency in this acquirement. This explains why, in some exceptional cases occurring in particular neighbourhoods, an extra class-room for teaching drawing has been thought necessary. As a rule, a special room for drawing is not necessary in elementary schools, such drawing as is there taught being easily carried on at the ordinary writing desks. A reserve class-room for many purposes of expansion in the working of a school may be, and is, desirable. It is, indeed, never omitted from a German school.

The Code forms the law for English elementary schools next in importance to the Education Act. Other subjects not set forth above, may be—and frequently are—taught under regulation. The "standards" may be regarded as the minimum instruction on which any grants of money can possibly be made.

The examinations take place annually, and are conducted by H.M. Inspectors, who are appointed by the Education Department. In addition to the examinations themselves, it is the duty of the Inspectors to pay occasional visits within their own districts, as far as possible, with the object of satisfying themselves that the conditions upon which the annual grants are made receive due attention throughout the school year. One anomaly in the work of the Inspectors is, that they inspect and report not only on the instruction—of which they are always excellent judges—but on the school fabric, its warming and ventilation, of the site and other matters,—in which, if not deeply versed, they certainly cannot be blamed. A very different state of things exists, as already shown, in Germany, where the inspection of the education and that of the building is each placed in separate, and duly qualified, hands with the most excellent results.

In theory, it is assumed that a child should be ready for examination in the subjects comprised under the first standard at the end of the first year after being transferred from the Infant school. In practice, it is often found impossible to present a pupil successfully until the next examination. Thus, attainments rather than age must always form the basis of classification. A child must have attended school 250 times, morning or afternoon during the school year, in order to be eligible for presentation at the annual examination.

It is not uncommon to find some of the standards taught together in the same class. The theory of instruction, however, clearly indicates that, in a perfect school, each subject of examination should be studied in a separate class if possible. Thus, following out the principles already laid down as to the proper numbers of a class, six classes of 40 each would appear to be the most perfect school, giving 240 as the total number of each department. In practice the numbers differ considerably, and the children in the lowest standards are more numerous than those in the highest. Any given number of children, entering a school at a given time, will have lost a certain per-centage by deaths before the sixth standard is reached. The higher classes, while containing fewer, consist of older and bigger children who require greater space than those in the lower. This equalizing or neutralizing tendency renders the planning easier. Perhaps the simplest solution is to make senior and junior class-rooms both for the maximum number of class adopted for the school (whether 30 or 40) and, in the case of the seniors, to fit them with desks for a somewhat smaller number. Another plan would be to abandon the division between the two senior class-rooms and adopt one room for a slightly reduced number.

The limitation of the number of class-rooms to the number of classes containable by the school-room itself has been alluded to, and it is clear that no further or more rigid general rule can be laid down. The size of the class-rooms should also be considered. Some hold the opinion that, although the double class consisting of 60 children may really be the best in point of size and facility of management for the certificated and the pupil-teacher working together, yet when this number comes in the ordinary course of school work to be divided—one half being placed in one class-room and the other in another—the rooms themselves (of course planned and built for 30 only in each case) are too small. In the plan suggested in a former chapter (woodcut No. 118, p. 174), it will be observed that the size of class in the *school-room* is only 30 in each instance, whereas in the *class-rooms* it is 40. This, providing good airy rooms for the classes when isolated, and limiting the numbers when in the school-room, has the manifest disadvantage that the classes are not of the same size, and, when moved for different lessons in the course of the day from school-room to class-room and *vice versâ*, the children will either be crowded in the one case or under the number in the other. Strict economy of building would therefore indicate that the class numbers should be the same in both cases and should be absolutely determined before the planning be commenced. The movement of the classes here alluded to forms one of the differences between German and English practice. In the former country, the child is taught entirely in one room until he is ready to be moved into the class next above.

The school-room should be so planned that the doors and fire-places are not on the same side as that along which the desks are ranged, otherwise space will be wasted and the room rendered draughty for the children who sit near the door. The desk side of the school-room should be to the N. or N.-E. The width should be 22 feet for rooms of large size, although in small schools 21 feet, or even 20 feet, may suffice. The length will be determined by the number of children to be seated in pair-desks, five rows deep, along one wall. The position of the fireplace should never be opposite the centre of the class, that being the natural place for the teacher. Perhaps nothing requires greater care in arrangement than the positions of doors and fireplaces. They must be convenient, and yet entirely out of the way. So disposed, also, as to afford the space necessary for the school cupboard.

Class-rooms should not be passage rooms, and should always be on the same level as the school-room. Yet, in the case of the *double* class-room, where,—the teacher and pupil-teacher working together,—the two rooms may in some sense be considered as one, the rule need not always be rigidly applied and one room may sometimes be treated as a passage to the other. While pointing out such an exception, it is hardly necessary to add that a separate access is, of course, better if it can be obtained without loss of space, inconvenience, or damage to other points of more importance. In the class-room the fireplace is most conveniently located in or near the corner.

As to the position of class-rooms in relation to the school-room, especially in view of the adoption of the grouping of the former in pairs—one opening into the other,—it may be remarked that the best plan would appear to be to place them across the ends of the school-room. The system of teaching by means of a certified assistant and a pupil-teacher working together then receives, apparently, its most convenient embodiment in the form of building. By this plan the expense and wastefulness of corridors is avoided, time in management is not lost because of the straggling and inconvenient positions of rooms, and easy supervision is secured, for the principal teacher can see at the same moment what his assistants are doing at both ends of the rooms, and each of these, in turn, can observe the progress of his pupil-teacher's work through the glass pane in the sliding partition. By it the whole school can be conducted with ease, economy, and with an *esprit de corps* and sympathy impossible for young children taught in the separate rooms of a German school. A further advantage is that it secures facility of "through" or cross ventilation by the windows opening at a high point, and thereby prevents stagnation of air during the hot days of summer when no other means of ventilation is of much use. When the partitions are drawn out between the two class-rooms, cross-ventilation by the principal windows is shut off. It therefore becomes necessary to insert a window (where it is certainly not required for light) high in the back wall and as near as possible to the partition. This window should be small and quite up in the corner of the room, or it may prove a nuisance to the teacher, especially if it happen to face the south.

One of the objections to the position of the class-rooms across the two ends of the school-room is that, in the case of the latter, it effectually cuts off the admission of light from the side, where —especially from the left of the children—it is of the greatest value. This objection has been minimized by placing the line of children with their backs to the north or east and elevating the windows as high as possible in the wall. The master, then, is not compelled to teach with a painful glare of sun in his eyes, or with a room darkened by blinds which he cannot dispense with.

In some cases the evil can be wholly removed by moving forward the class-rooms so as to leave space for windows at each end of the school-room. Where the site, and other conditions, will allow this, it forms an improvement on the plan already described. The arrangement which places the class-rooms parallel with the school-room, and in which the doors of the former are immediately opposite the line of desks in the latter, is not desirable. Ventilation and lighting are then both more difficult, and the teacher has always his back to one part of the school or another.

The plan advocated is not the only possible arrangement. Others will often be unavoidable from the conditions of the site. On the whole we consider it the best.

Another item of arrangement, which becomes of serious importance as the graded department approaches its maximum size, is the necessity for clearing any portion of the school easily and quickly without disturbance to any other portion still at work. The children in class-rooms placed across the end of the school-room should have direct power of egress without being obliged to traverse the school-room. This points to the desirability of the staircases being in the corner, accessible both to school-room and class-rooms separately, at the landing. In a department of the largest size, 240 for instance, there should be *two* staircases each so placed, if confusion would be avoided.

The position of every window, every door, every fireplace, every teacher's desk, every cupboard, and every opening for ventilation should be such as to facilitate the general work and comfort of the whole school. In each case, and under every fresh set of conditions, there is always one position and arrrangement for each detail better on the whole than any other. To determine each of these successfully, constitutes much of the success or non-success of school planning. If the black board cannot be well seen, if the teacher or children are exposed to draughts or cold, if the cupboards, teachers' desks, or fireplaces are in the way, if the doors are inconveniently placed, or if there is not sufficient space for proper gangways between the desks, then there is less or more fault to be found with the building arrangements, and the work of teaching—always arduous enough—is by so much impeded instead of assisted. If the difficulties be got rid of by the provision of excessive floor space and needless accessories—as they may easily be—the result becomes one of sheer extravagance. It is only when all the several points are contrived in the best and most skilful manner within proper economical bounds, when the union of perfect arrangements with strict regard to expense is found, that the school can be considered satisfactory. Then, and then only, do we arrive at good school planning.

The several matters alluded to are treated of under their respective heads in this work, and can be applied to particular cases as occasion may arise. In addition, specimens of Graded (and other) schools, shewing the arrangements necessitated by site and circumstance, are given by woodcuts of sufficient number and variety from among those erected recently in London.

CHAPTER XII.

PRACTICAL DETAILS.

Temporary Schools—Walls—Entrances—Staircases—Lavatories—Hat, Cloak, and Bonnet-Rooms—Their Arrangement in Mezzanines—Latrines—Playgrounds—Teachers' Rooms—The Master's House—The Caretaker—Windows—Sliding Partitions—Floors—The Dado.

Temporary Schools.—Before incurring the great cost of a permanent building, it sometimes becomes necessary to test the accuracy of the statistics previously taken, by planting in a neighbourhood a temporary school, so that the experience of a few months may indicate its future character, as well as the existing local demand for instruction. The power of enforcing children's attendance would seem to convert an array of carefully obtained statistics into an infallible guide for the numbers of a school in any neighbourhood. In reality, the facts and figures must be taken together, and the former as well as the latter tested in the most certain manner, if the erection of useless or unsuitable buildings would be avoided. The law of demand and supply deserves special observance when the creation of costly public buildings is involved.*

---

* For nearly two years after its election, the first School Board for London occupied itself in establishing temporary schools (principally in hired rooms) and collecting statistics. Apparently, time was wasted. According to the literature of that period, any situation and any building must be fit for a school, and the Board was unfit to cope with the stupendous ignorance of the Metropolis. Towards the close of their three years' term of office, the charge became different. Formerly they were doing nothing, now too much. A careful inquiry before a Committee of the House of Lords brought out the fact that *their sites had been chosen in the right places*. No greater proof of far-seeing economy could be adduced.

Temporary schools will sometimes, also, become necessary where a population is changing in character or numbers from local causes, or where existing schools, though sufficient in quantity, are unsuitable to the class of people in their neighbourhood.

The provision of a temporary building is not easy. In London, iron buildings are allowed, but they are found to be so pervious alike to the heat of summer and the cold of winter as to be unfit for school purposes. If lined throughout with wood they become expensive. Yet this is necessary so as to obtain a circulation of air between the wood and the metal, and to prevent heat or cold striking directly into the interior. The Metropolitan Building Act prohibits the erection of temporary buildings wholly of wood because of liability to fire. For country districts, they are best, as their use has proved in the damp outlying parts of Ireland. Framed in lengths or sections of ten to fifteen feet, and put together with screws, they can be taken down and removed from place to place. If plastered inside, and covered outside with cement or rough cast enveloping the whole of the wood, they are the driest and warmest buildings possible.

The pressing wants of the long-neglected population of London have necessitated temporary buildings in two or three cases. In view of the requirements of the Building Act, the outer covering is entirely of corrugated iron, though the construction is of wood and iron combined, and the wooden lining or match boarding has been found necessary. They do not remain long in one place,—being usually located on the playground till the schools are completed. Their cost is nearly as great as a common brick-building.

Whenever obtainable of suitable character, it is better and cheaper to hire existing buildings, warehouses, Sunday Schools, or other places, especially if not requiring much preliminary outlay to render them fit for occupation, until the permanent fabric can be reared not only worthily but of a kind suited to the wants of the people among whom it is placed.

Walls.—The external walls should be of sufficient thickness. A thick wall retains warmth better during the winter months, and secures a cooler school-room in hot weather. Proper warming and ventilation also requires not only plenty of chimneys, but also of air shafts for ventilation, and these cannot well be obtained in thin walls. In London, the Building Act settles the *minimum* thickness, and although an experienced architect will not waste his material or dispose it unskilfully there or elsewhere, yet a Public school comes within the category of public buildings, and is hardly a fit subject for

"The lore
Of nicely balanced less or more,"

especially in those matters which pertain to the solidity, strength, and durability of the fabric. The truest economy will generally point to the adoption of something more than the compulsory requirements of law.

The material should be selected according to the yield of the locality. In London it is naturally brick, and should there be selected in reference to good colour and to the hardness necessary for successful resistance to the climate.

Entrances.—The door of a school-room should no more lead directly from the street than the door of a room in a dwelling-house. There should always be a lobby or porch, because, while the children are arriving, the door is either fastened open, or is continually being opened and shut, and the warming of the room is rendered well-nigh impossible until all have entered. Again, it is well to have a place of shelter where a child can stand while waiting to be admitted, otherwise the general health of the school will suffer, particularly in the case of the youngest children.

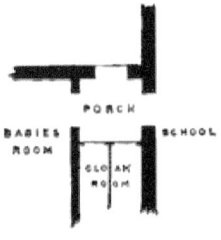

137.—PLAN OF ENTRANCE PORCH (INFANTS').

This porch or lobby should not project into the room, for it is then not only liable to be unsightly, but must always be inconvenient to the internal use of the room. The doors themselves should be so contrived that a child can easily open them. They must be sufficiently wide to admit of the passage of furniture, but to have doors of such size constantly in use is highly inconvenient and productive of an unnecessary amount of chill to the school-room: it is easy to have them made folding, so that one portion,—say of 2′ 3″ or 2′ 4″ wide, can be generally used, and a second portion, say of 1′ 2″ or 1′ 4″ wide, can be brought into play when the children are being dismissed, when furniture is being carried in or out, and at any other time.

The position of the respective entrances is also no unimportant point. Sometimes the absence of more than one street frontage, or the necessities of the general plan of the school as controlled by other peculiarities of site, will bring all the entrances near to each other.

Where it can be secured, the Girls' and Infants' entrances should be in one street and the Boys' in another. It is common for girls who are themselves going to school to take their younger brothers or sisters to the Infant school, and this reason alone is sufficient to indicate the advisability of these two entrances not being too far apart. Again, when the school is dismissed, the boyish element is apt to be hilarious somewhat at the expense of anything coming in the way, and children only just beyond the walking age are better out of range.

In rough neighbourhoods the lower part of the door should be covered with strong sheet iron, able to bear successfully the perpetual kicking which it is sure to receive.

The principal entrance should always have a bell communicating directly with the caretaker's rooms. Also a letter box.

Staircases.—Every department of a school, above the level of the ground floor, should have a separate staircase, and, if of very large size, even two separate staircases. Such arrangement will tend to avoid confusion among the children. It is also desirable that the separation shall be so complete as to render almost impossible any dispute or rivalry in jurisdiction between the various masters or mistresses in charge of the several departments. It is important in the case of large schools to be able to discharge the occupants of one set of class-rooms, or to allow a single child to go out, without disturbing the entire school. Perhaps it may be laid down, generally, that where the number reaches 240, a second staircase is desirable. Below this number, one staircase should be sufficient, if placed in a central position easy of access from both school-room and class-rooms. It should always approach the school-room on the side opposite to the desks, so as to avoid waste of space in the arrangement of the latter.

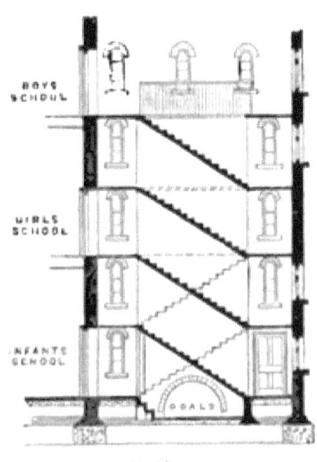

138.—THE DOUBLE STAIRCASE.

Considered in the light of part of a public building, and constructed, therefore, of stone or other fireproof material, it is natural for architects to plan school staircases with some degree of size and importance. What is the difference between this and any other staircase? First, the use is different. And this should be the first consideration. The staircases are to be used by *children*, hence the steps should be different from the usual proportion. No "risers" should exceed six inches. No "winders" ought to be used at all. The landings should be wide, but the staircases narrow. Both sides of each flight should be provided with a handrail, not of iron, but of oak or other hard wood. When the width is enough to allow several children to go abreast, there is risk of those in the middle being pushed down stairs, and hence it would appear that 3 feet 6 inches to 4 feet as a maximum may be regarded as an extreme width sufficient for the largest schools. Indeed, the numbers in the school ought not materially to affect the width and size of the staircase, but rather to suggest more staircases. The Germans provide very wide staircases,—sometimes more than twice the foregoing,—but they are careful to use very short flights, and broad low steps.

The *double* staircase (woodcuts Nos. 139 to 143) is economical of space, and also, somewhat, of cost. It has also the merit, while keeping the children in each section as distinct by means of a solid wall as though the two staircases were hundreds of feet apart, of facilitating supervision by managers from pass-doors on the half-landings, and of rendering easy the economization of space for lavatories and cloak rooms in mezzanines. Its chief objection is the unavoidable length of each flight in order to obtain headway. It has been much used in the new schools of London in localities where land is very dear.

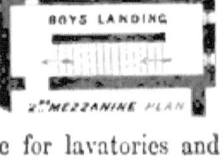

The *following* staircase separates the children completely, and used with an open well-hole, does not prevent them from at the same time seeing each other. It derives its name from the fact of two flights leading to different stories following one over the other in the height of one story. Pass-doors are not easily obtained, except at the top and bottom of the staircase respectively. And managers and teachers do not like the trouble and discipline it involves.

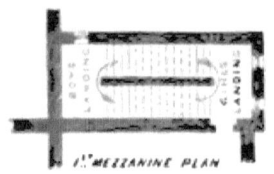

139 TO 143.—THE DOUBLE STAIRCASE.

It is so important that a staircase for children should have flights as short as possible, whether for safety or the rest obtained by landings, that, where the general plan will admit of it, or where it is not very important to economize land, the short-flight plan must be pronounced the best. If adopted, another complete staircase will always be required for a schoolhouse of three stories, and it is, therefore, not the cheapest. The height of the stories being 15 feet from floor to floor, *three* flights (woodcuts Nos. 144, 145 and 146), providing two intermediate landings, can be arranged in the ascent of each. In some cases, *four* flights can be adopted and space on plan saved. Of course the school of maximum size, providing graded departments of 240 each, would require four complete staircases, viz., two to the first floor and two to the second.

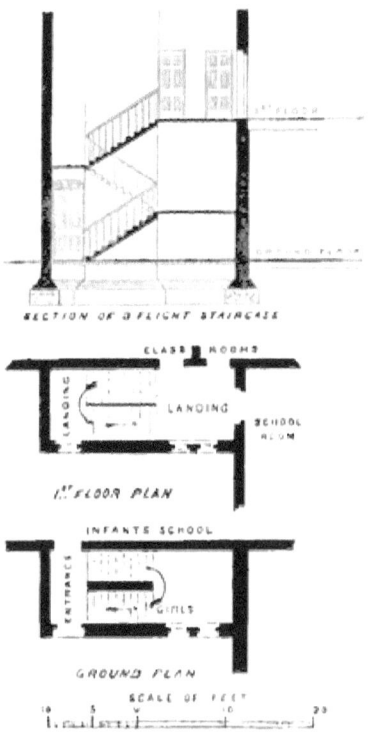

144, 145, 146.—THE SHORT-FLIGHT STAIRCASE.

Lavatories.— The washing-rooms for children should not be so placed as to involve possible cold or wet feet in reaching them, as when a yard or playground has to be crossed. Neither is it a good plan to utilise one or both sides of a porch or entrance passage with lavatory fittings. For wherever the washing-process is carried on there is sure to be more or less of sloppiness or untidiness, which is best placed apart and away from the eye in a separate, though small, apartment convenient of access from the schoolroom. The common method of placing the basins across one end of the cap and cloak-room should not be adopted where the general plan admits, without extravagance, of

P

a better arrangement, for the caps and cloaks are thereby liable to become splashed and wetted. If adopted, it should be with the provisions for separation shown in the woodcut, the lavatory really forming a separate room although reached through the cloak-room. The water should be cold only, for any attempt to provide hot water would only result in all kinds of evils, not among the least of which would be the constant state of disorder and want of repair. The schools in which, according to some theorists, children should be washed from head to foot, are not

147, 148.—LAVATORY AND CLOAK-ROOM COMBINED.

now alluded to, but simply those in which the ordinary cleansing of hands and face is *occasionally* necessary, as in the elementary schools.

Considerable difference of opinion exists as to the number of basins which should be provided. Some teachers aver that there should be a sufficient number to wash an entire class in two relays, as otherwise much loss of time and consequent delay in commencing lessons must ensue. Others contend that a good disciplinarian will compel each child to come clean to school, and that although the lavatory is unavoidable, its use is rather exceptional (as for instance after play, or at the commencement of a writing-lesson) and that therefore three or four, or at most half a dozen basins are sufficient for the largest school. Perhaps a mean may be drawn between these opposite opinions, and an arrangement giving two or three basins to each hundred children in the school may be found satisfactory. Let us not forget, that in the well-appointed schools of Germany lavatories are not provided. In the case of two infant schools side by side, the second being also under the general control of the mistress of the first, one lavatory, if conveniently placed, may suffice for both. Where, as in infant schools, a child cannot wash by itself, the number of basins is naturally limited by the number of teachers, and a large array would be absurd. The distance apart should also be greater, and

149.—INTERIOR VIEW OF LAVATORY.

the height from the ground less. The infant lavatory should be 1' 11" or 2' 0" from the floor, those for graded schools 2' 2". In workhouses, where children are all washed together, it may be proper to have the water turned on at every opening simultaneously, by means of one tap. In public schools this would involve great waste of water every time a child went to wash his hands. Each child should be able to obtain water separately at a single tap. The woodcut (No. 150) shews a kind of lavatory which has proved best under all the circumstances. Instead of being made to tip up on a pivot (a first-rate plan for a club), the basin is fixed, but the removal of a couple of screws is sufficient to release it for the removal of any stoppage. The plug cannot be removed, and lost, as when attached to a chain. For letting out the water, it is lifted half an inch, and turned half round. The water cannot be left running, for the removal of the child's finger lets drop the leaden (or iron) weight and turns off the tap.

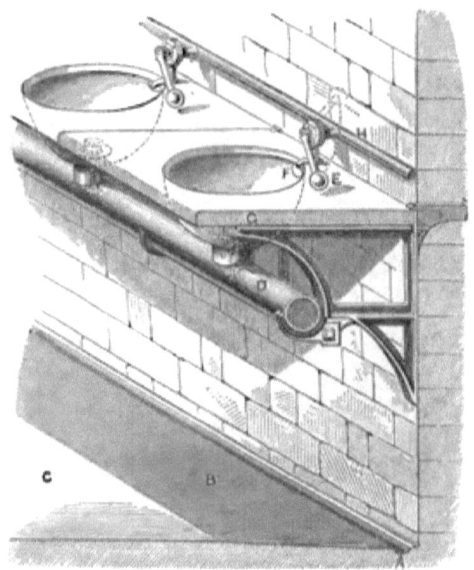

150.—SKETCH SHOWING MECHANISM OF LAVATORY.

References.

A. Gutter in floor, under lavatory.
B. Portion of floor sloped.
C. Level floor.
D. Iron drain.
E. Tap, with lever and ball of lead or iron.
F. Overflow in enamelled iron basin.
G. Slate slab.
H. Supply pipe.

This kind of tap has the disadvantage of tending to jar and burst the pipes if the ball be dropped violently. Another kind,— having a self-acting screw regulated to run off a certain quantity of water, and closing of itself when this has been discharged,— is in some respects superior. Instead of the enamelled iron, liable to lose its enamel in a few months, a thick slate slab is on the whole considered best for the tops, the former material being retained for the basin. An overflow should always be provided from the basin, so that in case of accident the room be not flooded. In the floor, and immediately under the line of basins, there should be a gutter to carry away the water always splashed on to the floor. The floor itself should not be of wood, for then the continual alternations of a wet and dry state would soon cause it to rot. It should be of stone, asphalte, or first-rate cement, and, for a certain distance back, should be sloped towards the gutter.

Cloak-rooms.—The common, indeed almost universal, practice of hanging up caps, cloaks and bonnets along the walls of the school-room, would appear (like some other arrangements) to be the result of a cheeseparing economy. Whether from the untidy appearance, the fact that at some seasons of the year many of the articles are wet and send out unpleasant odours, or other reasons, it is a practice not to be encouraged. On the other hand experience in some of the denser and rougher parts of London shows that when an outer corridor or entrance is used as a cloak-room many of the articles are stolen. Hence teachers will frequently be found to favour the use of the school-room for the purpose.

To a well-contrived little room, specially devoted to the purpose, no such objection can attach. If the upper part of the school-room door be of clear glass to enable the teacher to command the entrance, so much the better. The Germans never have separate hat and cloak-rooms, and invariably use two of the class-room walls for the purpose, yet Dr. Wiese says:—"Near the entrance to every department, in the most convenient place that the peculiar arrangement of the school building may afford, there should be a peg for the use of each scholar, and near at hand there should also be convenience for washing. In the Victoria School for girls (Berlin) there are wooden cases divided into small compartments, for holding overshoes, &c. Umbrella stands and other conveniences are also included in the admirable fittings of this superior establishment." The fittings of this school certainly form a remarkable exception to the general rule.

The cap and cloak-room should always have two entrances, and a wooden division not more than six feet six inches high. The caps, &c., can be hung along the walls and on both sides of the partition, and, on departure, the children can be filed in at one door and out at the other by the teacher, and order preserved to the very last. In girls' schools, whether elementary or secondary, the cloak-room should be provided with places for shoes, and a bench on which the girls may sit while changing shoes. In the case of Infant schools, the caps of the youngest boys are frequently collected in baskets. If this be preferred, the plan of the cap-room will be the same, minus the wooden division, for the elder infants will file in and out as before described, and the baskets will be deposited in the centre. It is better to inculcate habits of order as early as possible, and only in the babies' room should the basket arrangement be permitted.

Some method of warming the cap and cloak-room is desirable, so that in wet weather, while the children are at school, their clothing has some chance of becoming dry. This can be done either by a line of hot-water pipes (where warming by hot water has been adopted for the school), or by hot air sent in from the back of one of the school grates.

In point of general disposition, also, a word may be said. A school-room is, or should be, some fourteen feet or more in clear height. Such a height is absurd for a lavatory or cloak-room. Hence it will be found economical to adopt a system of *mezzanine* floors (woodcut No. 147, page 210), obtaining two stories of these little rooms in the same height as one story of the school-room, always taking care (in view of possible theft from the outside) to place the cap and cloak-room on the level of the school-room to which it belongs. The lavatory may, with little objection, be placed on the half-landing. The hat-pegs should always be of *wrought* iron, if intended to last more than a month. They should be fixed fully eight inches apart to avoid the caps or bonnets touching each other, and should be consecutively numbered. Several rows of pegs may be placed above each other, but the highest should not be too high to be reached by the elder children.

Latrines.—The latrines should be so located as not to be too far from the school, and yet sufficiently far off to prevent any possible smell reaching the windows. But for the fact that due attention is not always paid to the proper condition of closets by school teachers and managers, it would not be necessary to point out this. The approaches and doors from the schools should, of course, always be separate for the sexes, and the former should—imperatively in the case of infants—have covered ways, so as to obviate any danger of damp feet.

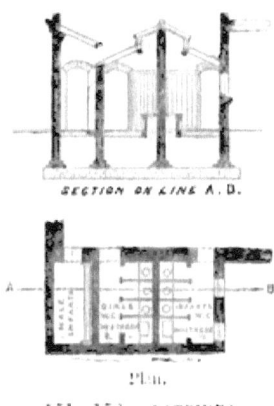

151, 152.—LATRINES.

In general plan (woodcuts Nos. 151 and 152), the water-closets and urinals should be masked. Without shutting any door, the

153.—INTERIOR SKETCH OF URINALS AND LATRINES.

whole should be perfectly enclosed from observation. Each closet should be separated by a partition, and each should have a separate door, with proper light and ventilation.

As to fittings, iron troughs have been found in practice to be the most economical, and the least open to tampering or liability to get out of order. They should invariably be thoroughly flushed out once every day, in summer twice a day. Neglect of this precaution frequently produces a nuisance. There are other sanitary inventions or appliances, equally good but more expensive.

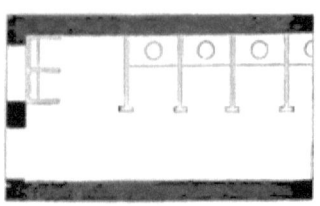

154.—PLAN OF URINALS AND LATRINES.

In the case of schools on upper floors it is desirable to have one or two closets approached from the staircase landings, so that a child may not be obliged to waste time by having a long distance to traverse during the hours of lessons. One closet, also, should be close to the babies' room of an infant school. The number should be, three to the first hundred, and one in addition for every succeeding fifty. The teachers and caretaker require also separate provision. The caretaker's should adjoin that for the children, to ensure the necessary attention to the latter.

Playgrounds.—The playground is a most important adjunct to a school, and, whether for fresh air, exercise, amusement, recreation, or discipline, is quite as necessary in the production of satisfactory educational results as a class-room, or any other portion of the school proper. Not inaptly did Mr. Stow call it the "uncovered school-room." A good teacher will often be found to regard it as but another place for another kind of instruction.

It should not be of a straggling, inconvenient form, but compact and without recesses or places where children can remain long out of sight. A northern or easterly aspect should never be wantonly provided when a southerly or westerly one could have been as easily obtained by no other outlay than that of a little common sense. A portion should be covered, so that in wet weather (which amounts in England to perhaps one-half the year) the children may not be compelled to play in their school-rooms. In the case of Infant schools, this covered portion is absolutely indispensable, as already shown, because marching forms so important an element in their preparatory instruction. It can generally be obtained in the form of a slight shed, open on one side, but, in some cases, and where land is dear, it may be convenient to raise the boys' and girls' schools on a low story of eight to nine feet high, and thus to obtain some portion of the covered playground underneath. In such cases, care will be required to prevent a cold draughty result.

As to the sizes of playgrounds for different schools it is difficult to be precise. On account of their more active outdoor games requiring space, the boys should undoubtedly have the lion's share, while the infants—too young to develop all the uses of a playground—will be happy in one much more limited. Perhaps a space of about twice the size of the school-room and class-rooms is necessary for the latter. Where land is dear, and in consequence limited, one playground may suffice both for the girls' school and the infants', an arrangement being made by the respective mistresses for its use at separate times. Without such arrangement there is risk of disorder, no *one* being responsible for the discipline of *all*. If there are two Infant schools or departments on the same site, the girls should be provided with a separate playground, because then the numbers are sure to be too great for one.

In point of arrangement it is found best to place the girls' playground next that of the infants, and to maintain separation without preventing vision by a division of wood or iron. That of the boys should perhaps always be isolated completely by a wall of, say, six feet high. Educationists, however, take as widely different views of these matters as of mixed schools. At the Home and Colonial Schools, for instance, boys and girls are separated when at play merely by a railing, and it is said that the arrangement is not found in practice to be objectionable. But at the Model National Schools in Dublin already alluded to at page 52, a much bolder course is adopted. There the *whole* of the children use the same playground. Lines in the pavement mark the portions devoted to boys, girls, and infants separately. And it is held to be a breach of discipline for any child to go beyond bounds.

Perhaps of all materials for forming the surface, gravel (unless of a hard-setting kind) is the worst. It affords endless ammunition for stone-throwing, and is apt in wet weather either to work

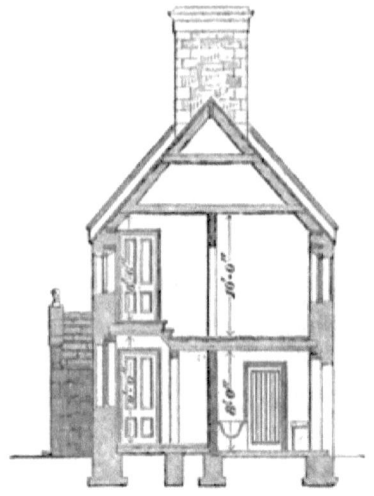

155.—SECTION A B.

up into a thick sludge, or to settle or wear into holes which stand full of water, while in dry weather the pebbles are all loose on the surface. Tar pavement, or asphalte, forms the cleanest and most generally satisfactory finish, the chief demerit being the presentation of a hard surface to a child falling from a swing or parallel bars, an objection more than counterbalanced by the advantages possessed. Wooden paving in blocks set endwise, is sometimes used, but it has no special advantage over the former, while being much dearer. A good tar pavement can be laid for half-a-crown per square yard in London, while the wooden block floor would be worth fully three times that sum.

All playgrounds should be laid with ample slope or inclination to carry off the water. The surface drainage cannot be too carefully attended to. They should also be supplied with proper gymnastic apparatus, which will be found separately treated of in Chapter XIV.

Teachers' Rooms.—Every large group of schools should have

156.—VILLAGE SCHOOLMASTER'S HOUSE.

one or more private rooms for the use of those engaged in teaching. In the case of those erected by School Boards, perhaps the most useful and economical plan is to provide, in a central

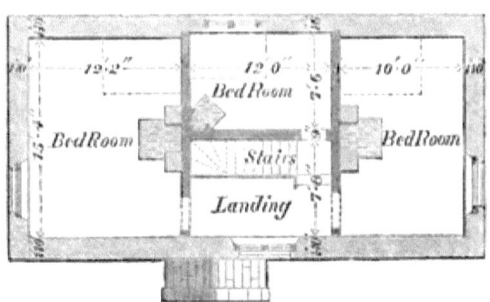

157.—PLAN OF FIRST FLOOR.

position, one room containing about 225 superficial feet or a size sufficient for the periodical meetings of the school managers. This would serve also as the common room of the masters and mistresses. Yet in establishments of the largest size it is

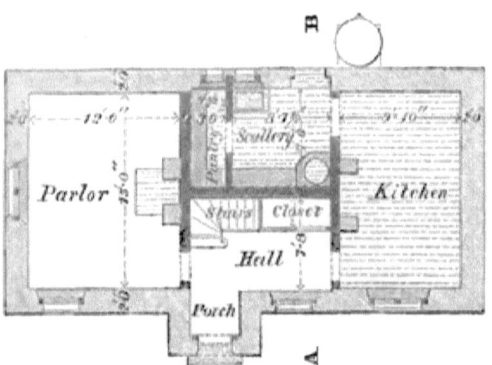

158.—GROUND PLAN.

preferable to provide a separate teachers' room of about 100 superficial feet for each of the three departments. It is usually one of the first items omitted on the score of expense.

Masters' Houses.—The practice of placing the headmaster's

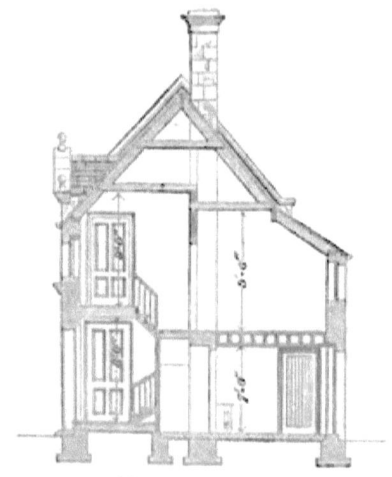

159.—SECTION A B.

house within the school precincts is not in vogue except in our large towns and there only in establishments of the largest size. It is considered better for the master of an elementary school to reside at some distance, sufficient to compel him every day to take a walk and to breathe fresh air. Attached to the

160.—VILLAGE SCHOOLMASTER'S HOUSE.

new Board Schools of London, no masters' residences have been erected. Country schools, on the contrary, are incomplete without these adjuncts. They are generally built detached from the

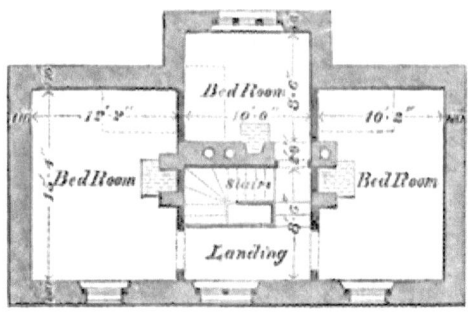

161.—PLAN OF FIRST FLOOR.

school-building. When close to it, particularly if a door of communication be added, the teacher is apt to forget the difference between work and leisure. There must be no internal communication between the house and the school. Appended are specimens suitable as adjuncts to small village-schools. They have been erected in Derbyshire for the School Board of Eckington.

If a town residence be wanted, it should contain a parlour

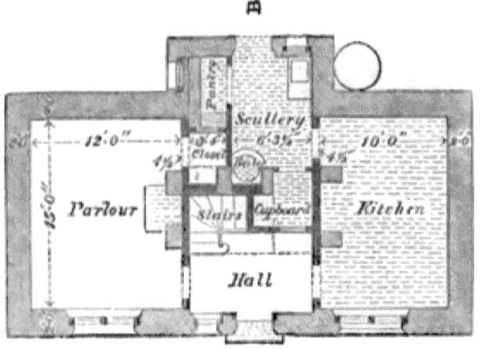

162.—GROUND PLAN.

kitchen, scullery, and three bedrooms of *not less* than the following dimensions, and none of the rooms should be passage-rooms.

		Ft. In.	Ft. In.	
A.	Parlour	14 0	by 12 0	
B.	Kitchen	14 0	,, 10 0	} or equal superficial area.
C.	One bedroom	14 0	,, 10 0	
D.	Two other bedrooms	10 6	,, 8 0	
E.	\multicolumn{4}{l}{The height to be not less than 9 feet to wall-plate, or, if ceiled at the collar, 7 ft. 6 in. to wall-plate.}

The Caretaker.—Where a master's house forms part of a small group of school-buildings, the master himself will usually take the general responsibility and care of the buildings, with such assistance as the extent of the buildings may render necessary. In the absence of a teacher's residence, and indeed almost always in large schools, caretakers' rooms, consisting of not *less* than two, and containing about two hundred and eighty superficial feet, with the usual adjuncts, become necessary. They should communicate easily with the principal door, and should be arranged, if practicable, like a porter's lodge, to command the entrance, and to have some outlook towards the rear. The caretaker, in small schools, may be almost any one fit to be trusted with such a general responsibility. He may be, as in America, a janitor. In large groups of schools, the accommodation provided should be rather more liberal than that above described, and the person should be a skilled plumber and glazier, able to take charge of the warming apparatus, to repair a gas or water-pipe, renew a broken pane, to regulate the ventilation, and to see to the daily cleansing and flushing of the water-closets and urinals. It will occasionally be found that such a person can undertake the minor repairs of several adjacent schools, and in other cases, that the repairs should be entrusted to some local tradesman. In all cases he should be responsible to some higher authority, and should have printed rules for his direction. One point, at any rate, is clear, viz., that in the larger schools, teachers cannot properly attend to any matters beyond their ordinary province, and are not always to be relied upon for keeping up sufficiently the open fires where that method of warming is in use.

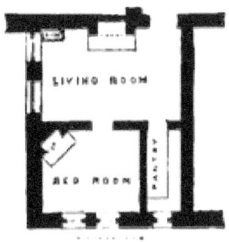

163.—PLAN OF CARE-TAKER'S ROOMS.

Windows.—The principal windows of a school-room lighted mainly from the back should face the north and east, these being the best aspects for ensuring a good and steady light for purposes of work, yet the importance of other windows on the sunny sides should never be overlooked. The position of the children in reference to the source of light, has already been discussed in Chapter IX., more particularly as to its effect on eyesight. It may again be stated, generally, that back-lighting alone is better than front-lighting alone, and that side-lighting is superior to both combined. The plan of an English school, as necessitated by the work, renders the invariable left-lighting, to which so much else is sacrificed in Germany, impossible. In the double class-rooms, for instance, if one room be lighted from the children's left, the other must of course be lighted from the right. Again, in the school-rooms, *all* the light cannot possibly be obtained at the sides of the classes, and then the back-lighting from the north or east, already described, should be adopted, but should be assisted, corrected, and diffused by other windows, highly placed in the opposite wall. The teacher, being thus made to face a cool, steady light, will not experience that common evil of having the sun in his eyes while teaching. This arrangement has the advantage of securing, at any time when required, a current of air through or across the room, and light both on the faces of the children and that of the teacher.

When a school-room or class-room has no story above it, top light may be introduced with advantage, but should never be sloped to the south or west, as the heat and sunlight will then be intolerable in hot weather, and difficult of regulation by blinds. The superficial area of light obtained from the top need not be so great as in the case of windows, because the rays are more direct, and the illuminating result more intense. It is better, also, that a "skylight" proper should not be made to open, because of its liability to admit rain and snow, but that some portion of the construction should be vertical in position, and arranged either to swing on pivots, or to fall back at the top.

The height of the window-sills from the floor should always be considerable, and the heads near the ceiling. It is a common error in new schools (in other respects excellent), to place the windows so low as to leave two or three feet of wall between their heads and the ceiling. Much of the cheerfulness of a school-room, especially in a town, depends on the amount of sky which can be seen from the windows. The height of the sills from the floor, therefore, should never be less than five feet, and may be even more with advantage. This will enable the top or head to be placed nearly, if not quite, up to the ceiling, and then the upper stratum of vitiated air in the room can the more readily be removed.

It will be seen from the foregoing, that any system, architectural design, or style which necessitates the adoption of windows having their upper portions permanently fixed, or half blocked up by stone, cannot be the best for the purposes of a school-house. Sashes, swing lights and casements are all admissible, and indeed each in their turn are better than the others, for particular places. A kind of window which has found some favour,

164.—METHOD OF OPENING WINDOWS BY MEANS OF IRON RODS.

Reference.

A. Sash hinged at bottom.
B. Sash hinged at side.
C. Sashes only moved for cleaning panes.
D. Wrought-iron rods.
E. Lever handles.

combines two methods of opening. The top portion is made rather shallow, say eighteen inches, and hinged at its base so as to fall backwards. The remainder is divided into two or more portions, of which the upper may be made to open inwards, as casements. Inwards, because, if outwards, the wire-guards often so necessary to school windows could not easily be added when the necessity became apparent. The casement opening outwards is more weatherproof, and the internal blinds are more easy of attachment. It may also be treated as an ordinary sash window, of which the bottom half would not generally be used, because of the draught coming too directly on to the children.

The window question has been shown to have so important a bearing on that of eyesight, that considerations in reference to the latter cannot be overlooked. It is not enough that a room be neither underlighted nor overlighted—in other words, that the children suffer neither from the straining caused by dim lights nor from the continuous glare of too much sun or light reflected from other surfaces,—but that the rays of light should not be distorted and rendered harmful by being made to pass through ground glass, rough plate, or any other of the numerous modern devices of similar kind. In the absence of some special reason to the contrary, therefore, school glazing should always be of clear glass. Where the windows are on the ground floor, so close to a public street as imperatively to require rough plate, or other thick and opaque glazing material in their lower portions, then such glazing should be limited to the lower portion, and clear glass used for the upper. Irrespective of the danger to eyesight, there is something peculiarly irksome in the feeling of a schoolroom entirely lighted by glass which imprisons the vision within the four walls, and does not allow even an unconscious glimpse of the sunshine and sky beyond.

In view of the rough usage to which all school windows are sometimes subject, it is not desirable to use panes of very large size. Without considering the artistic reasons in favour of small panes, the practical grounds are ample to ensure their adoption. They are better supported by the framework of the sash or casement, and from their strength less likely to be broken. In case of breakage, the danger to the children from falling glass is much slighter from moderately-sized pieces, while the ever-recurring glazier's mending bill is in each case less. Now that modern science has succeeded in making sheets of glass larger than a dining-table, we are in danger of passing with too rapid strides from the old-fashioned diamond quarries to the opposite extreme. There could hardly be a more complete mistake than to indulge in the taste (or want of taste) where a school is the subject.

For opening school windows, cords are undesirable, as being liable to give way from violence, and to involve, in the consequent breakage, damage to the window itself. A series of small iron rods of simple action is much better, moves the window with more certainty, is less liable to damage, and looks more orderly than the streams of endless cords sometimes seen hanging about school windows. Woodcut No. 164 shows the method adopted in many of the London schools.

Moveable partitions.—The occasional necessity for some species of moveable partition arises from the desire to isolate the children for class-work without losing the power of collective teaching. For some years sliding partitions running on small wheels fixed at the bottom have been used, but have been found so liable to get out of order, that heavy curtains moving either vertically or horizontally came more into favour. As a material the latter certainly have the advantage of a greater power of deadening sound, but they harbour dust and vermin, and are frequently merely a source of amusement to the children, who first make holes and then peep through at their fellows.

In point of fact there is no known moveable division which is at once simple, easily moved by a child, and perfectly sound-proof. Whether for this reason, or because of the additional expense, moveable partitions should never be adopted unless the arrangement and working of the school render them absolutely necessary. Where such necessity clearly arises, the value of an enclosed airspace should be remembered, and the partitions should be made double, with a clear intervening space of at least nine inches. Without this, and with school work in full progress on both sides, the division will be an obstacle to sight, but only partially to sound. Many of the partitions thought to be admirable, are so regarded because they stand between two large rooms of old-fashioned type, in each of which the noise kept up is so great that in neither can that of the other be heard. If framed wooden partitions be adopted, the reverberation or drum-like effect in the

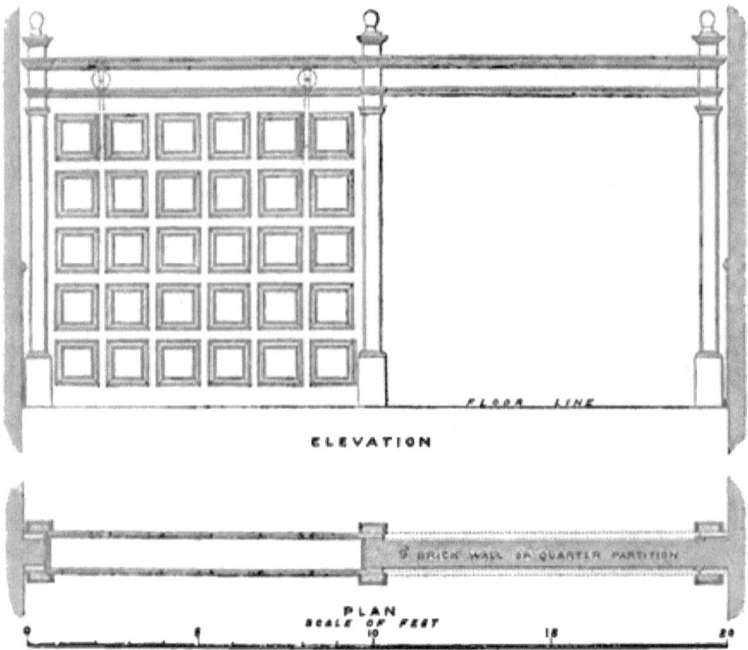

165.—SLIDING PARTITIONS.

room from the large wooden surface must be guarded against, and the surface limited in extent. All complicated machinery or any ingenious gearing likely to get out of order, and, when out of order, to interfere with the work of the school (as would certainly be the case), should not be admitted on any pretext. The simplest and safest plan, and that which has been found in the case of railway gates to involve the least amount of friction, is to hang the partitions from the top by wheels of not less than six inches diameter. This system, both with single and double partitions of framed wood, has been in use at the Home and Colonial Schools for some years. If thought desirable, small wicket gates can easily be made in the partitions, and one or more small pieces of plate glass inserted at a height of 4 feet for observation on the part of the teacher. Revolving shutters, whether of vertical to horizontal action, and constructed either of wood or iron or of both combined, have not met the difficulty so well as the simple piece of rigid framing made to slide horizontally. And the fewer specimens in use in any school, the better will probably be the condition of the school.

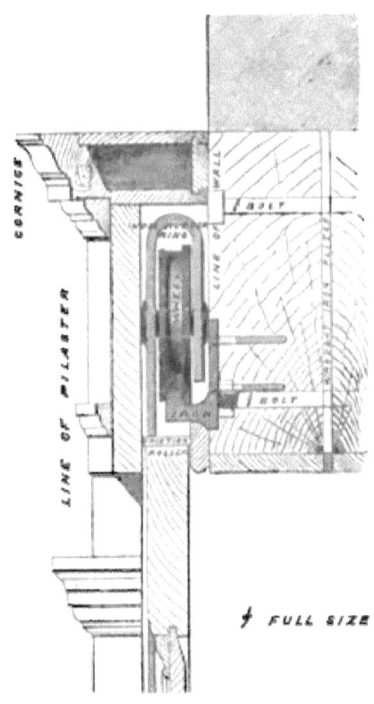

166.—SECTION SHOWING HANGING OF SLIDING PARTITION.

In view of the noise created by marching exercises, the floor of an Infant school requires precautions which are only necessary in degree for Graded schools. One method (woodcut No. 167) is to pave the whole floor, except perhaps that of the senior classroom, with wood blocks forming cubes of four inches set *endwise*, on a layer of asphalte spread on a concrete bottom. This is the most durable method. Any damp from below can find its way through the upturned end of the wood, and the capacity for wear in the end-grain is naturally much more than in any other. It attains the object in view, but has not the clean, fresh appearance preferred by teachers, because of the facility with which it receives and retains stains of all kinds.

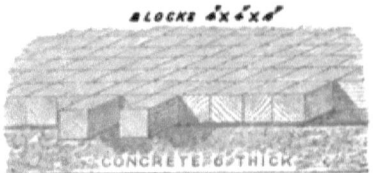
167.—WOOD BLOCK FLOORING.

If the cubes were set with their *sides uppermost*, no precaution would probably suffice to prevent rot, and the surfaces would tend to flake off in consequence of the numerous edges exposed, and soon render the whole floor ragged and dilapidated in appearance.

To obtain the advantages presented by the block floor without the disadvantages, a plank floor in short lengths is suggested

168.—PLANK FLOORING LAID ZIGZAG.

(woodcut No. 168). Pieces of wood $14'' \times 3\frac{1}{2}'' \times 2\frac{1}{2}''$ set in such a manner that the end of one piece always abutted against the side of another, would form a good floor having the side uppermost, and could be kept clean as easily as any other floor. It would be set dry, on a layer of asphalte above concrete, as in the case of the blocks.

Floors.—The floor above the level of the ground requires some consideration, as to deadening sound, in addition to that afforded by pugging. A strip of felt along the top of each joist, kept in place by means of a lath till the floor-boards are nailed down, has been found useful. Long lengths of floor should be broken near the class-room doors by a yard or so of boarding, laid in short pieces, zigzag fashion. Permanent durability and facility for cleansing are also worthy of attention. It is preferred in Germany to lay floors in the driest and hottest season, and to saturate them two or three times with hot linseed oil applied with a large brush, the operation of course being commenced at the point furthest from the door. A little stain or colour is added to the oil at the last saturation. At a later period the process is repeated. The floor can afterwards be easily cleansed by washing with a little lukewarm water, and does not require the deluge, and the mighty scrubbing to which English floors are periodically subject. This German system of oiling the floors has been tried, as an experiment, in some of the rooms of the new school-house erected in Gloucester Street, Stepney, by the School Board for London.

The Dado.—In schools of higher kind the rooms should be panelled in wood or match-board up to the window-sill line, and finished with a moulding. This is best for sound, health, comfort, and appearance. In elementary schools this dado should be of cement or hard-setting mortar, finished with a cement torus or other simple moulding. It cannot harbour insects, and is easily washed if well painted of a dark colour.

CHAPTER XIII.

MIDDLE SCHOOLS.

The Old Foundation Grammar Schools—Object of Middle Schools—Desirability of State Control—Sources of Inspiration for new Plans—German Higher Schools—Absence of recognised Code for English Secondary Schools—Girls' Schools—Boarding Schools—Milton Mount College.

THE old Grammar Schools of England, many of them dating from the time of King Edward the Sixth, and some—like Eton, Westminster, and Christ Church—being of royal foundation, were originally formed for a kind of education in which Latin and Greek largely predominated. Their endowments and scholarships, often intended for the meritorious youth of humble rank, have fallen into the lap of others better able to pay. This may in great measure be ascribed to the kind of subjects taught, many being comparatively useless to those who have afterwards to earn a living by the exercise of some modern trade. The simple fact of their possession of endowments having determined the position of these schools as clearly distinct from those which may be termed private enterprises, Government has claimed and exercised its undoubted right of supervision, and even, in some cases, of a rearrangement of the endowments. As the Endowed Schools Commissioners proceed with their work, many funds either lying useless, misapplied, or not applied to the best advantage, will be diverted into a new channel. From this source as a nucleus may be expected the creation of good Middle Schools.

The object of Middle or Secondary Schools is to impart a higher education than comes within the scope of Elementary Schools, and to lay a foundation for still higher studies to be afterwards pursued, either at technical schools or universities. The Germans call them Higher schools, the French Secondary, while among ourselves they have no clear and descriptive title. Much greater difficulty may be anticipated in the establishment of a complete system of these, than has been experienced in the case of schools for elementary training. People easily understand the practical use of that primary education which assists them in every transaction of their daily life. Reading, writing, and arithmetic, and even some further acquirements, are known to have their price in the market. For every step taken in knowledge beyond these necessary acquirements the apathy with which it is regarded increases. Secondary and higher education labours under the difficulty that when not possessed it is not much missed. Mental culture, producing enlargement of mind and a wider horizon, is seldom steadfastly pursued from a comparatively early age or in the lower walks of life. Its light comes later, or belongs to another class. A law compelling attendance can never be applied to even a lower kind of Secondary School. The increase of higher education must to some extent await the national appreciation of its advantages. But the public mind may be awakened and stimulated by wise legislation on the subject, taken in stages, especially if kept apart from the miserable squabbles of party politics.

On this subject Mr. Matthew Arnold says: "The middle classes in England have every reason not to rest content with their private schools; the State can do a great deal better for them; by giving to schools for these classes a public character, it can bring the instruction in them under a criticism which the knowledge of these classes is not in itself at present able to supply; by giving to them a national character, it can confer on them a greatness and a noble spirit, which the tone of these classes is not in itself at present adequate to impart. Such schools would soon prove notable competitors with the existing public schools: they would do these a great service by stimulating them, and making them look into their own weak points more closely: economical, because with charges uniform and under severe revision, they would do a great service to that large body of persons, who, at present, seeing that on the whole the best secondary instruction to be found is that of the existing public schools, obtain it for their children from a sense of duty, although they can ill-afford it, and although its cost is certainly exorbitant. Thus the middle classes might, by the aid of the State, better their instruction, while still keeping its cost moderate. This in itself would be a gain; but this gain would be nothing in comparison with that of acquiring the sense of belonging to great and honorable seats of learning, and of breathing in their youth the air of the best culture of their nation. This sense would be an educational influence for them of the highest value; it would really augment their self-respect and moral force; it would truly fuse them with the class above, and tend to bring about for them the equality which they desire."

The wealth and perseverance of the nation has enabled us, hitherto, to hold our own in most branches of science and art, against the foreigner, though generally after much blundering and consequent waste. Foreign nations, conscious of their comparative poverty, have steadily set themselves to know thoroughly their business in different branches, and so to avoid the extravagance and wastefulness attendant on spoiled goods and costly failures. It may now be plainly stated as an indisputable fact, that in more than one direction they are ahead of us. The time, therefore, has gone by when England can afford to allow its professional, scientific, and mechanical men, or even its skilled artizans, to be educated merely by rule of thumb. The medical profession has its schools, because without them it would be dangerous to allow any tampering with human life. A wise economy would dictate the provision, also, of schools for the various professions which have necessarily considerable control over the expenditure of the individual and of the nation.

Our old foundation Grammar Schools furnish us with few ideas as to the future planning of Public Middle Schools. Their sole provision was usually a single lofty and noble hall of oblong form, in which the whole of the boys might be seen engaged in their various lessons—learning by heart, or carefully plodding with grammar and dictionary—within sight of the master, who was placed on a raised platform. No class-room ever, until during quite recent years, spoiled the simple dignity of these architecturally excellent school-houses. And their fittings were of the rudest and simplest kind.

It would be possible to collect specimens, possessing good features, from among the superior English schools erected at a much later period—say during the last thirty years. It has been found better for our purpose to go to other countries, of greater experience as to buildings constructed with clear meaning and forethought, for a study of plans and arrangements. The school-houses of Germany have been found much more suitable for Secondary than for Elementary Schools. Considered simply as day-schools for boys, it would be difficult to point to any models more worthy than some of these establishments to receive the careful attention of an architect entrusted with the arrangements of a new school-house of the secondary kind. In Germany, the public school system does not include in its building programme any arrangements for taking boarders. Boarding-schools are usually a different kind of establishment, the creation of private enterprise, and do not hold quite the same rank in general estimation as the public day-schools. Several examples already given—among others that of the King William Grammar School at Berlin—have furnished evidence in detail of what is here meant. Pupils from a distance in schools of this rank usually board out, either with one of the professors, with friends, or in some other way, and there is not the same kind of connection between school and the daily life which is found in some English Public and Proprietary Schools, fitted, however imperfectly, with dormitories, kitchens, dining-rooms, and other adjuncts of the boarding-school. In this view of the subject the Collége Chaptal at Paris, already given in a preceding chapter, forms an instructive specimen, combining as it does the necessary arrangements both for instruction and daily board and lodging. While admitting the existence of English schools of secondary kind, wherein the education is excellent and the boys are well taken care of, it is necessary to maintain that there is nothing sufficiently perfect which we can select as the type of a system.

Owens College at Manchester, founded more than twenty years ago in accordance with the bequest of the late Mr. John Owens, is perhaps a typical illustration of what, in their teaching and arrangements, English Secondary Schools might aim to become. The course here pursued was eminently wise. Instead of immediately erecting a building, and finding out afterwards what it ought to have been, the trustees of the founder have spent some years in carefully developing the educational aspect of the college. The new building which they have recently completed, ought therefore to be the embodied result, in building form, of long and varied educational experience. Its cost has been upwards of 100,000*l.*, and an endeavour has been made to render the chemical laboratory superior to any corresponding department in the schools of Germany, Austria, or Switzerland. We have, as yet, no Code, no Act of Parliament, no information of any authoritative kind, as to the method of teaching to be pursued in the English Middle School of the future, and in consequence no theory of the building developed from stated educational conditions. To design buildings for an undefined use would be a resultless labour. To point out what may be done in the future is quite within our scope. Our German examples have the advantage of being the embodied results of mature study, and of planning intended to meet a distinct programme of requirements. On them may easily be grafted the further requirements of an English school.

When the organization of higher schools has fairly begun, the principle of separate class-rooms with a properly qualified teacher to each will probably be found best. The English school-room, always available after the opening of school for some few classes, is not likely to be abandoned in favour of an Examination Hall, only useful two or three times a year and never used at others. The lighting is not likely to be better than in the German class-rooms, but the ventilation may easily be made superior. More variety is likely to be seen among the plans of English architects than we have found among those of other countries. The use of a gymnastic hall may be expected to grow in favour in this country exactly as it has done in Germany, notwithstanding prejudice and the national love of out-door games. It has already been pointed out that attention to the thoroughness and efficiency of education in any school would soon react upon the buildings in which the education is given, and produce good school-houses. The education, to some extent, must come first.

Schools for girls involve the consideration of boarding-schools, and, in view of their greater neglect, require even more attention than those for boys. Boys can easily walk a distance to the school in which they are being educated, or they may be boarded at, or near, the building under the care of one of the masters. Competent judges tell us that the life of a large English public school is the best preparation for after public life. The boarding with other boys belonging to different families of different mental faculties and diverse home training may perhaps be an additional element of training for the world's wide stage. It can never be considered a moral or practical necessity, and is nowhere so regarded on the Continent. In point of building result we need hardly press the inquiry further. If in planning a new school-house we embody the principles of a German example, and do not forget English wants;—if we translate it, so to speak, into good English, we may easily arrange the boarding portion. Unlike the arrangements of the girls' school at Gravesend, the dormitories need not all be subdivided into little cells. Large sleeping rooms will be found best, except perhaps for some of the oldest boys.

When the education of girls is concerned the principle is further open to consideration, because the results to be attained are somewhat different. The ordinary course of studies, aiming at an education equal in quality to that of boys, must be supplemented by an amount of domestic training and home influence not so necessary for the sterner sex. Secondary day-schools for girls are sufficiently common in other countries to need no special advocacy here, and the diminution of attendance in ordinarily bad weather is not so great as might be supposed. The prejudice among us in relation to them has arisen mainly because of their universal inferiority in England.

Public Boarding Schools, unless of such a size as to be economical in point of teaching and managing power, must always be rare. And it is clear that if girls are always to be boarded and lodged as well as taught, secondary education will continue at a low ebb. Roman-Catholic countries prefer the boarding-schools attached to such convents as the Sacré Cœur, for the sake of the religious hold obtained over the minds of the girls. The Englishwoman sends her daughter to a private boarding-school, in the vain hope that to a sound education may be added some of those external graces and accomplishments which add to the attractions of the sex. As to the former she is in most cases miserably deceived. Too often the solid instruction has been neglected when the acquisition of more showy acquirements stood in the way. An unaffected home-bred girl is, in one of these "establishments," sometimes transformed in a few months into a fashionable fine lady, with little knowledge and many airs. The proprietors of this class of school are perhaps less to blame than the public, which prefers the shadow of mere accomplishments to the substance of mental culture. If the teaching be entirely wrong, the buildings are almost invariably unfit for school purposes.

While touching on the relative merits of day-schools and boarding-schools for girls, the opinions of those who have long studied the question are of value. Miss Wolstenholme, in an essay on the education of girls, when contrasting the small boarding-school with the large day-school, writes as follows: "The experiment of large day-schools has been successfully tried, and the results are conclusive as to the superiority of the system from whatever point of view we regard it. Their superior economy is obvious. Morally, we believe the gain to be also great. We want in every considerable town in England a high-school for girls, which should offer the best possible education on moderate terms,—one which should serve as a model to all those private establishments for which in future, as at present, there will no doubt be abundant room. To such a school as this it would be very easy to attach all manner of appliances and apparatus in the way of lectures and special classes, which might be attended from private families or smaller schools."

The efforts of Miss Buss, who for twenty years has been the lady principal of the North London Collegiate School for Girls, Camden Town, have not failed to make an impression on this branch of the subject. The North London establishment, as well as the Camden School for girls, have recently been handed over to trustees. And, as in both the demand for admission always exceeds the accommodation, a considerable enlargement on an improved plan, or an entirely new building, may be looked for at an early date. These schools, in their education, though not at present in their buildings, represent the want described by Miss Wolstenholme. A good education having been secured, good buildings suitable for the purpose only await sufficient funds.

The College, formerly at Hitchin, for the higher education of girls, has been removed to Cambridge, where under the name of Girton College, and in a new building, it will continue under the auspices of Miss Emily Davies. It is really as much beyond the scope of our inquiry as a continental Polytechnic or a University, yet it cannot be omitted where the higher education of women is the subject. Its establishment forms the first college for women at one of the great seats of learning, and marks an epoch in the higher education of the sex. Its plan is similar to that of the colleges for men, common to Oxford and Cambridge, consisting of an open quadrangle or "quad" surrounded by buildings. The sets of rooms are always a sitting-room 13ft. 6in. × 10ft. 6in. and a bedroom 13ft. 6in. × 8ft. 6in. This goes far beyond the ideas of a boarding-school. The plan of the building is an evidence of clear intention to educate women in the same manner as men. The waitress is designated a "gyp." Why should not the mistresses be styled "professors"? As a commencement, only one side of the quad, together with the kitchen offices, is built, the erection of the other three sides being deferred for the present. The external architecture is a kind of modern Gothic. The building is of two stories, together with an attic storey, lighted by roof-dormers.

Milton Mount College, Gravesend (woodcut No. 169), intended

169.—VIEW OF MILTON MOUNT COLLEGE.

for the daughters of Congregational ministers, and erected from the designs of Mr. Robins, forms a rare specimen of a carefully considered schoolhouse for 150 girls on the boarding principle. The sleeping apartments on the first floor (woodcut No. 170) are planned on the model of the ancient monastic dormitories, a central corridor down a large hall or room leading on right and left to cells, chambers or cubicles formed by wooden panelled or match-boarded partitions, 6ft. 6in. high, arranged along both sides. The monks' dormitory at Durham, erected in the 15th century, and now used as a library, was identical with this in plan. The

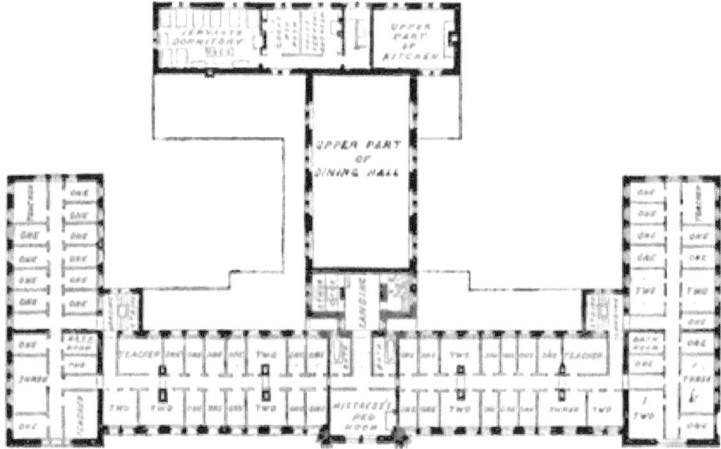

170.—PLAN OF FIRST FLOOR.

advantage of the arrangement is that each scholar has a room to herself, while ventilation is promoted by the upper part of the room being entirely unencumbered by divisions. The cubicles measure 9ft. 6in. by 5ft. 6in., and occasionally two are thrown together by the omission of the partition for occupation by two sisters or friends. Woodcut No. 171 shows an interior view, supposing the partition to be removed for the purposes of a sketch.

On the ground floor (woodcut No. 172) are the school-rooms, class-rooms, teachers' and visitors' rooms, library, and music-rooms. The last are fitted with shelves and portfolio lockers. The little practising-rooms each contain a piano, and it is found that sound from room to room is sufficiently deadened by the plan for the purpose in view. In each of the school-rooms, book-closets are formed against the wall-dado, and surround the room. One school-room is fitted with a raised gallery and desks. There are four class-rooms, each for twenty pupils. The bonnet and cloak-rooms are fitted with separate closets for every three girls. The dining-hall is capable of accommodating 150 girls and eight mistresses.

No plan is given of the second floor, the arrangement of dormitories being almost identical with those of the first floor (No. 170).

The plans of the future secondary schools for girls, if intended

171.—INTERIOR OF SCHOOL DORMITORY, SHOWING SUB-DIVISION INTO CUBICLES.

as day-schools only, will probably not differ materially from those intended for boys. The idea contained in German secondary schools will be moulded to English uses. And the adjunct of the gymnastic hall is not likely to be wanting.

The relative merits of day-schools and boarding-schools in reference to the education of girls may be left for the discussion of others. Whichever kind be preferred, the teachers and the school buildings should be under the regulations as to qualifications and fitness which must ultimately prevail concerning boys' schools. None should be allowed to exist without a license from Govern-

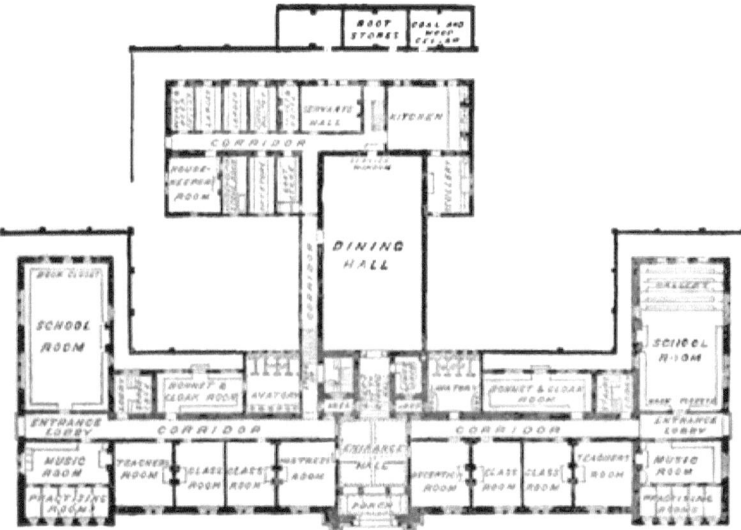

172.—GROUND PLAN.

ment. And the main examinations should be public, in the same way as in Germany.

Some legislative enactment in favour of securing good teachers and good school-houses for secondary and private schools appears to be the next important item of progress required as a step towards the establishment of a complete national system of education. It would not be very difficult to obtain Parliamentary sanction to such a measure, if once the nation understood what it meant.

CHAPTER XIV.

PHYSICAL EDUCATION.

Its Rise and Progress in Germany—Desirability in England—The Turnhalle and Turnplatz attached to Berlin Elementary Schools—Difference in Fittings respectively—Apparatus for the Playground—Description of its Construction —Extension of Turnhalle to Elementary Schools—Turnhalle at Hof—Apparatus for the Gymnastic Hall.

AMONG the ancient Greeks, physical education formed an important feature. Large and suitable buildings, designed with characteristic architectural beauty, were devoted to its cultivation, and great skill was attained in the various exercises. English athletic sports are not pursued in the same manner. In the schools, outdoor games, such as cricket and football, take the place of more scientific exercises having for their object muscular development. At the present moment, there is probably no single example among us of a school possessing a hall specially set apart for gymnastics. Still less is the subject taught as part of a school course. Germany has been taking to it in the earnest way usual with the nation when prosecuting anything from which important results are expected. The commencement in that country was probably in the year 1776, when Basedow at Dessau, and Gutz Muths at Schnepfenthal introduced it into their respective schools. First, by conducting exercises in the open air, and afterwards by erecting covered halls in which the system could be further developed and practised, the various professors gradually advanced its cause and their own, until people in different parts of the country became well accustomed to gymnastics, and learned to regard them as necessary. In 1811, Jahn, a high authority, established a hall at the Hasenhaide near

Berlin, from which may be traced the foundation of the German school of gymnasts. Through many vicissitudes, sometimes being regarded with suspicion or dislike by the State, and after being entirely banished from public institutions by the reactionary movement of 1819, gymnastics grew in importance until they finally became popular everywhere. The regulation of June 6th, 1842, formally recognised in Prussia, by a Cabinet Order, the principle that physical education had become necessary and useful for boys, and it has since been a part of the national system. At the present moment, no public school in Germany is regarded as complete without its Turnhalle, and the fittings and apparatus are often of the most elaborate and expensive character. The exercises are conducted by a regular master, are adapted to every stage of muscular development, and are studied with as much precision as any of the other school lessons. The same ideas have spread in Austria, Switzerland, France, &c., till the teaching of scientific gymnastics may be considered to be the general rule among continental schools.

At this time, when the foundations of a great system of national education are only just being laid, it is impossible to say whether similar gymnastics will ever attain in England the popularity they have long enjoyed abroad, or obtain recognition as a part of public instruction. The national love of field sports, and the delight of English boys in outdoor games, may be said to mark an idiosyncrasy different from that of the taciturn German. Yet the subject cannot fail to receive more and more attention as the educational system gradually becomes more matured. Its application to new establishments is worthy of careful consideration. At our public schools all the boys do not join in the rough play. Some hold aloof from shyness, want of interest in the games, or from a more studious disposition, and these form a large percentage in every school. The hard-working, studious boy would derive more benefit from half-an-hour on the trapeze or parallel bars than from his "constitutional," which is often lost altogether when the weather is bad. It is precisely this kind of boy to whom really well-taught gymnastics would bring the greatest boon, and who also requires a special measure of consideration. As a rule, the German boy works harder and has more to go through than the English school-boy. In place of a constitutional, or of those outdoor games which he never plays, he has his gymnastics all the year round either in the open air or in the hall devoted to the purpose. Physical education of all kinds is in Germany pursued so seriously that, in many of the towns, large tanks are provided for instruction in swimming, the child being supported by the teacher from a stick and a cord very much like a fish at the end of a rod and line.

Attention has already been called to the playground fittings suitable and safe for the use of infants. And the subject of playgrounds for boys and girls has been discussed at length on the supposition that no fittings were to be provided. It remains to be shown what playing apparatus might be with advantage used in the open air for the graded portion of elementary schools, and for all higher schools; and also the kind of hall which might be provided for indoor instruction in athletics.

In endeavouring to show this, no English specimen can be

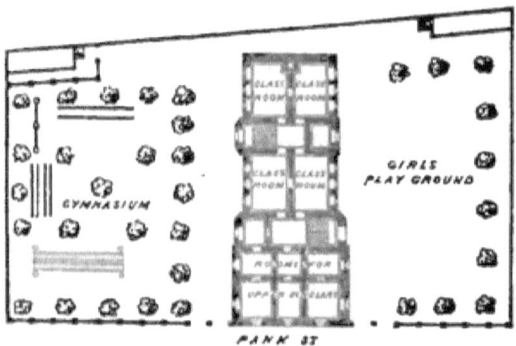

173.—PLAN OF THE PARISH SCHOOL IN THE PANKSTRASSE, BERLIN.

given which would afford any illustration of our meaning. English schools, having neither gymnastic halls nor playgrounds properly fitted with suitable apparatus, furnish no examples. We must turn to Germany and, amid some of the lowest schools of that land, seek for a plan of a playground.

While discussing the Parish School-house in the Frankfurterstrasse at Berlin one of the features of the general plan (woodcut No. 52, page 82) was shewn to be a well-appointed and furnished playground. Another specimen of similar class, in the Pankstrasse at Berlin (woodcut No. 173), is so placed as to divide, by its position, the girls' playground from that of the boys, and the latter is supplied with complete apparatus for outdoor gymnastics. Even in connection with the higher English schools, it is extremely rare to find any isolated playground fittings, still less to find them carefully and scientifically considered and arranged, while in elementary schools they have simply not yet been thought of. The playgrounds of none of our Board schools are as yet properly furnished—a circumstance arising principally from the engrossing labour entailed in creating the buildings themselves and organising the instruction. In Germany those of the corresponding schools are hardly ever entirely deficient.

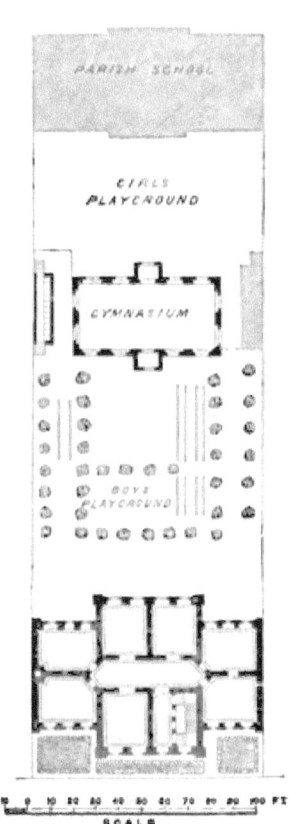

174.—PLAN OF PARISH SCHOOL IN NAUNYN-STRASSE, BERLIN.

It should be remembered that the *Gemeinde*, or Parish School, of Berlin corresponds to a London Board School for the poorest kind of children. A third example from this class of establishment may be quoted, and it advances yet a step further than those preceding. The Parish School in Naunynstrasse (woodcut No. 174) has not only a well-provided outdoor playground for boys, but in addition a *Turnhalle*, or hall for gymnastics, admirably fitted for imparting physical education at every season of the year.

In the Gemeindeschulen, Volkschulen, and the lower schools generally, the turnhalle is not commonly found, and the apparatus of the *turnplatz*, or playground, is there more elaborate than in that of the higher schools, yet not so complete as in the turnhalle of the latter. It is studied in reference to economy, durability, and open-air use, and its construction and provisions are of necessity different from those followed in the turnhalle. For instance, instead of cushions, the parallel and horizontal bars are provided with a layer of sand or bark 12 inches thick, laid level with the surface of the ground. A jumping ditch is formed, about 3 feet deep at one end, and gradually sloping up to the surface at the other. Its sides are lined by a wooden border, and it has a floor of washed river sand 2 feet thick. The whole playground is always carefully drained. Where there is also a building for gymnastics, the outdoor apparatus is purposely kept as simple as possible, because it is found better to give the lessons indoors, even in summer, on account of the liability of the boys to take a chill after the heat of exercise. The professor of gymnastics at Emden, Herr Droop, has described the apparatus which, from his experience, he considers to be most suitable for the playgrounds of the Prussian Volkschulen. Some portions of it may be found equally well adapted to those of the people's schools in England; if not immediately, perhaps at no distant day when the necessary buildings have been erected, and time has been allowed for a full study of the subject apart from prejudice. We therefore extract some portions of the description. Herr Droop advises that, for an open-air class of 40 to 50 scholars, a space of at least 3,230 superficial feet is necessary. The best shape is a quadrangle with the sides in the proportion of 2 to 3. The ground must be near the school-house, flat, of a sandy soil, and fenced in by boarding or wall, especially high towards the N. and E. The apparatus is to be made of dry, well-seasoned wood, having all sharp edges removed, and is to be well saturated with hot linseed oil before being used, for purposes of preservation. All movable parts of the apparatus, such as climbing poles and ladders, are not to be oiled, but are to be kept under cover in a dry, airy room when not in use. In his recommendations Herr Droop has studied economy, both of cost and space, by using the standards of the climbing-frame as posts for the horizontal bars and as supports for the leaping string, without interfering with the free use of other parts of the apparatus. He thus saves a couple of posts for the horizontal bar, and a pair of standards for the leaping-string, besides gaining space. The following is his list, viz.:—

    1. The Climbing Frame, of two kinds, single and double.
    2. The Rung Standard.
    3. The Ladder.
    4. The Climbing Poles.
    5. The Climbing Rope.
    6. The Horizontal Bar.
    7. The Jumping Frame.
    8. The High Jumping Frame.
    9. The Swings.
    10. The Staffs or Bars.
    11. The Fixed Parallel Bars.
    12. The Adjustable and Fixed Parallel Bars.
    13. The Adjustable and Movable Parallel Bars.

1. *The Climbing Frame.*—The specimen given in the woodcuts (Nos. 175 and 176) is of the single kind only sufficient for a class of from 20 to 25 scholars. As the gymnastic classes are usually double that number, or from 40 to 50, two such frames become necessary. They should then be placed together, with the middle post in common, and arranged to receive two horizontal bars as hereafter described. When used in its single form the climbing frame comprises the following apparatus, viz.:—

    (*a*) One sloping Ladder (fixed perpendicularly, if preferred).
    (*b*) Four Climbing Poles.
    (*c*) Two Climbing Ropes.
    (*d*) One Horizontal Bar.

(e) One Jumping String.
(f) Two Step or Rung Standards.
(g) One Arrangement for Deep Jumping (fixed to the Rung Standards).

The general construction consists of two fir posts or standards (A) (woodcut No. 175), of 14 or 15 feet clear height, connected with, or tenoned into a cross-beam (B), and extending 3ft. 9in.

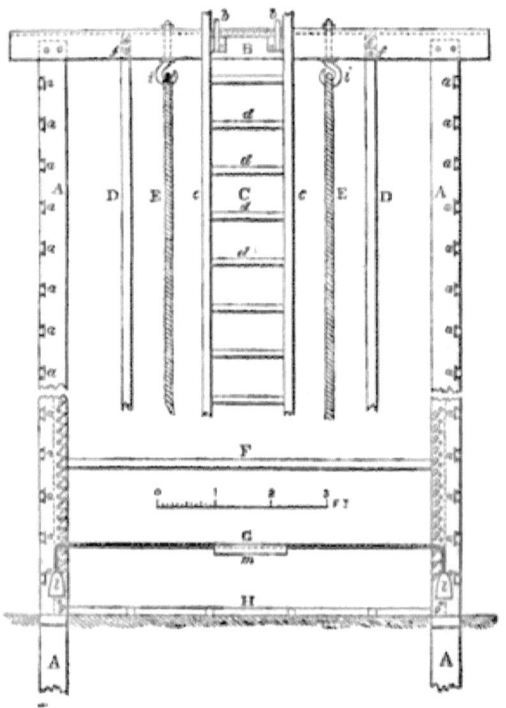

175.—GYMNASTIC APPARATUS FOR OPEN-AIR USE. FRONT ELEVATION.

into the ground. The standards and cross-beam should all be about 8in. × 6in. in section, the former having the broader faces towards each other. The underground construction is shown by the end view (woodcut No. 176). The standards are fixed by a tenon (d) in a horizontal beam or sleeper (b) 5ft. long, and maintained in position by two struts (c c) 4ft. long, firmly connected with standard and sleeper. This footing is found to give sufficient firmness to the structure, which is thus enabled to withstand the constant tendency to overturning caused by the strain of the gymnastic exercises to which it is subject.

When used for a class of 40 to 50, the climbing-frame in its double form comprises exactly twice the apparatus set forth in the list above, except that there are still only two step or rung standards. The middle standard should measure 8 inches square, and should be fixed in a double frame because of the necessity of having on both sides a groove for the horizontal bars.

The two outer posts (A A) are arranged as step or rung-standards. On the outer faces are cut horizontal grooves (*a a*) of dove-tail form, six inches apart, into which are fixed the steps or rungs 2ft. 3in. long, and projecting nine or ten inches beyond the standards, as shown by the end elevation.

The rung should be made of English ash or white beech, with edges rounded, causing it to show in section (at *a a*, woodcut No. 175, if the scale be not too small for the reader's eyes) an oval of 1¼in. × ⅝in.

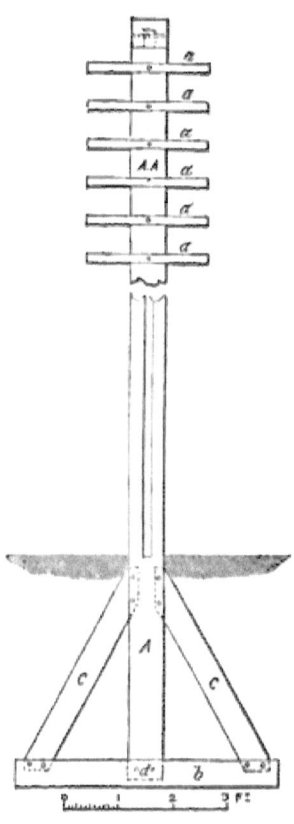

176.—END ELEVATION.

In the centre of the cross-beam (B) are firmly screwed two iron hooks (*b b*) about a foot apart, which serve to fix the ladder either for the sloping or the perpendicular position. The ladder differs little from other ladders, but, to make our notes complete, it may be described as 18ft. long, 14½in. in clear width, and with rounds nearly 8in. apart, connected with the sides by *square* heads passing through and securely wedged.

The method of fixing will vary somewhat, just as the inclined or horizontal position is preferred. If the ladder be sloped, its sides will rest upon the upper edge of the cross-beam (B) while the topmost round will rest in the hooks at *b b*, and the result such that no lateral movement or downward sliding can take place. If it be fixed perpendicularly, the shorter space will necessitate its being hooked to the beam by one of the lower rounds, and the bottom must be fixed by tenons of 4in. long into holes in the footboard.

The climbing poles (D D) are 14ft. to 15ft. long and 2in. diameter, made of perfectly clean fir. They are fixed into the cross-beam by the mortices at *e*, and into the footboard H by the mortices *i*, while an iron pin *f*, passing through both cross-beam and pole, really carries the latter suspended, and causes the weight of the climbing gymnast to pull upon the former. The outer poles are about a foot from the standard; the middle poles (if any) are placed exactly opposite the sides of the ladder; and the poles, when four are used, are kept about 14 or 15in. apart.

The ropes for climbing are of the length of the standards (14 or 15ft.), and are made of rough finish but great strength, 1½in. in diameter, the upper end of each being provided with an eye of horse-shoe shape lined with iron, by means of which they are attached to the hooks *i i*, fixed by screw and nut into the cross-beam B. At their lower ends the ropes are covered with strong leather to prevent untwisting. In the event of a scholar desiring to use two ropes at the same time, the suspension should be from the ladder hooks at the points *b b*.

The horizontal bar, 1¾in. diameter, is made of long-grained ash free from knots, and, although in itself round, the ends are 3in. wide, fitted with iron heads covering both sides and firmly screwed on. Grooves 1¾in. wide and 2¾in. deep, for receiving these iron heads, are provided up the inner faces of the two standards A (woodcut No. 176) to a height of 8ft. 6in. In the opposite direction a number of ⅞in. holes are bored at intervals of 2¼in., into which, and through the head of the bar, is passed a small iron bolt secured by a wedge. Bolt and wedge are both attached to the standard by a small chain.

For jumping, the climbing-frame may also be used. Two bolts, of the kind just alluded to, but without the aid of the wedge, will serve to hang up the jumping-string as at G (woodcut No. 175). The string, as there shown, is 7ft. 9in. long, and ⅜in. thick; both ends carry a leather bag *l l*, filled with sand; and to the middle of the string is sewn a piece of strong canvas, two or three inches wide, to render the line more clearly visible. The length of the string is sufficient for two scholars practising at the same time, but each will require a separate spring-board. The spring-board shown in woodcut No. 177, by B on plan, and by B B in side view, consists of two 1½in. deal boards *b b*, having along one edge a plate *d*, measuring 2¾in. × 3½in. fixed by means of four bolts *c c*. A strip of oak, *f f*, is also bolted across the board under the centre for the purpose of giving greater strength. The other edge *h* is bevelled off so as to terminate in a thickness of ¼in. and covered with sheet iron on both sides for protection. The upper face of the board must be left rough, and all nails avoided as being likely to work out and cause danger to the gymnast.

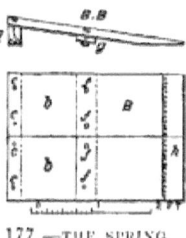

177.—THE SPRING BOARD.

The arrangement for deep jumping is shown in woodcut No. 178, E representing the side view and F the plan of the underside. It consists of a board hooked on to the rounds of the rung-standard by means of the angle-irons *n*, and supported by the bracket *o*, which is fitted into a groove on the inner face of the standard. As in the former case, the upper face of the board must be rough.

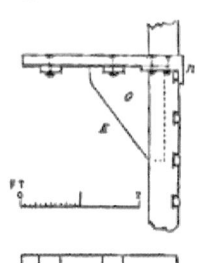

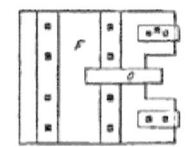

178.—APPARATUS FOR DEEP JUMPING.

The swing-trees are shown in woodcut No. 179, G being the

side view and GG the view as seen from the end. The construction of this item of gymnastic apparatus is very simple. A fir-tree 16ft. long, wrought completely round to an equal diameter of 6½ inches, is sawn lengthwise down the middle, each

179.—THE SWING TREES.

half thus forming a useful swing-tree after having received the addition of two pairs of legs. These latter are of wrought white beech, 17 inches long, fixed at a distance of 15 or 20 inches from the ends, and at such an angle that, on the ground, they spread out to 15 or 16 inches.

The staffs are always of wrought fir, perfectly free from knots, 4ft. 9in. long by 1¼ diameter, and with both ends rounded. Each scholar of the gymnastic class requires a staff and therefore a sufficient number should be provided in arranging the details of the apparatus.

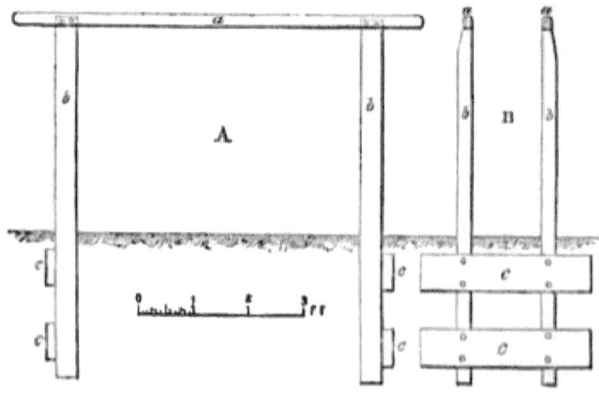

180.—FIXED PARALLEL BARS.

In teaching gymnastics, parallel bars are much used, and should either be provided in separate articles of different height and width, or in one so contrived with adjustable bars as to allow the height to be changed when required. The best and most convenient kind, though also the most expensive, is the movable apparatus with adjustable bars.

The simplest and most ordinary kind of what we term *fixed* parallel bars, is shown in woodcut No. 180, A being the side elevation and B the end view or section. The two bars, *a a*, are each supported by two posts, *b b*, measuring 4½in. × 3½in., fixed in the ground and connected underground by two planks, *c c*. Both posts and planks are best when made of oak, but the bars themselves, *a a*, should be of English ash wrought to an oval section of 2½in. × 1¾in. The apparatus should be 3ft. 9in. from the ground and 15 inches wide between the bars, if intended for boys of 12 to 14 years of age.

The construction of the *adjustable and fixed* parallel bars is set forth in woodcut No. 181, A being the side view, B the end

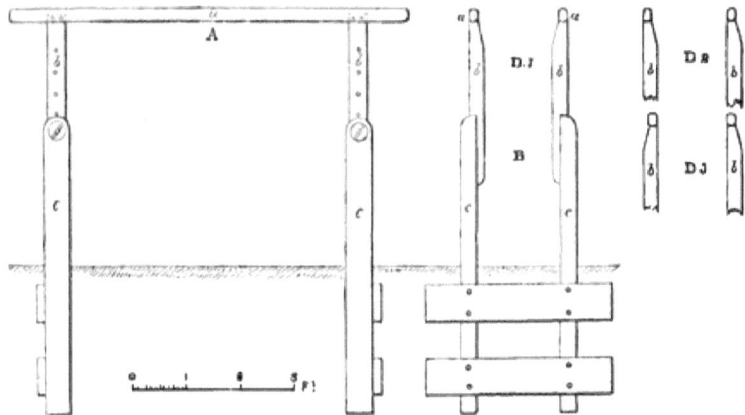

181.—ADJUSTABLE AND FIXED PARALLEL BARS.

elevation, D1, D2, and D3 being three different widths of bars suited to the different ages of the pupils. The bar *a* is fixed upon two supports, *b b*, oval in section, which fit into a hollow wrought out of the post *c* in its whole length. In the supports are bored 6¾in. holes, 4¼in. apart, the uppermost being 6½in.

below the bar. The upper end of the post contains one such hole, and a bolt passed through it and through any hole of the support (according to the height required) can easily, by means of a nut, be made to connect both pieces firmly together. In this apparatus the posts should be made, as in the former, of oak, but the supports should be of fir. The same lower construction and fixing in the ground will suffice. It will be noticed in the end elevation B, that the bars are fixed *flush*, or in a line, with one side of the supports, the latter being sloped away at the other side to the thickness of the bars. This renders it possible to obtain the three different widths between the bars, viz., 17½in. 16in. and 14½in. by fixing the supports in three alternative positions shown at D1, D2, and D3 in the woodcut.

The last remaining kind of bars to be noticed, as well as the last item of gymnastic apparatus for out-door use, is the *adjustable and movable* parallel bars. These are capable of alteration in height and width like the second, but differ in construction from both the two kinds already described, in being made with the view of removal from place to place, instead of being fixed into the

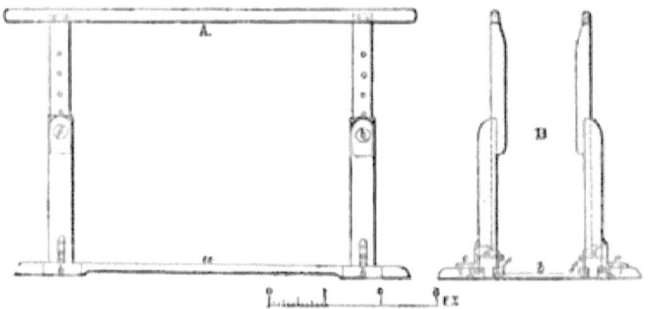

182.—ADJUSTABLE AND MOVABLE PARALLEL BARS.

ground. The position in the playground can thus be changed with ease, and during the more severe winter weather the apparatus can be taken indoors if thought desirable. A (woodcut No. 182) shows the side elevation, and B a section. The dimensions need not be varied from those last described, but the

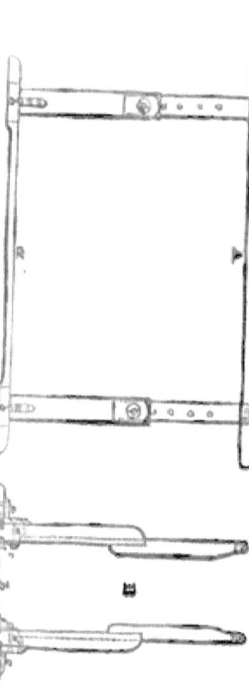

supports must of necessity be shortened according to the length of hollow in the posts, which in this case cannot reach the ground. The posts are tenoned into a framed footing, *b b*, and fixed by angle-irons, *e e*. The bolt, *d d*, is screwed to the bolt *f f*, which reaches through the framed pieces *a* and *b*.

It remains to consider the position of the gymnastic apparatus in the open playground. The climbing-frame should be so placed that as much free space as possible is left for other purposes. If placed at one side of the ground, the standards of the frame may be within 10ft. of the boundary wall. Parallel bars, when permanently fixed, are always best placed near and parallel to one side of the playground, with a clear space of 7ft. all round. Between the standards of the climbing-frame the soil is usually taken out to a depth of one foot and filled in with refuse oak-bark, which ought to be frequently turned over, and entirely renewed once a year. This forms a soft ground for jumping, but no such preparation is required around the parallel bars, for here it is better to accustom the scholar to sure and firm jumping.

The open-air playing ground, however well fitted, is not regarded by German educationists with the same favour as the turnhalle, from the liability of lessons being interrupted or entirely prevented by bad weather, from the greater difficulty of giving them in the open air, and from the liability of the boys to catch cold. The turnhalle has therefore steadily grown in favour until it is considered inseparable from the higher schools. It is now being introduced in connection with all the best *elementary* schools of new erection through Germany. Although long regarded as a part of a higher school, no more to be omitted than any other class-room devoted to a special subject, this last fact indicates a movement of great importance. We have already shewn how Prussia and Saxony especially favour it, and the lapse of a few more years cannot fail to render the turnhalle almost universal in connection with the primary schools of those countries. It is, therefore, desirable to give a representative specimen of the kind of building commonly used in Germany for gymnastics, so that its shape, size, and furniture may be better understood by the English reader.

A good specimen is that attached to the *Gewerbeschule* or

183.—GYMNASTIC HALL AT HOF. WEST ELEVATION, LOOKING TOWARDS PLAYGROUND.

Trade School at Hof, which has been well described by Herr Thomas in his pamphlet on the subject. The external architecture, as shewn by the two woodcuts (Nos. 183 and 186), although

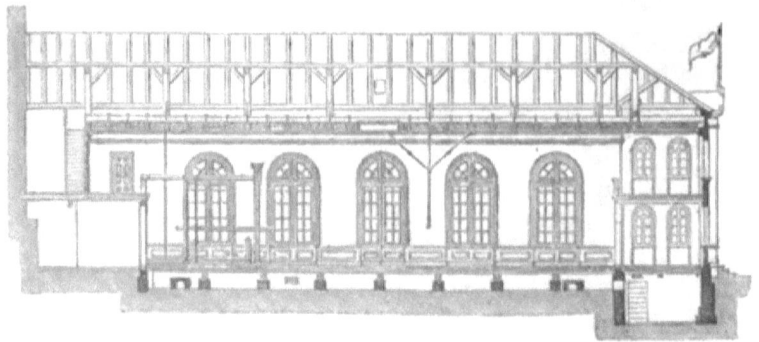

184.—LONGITUDINAL SECTION.

without any pretensions, is yet not treated as though inferior to, or in some sense apart from the school buildings themselves. On entering the building (woodcut No. 185), there is to the right a room for the committee and the master, to the left a dressing-room, while before us is the hall. A staircase leads to a gallery formed over the small rooms. At the opposite end of the hall is a room for requisites and apparatus, with a staircase leading to another gallery.

The floor of the hall is carefully laid of clean boards, nailed through the feather or tongue so that no nails may be in the

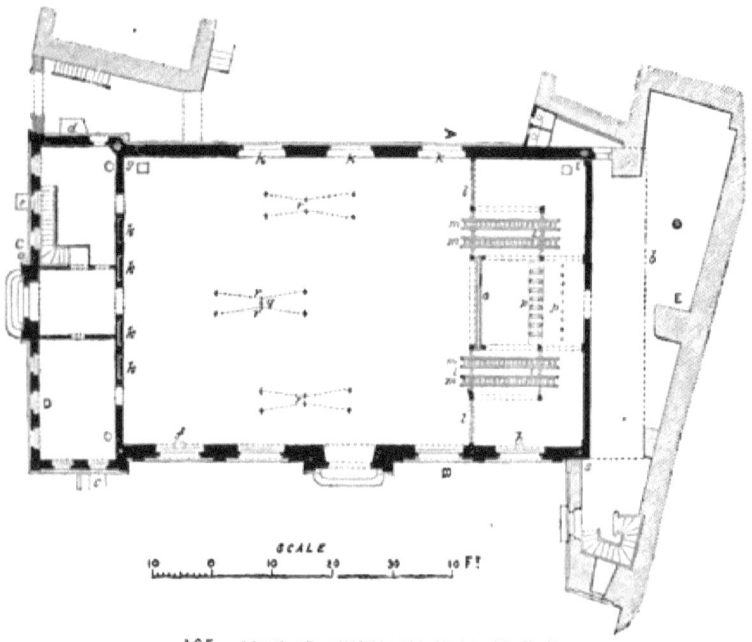

185.—PLAN OF GYMNASTIC HALL AT HOF.

way. The walls are match-boarded to the height of the window-sills, and whitewashed above. The hall contains about 111,000 cubic feet of air, and is warmed by slow-combustion stoves having a heating-surface of 40 superficial feet, the cooling surface of the hall being about 12,700ft. For ventilation there are provided sheet-iron flues or tubes which run through the roof. Around these are three spaces of $2\frac{1}{2}$in. diameter, which are continued in the wall downwards below the floor line, and enter the hall at F and H. According to the degree of warmth prevalent near the ceiling, and the consequent exit of air by the tubes, a more or less strong current of fresh air will be produced and drawn in at the bottom at F and H. The hall is lighted by gas, having in all 120 jets. The furniture is so arranged that two classes of sixty boys can each exercise without interruption to the other.

The apparatus necessary for the turnhalle is much more complicated than that intended only for use in the open air. It consists of two principal kinds, the *climbing* and the *movable*, as set forth by Herr Thomas in the following list, viz. :—

CLIMBING APPARATUS.

(A) *Fixed to floor and ceiling.*

Four horizontal bars in front.
Two ditto behind.
Four ladders.
Sixteen swarming poles.
One beam (removable).

(B) *Fixed to ceiling only.*

One giant stride, with six ropes.
Three pairs of swinging rings, or, if preferred, flying trapezes.
Four (or more) swarming ropes.

MOVABLE APPARATUS.

Six portable parallel bars, of equal size.
Two upholstered horses of movable height.
Two ditto   ditto   ditto   (for leap-frog).
One jumping table.
Eight small spring boards.
Eight large ditto.
Two pairs of jumping boards, without springs.
Two long rope swings.
Four round trees, 3 in. in diameter, for laying across horizontal bars.
Six iron bars, 4 feet long, of different thicknesses.
Sixty wooden bars.
Sixty pairs of dumb-bells.
Eight weights.
Two poles, 2½in. in diameter.

The climbing apparatus placed at the back of the room comprises all stationary apparatus not fixed to the ceiling. The four horizontal bars are placed in front between six posts, for adults and boys. The two others are behind, between four posts.

Each set serves as a rest for the horizontal ladder, which is lashed to the iron bars with iron thongs. In the section (woodcut No. 184) all four ladders are shewn horizontally, which would rarely be the case in reality, one pair being generally slanting, the other

186.—NORTH OR PRINCIPAL ELEVATION.

perpendicular. In order to obtain for the bars a permanently smooth slide, upright guides of cast iron with smooth edges are screwed on to the supporting posts. In the back posts this guide is made to reach from floor to ceiling, and in the front to a height of 9ft. To secure the horizontal bars in position, small holes are drilled at intervals of $3\frac{1}{2}$in. for the insertion of iron pegs. The front bars serve as supports for the spring-boards, and the pegs for securing the cord for the jumping line. The vertical posts are $7\frac{1}{2}$ft. apart, to allow two to exercise simultaneously. Some of the bars are of iron, others

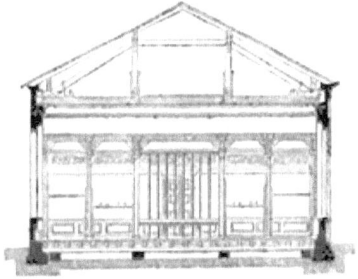

187.—TRANSVERSE SECTION.

of wood. Behind the front middle posts of the climbing apparatus, are fixed in the floor two smaller posts, 5ft. high and from 9 to 10ft. apart, supporting a beam $15\frac{1}{2}$ft. long and 9in. square, at different heights as required.

Besides these horizontal bars, are the horizontal, inclined, or perpendicular ladders. Also, the beam.

The sixteen swarming poles are placed in pairs, one behind another, down the middle of the apparatus. The whole are carefully secured, to ensure safety.

All round the apparatus there is left, next the wall, a clear path of 7ft. for marching. Other climbing apparatus is fixed to the ceiling and does not impede passage.

Nine hooks are fixed between the swarming-poles and the gallery, near p (woodcut No. 185), to which swarming-ropes can be fastened.

In the middle of the room is the giant-stride, and to right and left, near r, two pairs of swinging-rings. A third pair of swinging-rings, r, r, can be added if the giant-stride be removed. The height, 23ft., renders it necessary to lower the fulcrum of the swinging-rings by 7ft., which is done by fixing in the ceiling four additional slanting hooks from which four thin iron bars of different lengths run down obliquely towards two perpendicular suspended irons, each 7ft. long, joined to a horizontal iron bar to which the swinging-ropes are hooked.

The whole of the above apparatus is for use indoors.

Having thus called attention to physical education as carried on among continental nations, both in open playgrounds and in buildings specially provided, and selected good examples of the apparatus used in both and considered necessary for its successful study, the subject of its introduction to English schools may be left to the consideration of all who take an interest in the instruction and well-being of the young. The practice of gymnastics in properly fitted buildings would not in any way lessen the zeal for outdoor games and athletics of the national sort. And it might be the means of preparing many youths for joining in the latter who would, otherwise, never be able to do so.

CHAPTER XV.

WARMING AND VENTILATION.

Their inseparable Nature—Importance to Schools—Application in relation to chool-work—Cooling power of Glass—Methods of Warming to be avoided—General Principles to be adopted—Demand and Supply—First Cost and annual Maintenance—The presence of a Caretaker as affecting System to be adopted—The open fire—Different kinds—The Gill stove—German Methods—The Hot-water low-pressure System—Cases where it may be economically adopted—Practical Usefulness.

ATTENTION has already been called to the influence of proper lighting on the eyesight of children, and to its alleged effects on future physical development and general health. The subject which we now begin to discuss possesses even greater importance in its application to schoolhouses. To treat it as two, capable of usefully existing in separate forms each apart from the other, would lead to no practical result. For our purpose, warming and ventilation must be treated as inseparable, or, at least, in treating of one the other must be always present to our mind. On any other principle the arguments would lead us to conclusions of a contradictory nature. For instance: The mere warming of a room is, in itself, a very simple matter. A brazier of live coal placed in the middle of the floor would speedily fulfil the task, but we should find ourselves half choked by the smoke and other productions of combustion. In the same way, ventilation, considered alone, presents no great difficulty. The omission of doors and sashes would probably be found sufficient in practice, but we might in consequence suffer from cold. These extreme suggestions, however absurd, are only in degree less compatible than any more mature devices having for their object the separation of one part of the subject from the other, especially when intended for the use of schools. Except during summer weather, when the natural temperature of the external air requires little or no artificial increase, or may even be higher than the air indoors, and when in consequence the element of movement,—the prevention of stagnation—alone is required, the two are so interwoven as to form at most only separate heads of the same department of school hygiene. Warming is, in the first instance, the more important. The cold nature of our climate must always cause it to be regarded as the greater necessity, and, in past years this has been conspicuously the case. Yet the mere liability to sudden changes of temperature, sometimes causing in a few hours the difference between winter and spring, would alone indicate the power of regulating the air indoors to be of no slight importance. Added to this is the necessity of removing foul gases, animal exhalations, and the impurities mingled with the air by the process of respiration.

We may commence by the assumption that it is unreasonable to expect good ventilation without effectual means of warming, and that the best and most complete methods of raising the indoor temperature cannot fail to produce offensive and unwholesome results, when used without complete power of control and without provision for the continual renewal of the air. "Ventilation" is one of the words frequently heard in the present day. Those who use it should remember that it cannot be satisfactorily effected save by differences of temperature. Hot air rises by reason of its greater levity, while cold air takes the lower place from greater gravity. Thus warming is the motive power of ventilation. Ventilation is the safety-valve which, by one of its functions, prevents overheating, and, by another, provides the means of freshness.

A high state of civilization has not tended to induce greater hardiness of race. We are perhaps less able than our forefathers to inhale with comparative impunity some kinds of impure air to which they were strangers. The products of the combustion, in modern dwelling-houses, of carburetted hydrogen, for instance, are so noxious that the use of this gas should always be accompanied by thorough ventilation. It should, if possible, be religiously excluded from all sleeping-rooms, or, if used, should be accompanied by ventilation more complete than ever.

The age of the persons occupying, and the length of time during which the buildings are occupied, are not without some degree of importance. A church may be warmed on the most mistaken principles, and may be entirely, or nearly, deficient in ventilation with less disastrous results than a schoolhouse. Most of the ancient fabrics were so. If warmed at all, the means was probably the brazier already alluded to, carried into the building when the fuel had become red hot. Modern specimens are little better. The respiration of the same air again and again by adults for a couple of hours on a Sunday is never so mischievous in its consequences as when experienced every day for five or six hours by children of tender years. If the freshness and purity of the air breathed by older persons has an important bearing on health,—as witness the difference between men whose daily work confines them much indoors and those who follow an outdoor occupation,—how much more in the case of a child kept long in a school-room with many others! If improper or insufficient lighting can be held to affect injuriously the general health, how much more the temperature and quality of the air which is to enter the lungs, to contribute by its purity or impurity to the healthiness of the blood itself, to affect the skin, and, in short, the whole human frame, directly or indirectly. The process of breathing directly tends to vitiate the air, carbonic acid being given out with each successive expiration. When confined and accumulated within the four walls of a room, it must soon render the air poisonous unless carried off by ventilation. The historical case of the Black Hole at Calcutta is only an extreme example of this principle. Dr. Theodore Beckar, in the programme of the gymnasium at Darmstadt, says that each boy emits, in the process of breathing, two-thirds of a cubic foot (the German foot differs little from the English) of carbonic acid gas every hour. Thus, in one of our single class-rooms, containing 40 children, poisonous gas would be produced at the rate of nearly 27 cubic feet per hour, even supposing the air to have been fresh at first! The respiration of the same air a second time is always unpleasant. Nature, even at this early stage, warns us that a deficiency of oxygen may soon become of vital moment. And the warning is given in an increasingly marked manner at every further stage of deterioration. The constant breathing of a vitiated or overheated atmosphere will not only affect the whole general health of a child, but will render it liable to catch cold from sudden change of temperature when emerging into the open air on its way home. Even when occurring in less degree, as in rooms where partial ventilation exists, much of the restlessness, inattention and apparent stupidity, often observable among the children, is due more to want of freshness in the air than to dulness in the scholar. A teacher will find his task materially facilitated if carried on in a light, cheerful, warm and airy room. However important in all rooms and collections of rooms, let us understand once for all that proper "warming and ventilation" is seldom of such vital moment as in a school-room, and that education cannot properly be carried on in its absence, or if carried on, suffers under the most serious disadvantage.

In considering the means to be adopted for producing and maintaining a suitable temperature and quality of air both in school-rooms and class-rooms, there is a preliminary condition to be observed which is not without considerable determining influence. The whole of the arrangements must be such that the proper and full use of the room for teaching purposes cannot be in any way interfered with or incommoded. The fireplace, the coil or line of pipes, the warm-air gratings, and other agents for warmth and ventilation must be entirely out of the way, and yet in proper positions for efficient results. If the teacher be compelled to pursue his work all day with the fire close to his back, or with a current of hot air whiffing across his face or steadily directed against his legs; if one part of the class be overheated and the other left shivering from cold; or if a coil of pipes, a fireplace, or an opening for admitting warm air be placed exactly where the school cupboard ought to stand, the result cannot be regarded as particularly satisfactory. Like the positions of the desks, windows, and doors, which have already been shewn to involve in their importance skilful or unskilful planning, those of all other appliances or necessities must be the result of care, thought, and study.

Bearing this preliminary condition constantly in mind, we may next proceed to consider what are the safest and most wholesome methods of generating heat, what amount of heating-power is necessary under given conditions, and in what manner, and at what rate, the constant movement and renewal of air shall be effected and maintained.

The quantity of glass contained in the windows, or skylight, has a direct voice in the amount of warming-power required. The general request for "plenty of light" in school-buildings is too often answered by the introduction of windows anywhere and everywhere. Not only is an unpleasant and trying glare of complex lights and shadows thus produced, but in the severe weather of winter it is found almost impossible to warm the rooms. The power of glass in cooling the atmosphere of a room heated to a higher point than the external air is so great that, unless we are prepared to adopt a system of double windows, as used commonly in the class-rooms of Germany, we must not introduce windows quite so lavishly as in a conservatory, nor without due consideration. If we place them exactly in the proper places, we shall find that a less quantity than is generally supposed will afford abundant results. We have already quoted, in a previous chapter, the German opinion as to the quantity of window-surface necessary for work. As to the influence of this surface on the temperature of the room, Mr. Hood, in his admirable work on Warming and Ventilation, tells us that experiments have shewn that one square foot of glass will cool $1 \cdot 279$ cubic feet of air as many degrees per minute as the internal air exceeds the external in temperature. Calculating the cubic content of the room and the superficial area of window glass, we shall easily find on this basis the total amount of cooling-power at work, and the corresponding increase required in warming-power. The more window there is, the greater the warming-power must be. To over-light a room is nearly as bad as to under-light it.

As to the amount of heating-power practically required in buildings, Mr. Hood further tells us that we should calculate for warming three and a half to five cubic feet *per head per minute*, and, in addition, one and a quarter cubic feet for each square foot of glass.

Among the many methods of warming practically known among buildings, are a series which, in their application to schoolhouses, must be condemned in the most unequivocal manner. Of the most deleterious, or dangerous, and the most carefully to be avoided, are those which may be classified under six general heads, as follows, viz. : —

1. All warming by means of ordinary gas stoves.
2. Every kind of stove not provided with a flue for the escape of smoke or products of combustion.
3. Any method which merely warms the same air again and again.
4. Any system by which the air is liable to be vitiated by direct contact with overheated metal surfaces.
5. All methods in which warmth is obtained by water pipes heated at high pressure.
6. All heating by steam.

The first and second of these require no comment, their objectionable nature being now pretty well understood and condemned by experienced school promoters and managers.

The third refers to methods which have been, and still are, almost universal in their application to churches, public halls, &c., and which are unfortunately not unknown to schools. One of these consists of a coil of hot-water pipes placed in a corner, or of lines of pipes carried round the walls, without any provision for a renewed supply of air. The coil, or line of pipes, heats the particles of air with which it is in contact, and thus transmits warmth; but the principle is merely to heat and re-heat the air which happens to be in the room, and which is being breathed by the children again and again with the certainty of becoming more impure at each respiration. Another system, although having its apparatus of heating surfaces placed in the basement, yet draws its supply of professedly fresh air from the *interior* of the room or building, and is even worse. (See woodcut No. 195.) The heated air rises to the ceiling, and, as it descends by cooling, is again drawn down to the basement to be re-heated, and to perform the same process as before. The heat is certainly economised, and the process may, by some, be ranked as "cheap," but the principle is eminently vicious, and the effects on health disastrous in degree according to the length of time during which the air is breathed.

This method, the result of over-zeal for economy, usually aggravates its evil by adopting as its means of obtaining heat the principle of the cockle-stove in some form, which we have condemned under No. 4. The radiation of heat from iron plates placed in contact with a furnace (almost always fired violently when warmth is quickly required), is also cheap, and also dangerous to health. The air is deteriorated, and numerous instances could be given, from actual observation, of the bad effects produced.

Water pipes heated by high pressure do not warm the air equally. Their average temperature may be taken at 350°, but is often, in parts next the boiler, considerably more. The steam is prevented from escaping. These three objections point to uncertainty of results, deterioration in quality of air, and danger of explosion when managed by the ordinary caretaker. If the bulk of water contained in the pipes at a temperature of 39° be increased by no less than one-twenty-third when raised to 212°, or boiling point, the force must be enormous at higher temperatures, with no provision for the escape of steam. In addition to its other serious objections, sufficient to exclude it from all school-houses, the high-pressure system is rather more costly in annual maintenance both as to fuel consumed and as to repair constantly required.

Steam warming, however suitable for manufactories where the waste steam from the boiler can be turned into iron pipes placed around the workshops with provision of final exit to the open air, cannot pretend to be of use for schools, if only from its troublesome character, and the circumstance that until steam is up no particle of heat can be obtained. If heating by hot water at high pressure can be regarded as somewhat dangerous, steam is, from its great expansive power, eminently so. Both require the constant charge of an experienced machinist or engineer.

Generally, all methods are objectionable which deteriorate or render too dry the air, which in any way tend to prevent a copious supply of oxygen, and which are not capable of simple and easy management.

It is much easier to point out the various systems which are bad in principle or in practice, and to determine what we ought not to do, than to draw final conclusions as to the course best for adoption in each case. The subject of warming and ventilation is perhaps the most difficult among all those questions which arise in connection with School Architecture. The conditions of the problem may easily be stated. The building requires not warmth only, or ventilation only, but the two in combination, each efficient, thorough and ample. The air for respiration must be perfectly fresh, comfortably warm, yet never too warm, always in movement so imperceptible as never to be productive of draughts. The system should be so entirely under control that, when the temperature of the external air changes suddenly, that of the internal air may be regulated accordingly, and on no account allowed to become stagnant or unwholesome. The apparatus should be capable of warming the building within two hours of the lighting of the fires, so that when the children first arrive the effects may already be at their maximum. A warming system is frequently pronounced inefficient because the school-room is chilly at 9 A.M. At that hour the warming-apparatus should be exerting its full power, to be reduced somewhat from time to time when the cold air has been thoroughly expelled.

These desired results may be attained in any one of several different ways, according to the building to be treated, to opinion on the subject, to proximity to a coalfield, or other conditions; but, whatever the course pursued, there is one great principle applicable alike to every system possible to be devised which we must strongly insist upon at the outset, viz., that of *demand and supply*. The removal (or attempted removal) of heated or vitiated air from a room by means of an aperture or flue is often supposed to be ventilation. It is only a part—a necessary part—of ventilation. To be of use, indeed to act at all in the manner intended, such a flue or aperture requires that fresh air of at least equal volume shall, by some other source placed at a lower level, be admitted to the room at the same time. The *demand*, set up by the outlet flue, requires the *supply* which can only be met by the provision of an inlet flue. In all those systems which attempt to warm the air of a room without allowing any of the warmth to escape, ventilation is entirely lost sight of. None can be really good which do not contemplate a continual removal of foul and supply of fresh air to the room. To provide such an amount of heating power as will admit of a constant movement and renewal of the air, the warm fresh supply being admitted at one place and the vitiated air being carried out of the building at another, involves sometimes a cost so considerable that sound principles are in danger of being sacrificed, and school hygiene forgotten, in a mistaken zeal for defending the purse-strings. A large outlay at first cannot be avoided if we would have thorough warming and thorough ventilation effectually combined, for, if the foul air be continuously extracted, and fresh air continuously admitted, the arrangements for warming the latter must be of great power, and for removing the former of great extent. True economy dictates that only of such methods as are sound in principle, healthful in practice, easy of management, and therefore suitable for a schoolhouse, should we take note.

Here, for the first time, occurs the question of *relative* expense. Among methods all to be classed as sound and good, we have to consider and judge between the respective economies of first cost as it appears in the price of the building, and of continuous expenditure as set down in the account of annual maintenance.

Warming by open fires, whether of wood or coal, has in its favour a strong prejudice in the mind of the English people. Some part of this may be due to the poetry of long association with the hearth and home, but the undoubtedly cheerful appearance of the blazing grate and its tendency to promote ventilation by causing a current of air, form strong arguments in its favour. The last reason, though, in favour of the open grate considered as a ventilator, must tell against it if regarded as a means of scientific and economical warming. The draught which naturally sets towards the fire, carries all the heat up the chimney except that derived from direct radiation, of which there is usually far less than from a stove standing out from the wall. Again, exact results and precise measurements of the warming power of open fires are not easily obtainable. The use of any particular grate in a given size of room does not necessarily carry with it a definite amount of heat, for one teacher will become so absorbed in his work as to allow the fire to die out, while another may be so susceptible to cold as to maintain it constantly in a state sufficient for the roasting purposes of a kitchen.

The erection of schools with command of funds sufficient for proper maintenance, occurring simultaneously with an enormous rise in the price of our staple fuel, marks a fitting time for the prosecution of further inquiry into the warming of the new buildings with an eye to their cost of annual maintenance. The use of coal, not very common before the 17th century, has now become so universal as to cause some alarm concerning the endurance of the supply. In any event the great price must prove a sufficient inducement to study economy under this item most strictly. In the Voluntary schools, which in the past have done great work for education, the annual cost was often obtained with greater difficulty than the capital sum for the first erection of the fabric. Limited funds often, in the smaller schools *always*, forbade the employment of a caretaker permanently, or for the whole of his time. He was expected for a small sum to take charge of the building, and to earn his main living during the day in other ways. The absence of this official became, alone, the cause of much prejudice against all methods of generating heat artificially. Among other causes operating in the same direction were, the adoption of systems bad in principle which proved equally bad in practice, and the use of others which, although good in principle, had been carried out in an ignorant way, so as to become bad in practice. The main cause, however, arose from the absence of an officer, whose appointed duty it was to look after the apparatus. The school promoter and the teacher united in saying, "These artificial systems never act well; there is nothing like the open fire after all." The open fire thus took possession, and would in all probability have continued in undisputed sway for another generation but for the price of coal, the increased size of the new schoolhouses, and the reflection that it is easier and cheaper to warm a *large* building by one fire than by twenty.

The presence or absence of a handy man in charge of the school fabric, able, perhaps, to mend a pipe or a window pane, to see to the cleanliness and daily flushing of the lavatories, waterclosets, and urinals, and to take charge of the heating apparatus, cannot fail to make an important difference in our estimate of systems of artificial warming as suitable to schoolhouses. The hot-water apparatus, with no one to look after it but the teacher—who never treated it as a part of his duty, and in fact, never quite understood the mechanism—was naturally liable to disorder at frequent intervals. During the Christmas holidays, the water being left forgotten in the iron pipes, and having ceased to circulate from absence of the accustomed fire, naturally froze and burst its

T

iron bonds if by any mischance a hard frost set in before the fires were again lighted. The new era of Public Elementary schools, in which the building invariably has a caretaker, enables us to look on questions of warming by hot water, hot air, or other artificial means, in the new light afforded by these new conditions. For the first time in the history of English Elementary schools, one or other of these systems can be adopted with absolute success in action, without fear of accident to teachers or children, and with certainty of results as to quality of air, temperature, supply, and power of extraction, *if only it be well considered at first and properly carried out.*

In warming by open fires, the first point is to render as much of the heat as possible available for the warmth and comfort of the room. To this end M. Gauger had applied himself so far back as the year 1713, with the result of recommending several points which have since become well known. The hollow chamber provided to the hearth and back of fireplace, for admitting air to be warmed in transit; the angle of 45° as being most scientific for reflecting the heat from the jambs; and the necessity of contracting the throat of the chimney, were all urged by him at that comparatively remote period. In more recent times many kinds of grate have been invented, all having for their object the supply of fresh air warmed in its passage to the room by contact with some portion of the grate. These agree, generally, in warming the fresh air by passing it up through chambers formed at the back of the grate and admitting it to the room at a higher or lower place. The point of admission is, with some, at the front of the grate, with others, at heights varying from 5 to 7 feet from the floor, and with yet others, nearer the top of the room. On this it may be remarked that the point of admission must depend on two things; (*a*) the volume of air to be admitted, and (*b*) the method of extraction employed. If there be no other means of ventilation than an inlet from the back of the grate, and an open fire, and the inlet be placed just above, or at the sides of the latter, the fresh air will be drawn up the chimney at once;

and therefore, in this case, it must be kept sufficiently high to be able to rise through the room before being drawn towards the fire. If, however, there be a separate flue provided for carrying off the vitiated air, the inlet may be at the front of the grate, and it only remains to provide by its means a volume of fresh air equal to the aggregate capacity of the extraction flue and the chimney. Where the ventilation depends only on the provisions connected with a grate having air admitted in the manner described, some provision for summer is necessary also. It is true that, in cold weather, the warmed air will ascend to the ceiling, pass across to the other side, descend as it cools, and, as a vacuum below is created, will finally be drawn up the chimney. In winter, this means of extraction will generally be ample, and the heat will be well utilized. In summer, however, the motive power of the fire is absent. Cold, fresh air will be admitted, instead of warm, through the chambers at the back of the grate, but the vitiated air will not be drawn up the chimney. It will be necessary, therefore, with this system, to provide extraction flues for summer use only, capable of being kept closed in cold weather. Where the use of extraction flues is contemplated for the winter months also, the ventilation will be more perfect, but, as a greater volume of fresh air must be admitted, so a more powerful stove must be used. The material of which the back of the grate is made, and with which in its hot state the air comes in contact, varies with different manufacturers, some preferring iron and some fireclay, but all adopting flanges or gills to increase the amount of heating surface. Fireclay has an advantage over iron in not readily parting with heat. Thus, air heated by contact with surfaces of hot fireclay or brick, is not so liable to be impaired in quality as when passed through a nest of hot iron plates, especially if the latter are so arranged that overheating is possible. This danger is sometimes avoided by making the iron back of considerable thickness, and sometimes by admitting the air in large quantity, according to the size of the grate and the room to be warmed. The point of temperature at which air becomes deteriorated for purposes of respiration by contact with heated surfaces of any kind, has never been clearly established. Some have placed it as low as 170°. Others hold that until boiling point, or 212°, be reached, there can be no possible danger if the original air be free from particles of animal or vegetable matter liable to decomposition.

Among grates which appear to be suitable in greater degree

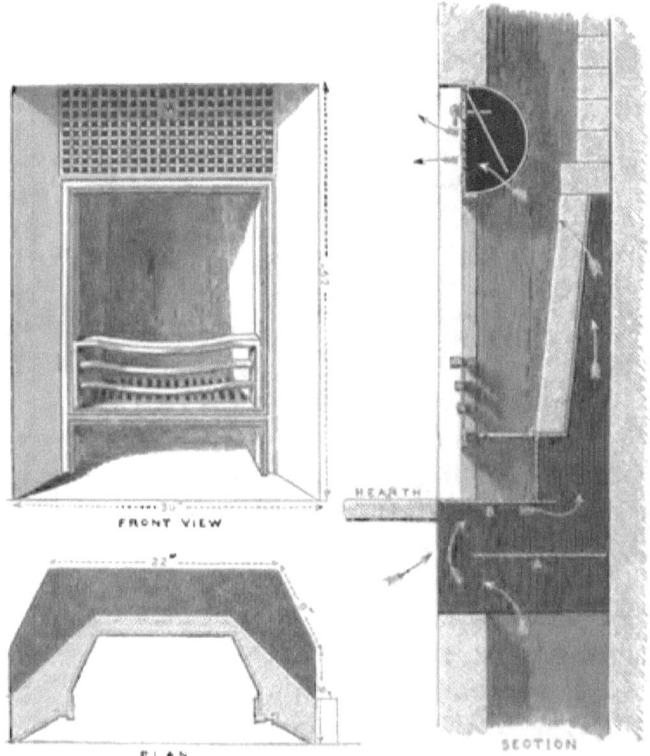

183.—THE BOYD GRATE.

than others for school purposes, which have been prominently before the public, and which are all designed in reference to the warming of air in its progress to the interior of the building, may be instanced the Galton, the Manchester, and those bearing the names respectively of Boyd, Pierce, and Longden. The two first have the same general aim; the former is more complicated and is said to be liable to smoke, while the latter has the merit of simplicity. The Manchester grate also aims at considerable results by the bold use of iron as a heating-surface, combined with the precaution of such a thickness of back as greatly to diminish the possibility of over-heating and deterioration of air in its passage to the room.

The Boyd grates are of several kinds, all remarkable for their great power and the large volume of warmed air which, by their means, can be brought into use. Woodcut No. 188 shews, of

189.—THE BOYD SCHOOL GRATE.

the medium size, the appearance and construction of one kind, which is manufactured of three different proportions and degrees of power. In every case the warm air is admitted to the room immediately above the fire, and experiments have shewn that owing to its considerable volume it is *not* immediately drawn up the chimney. Where the position of the fireplace does not occur in an external wall, a trench leading to the back of the grate must be formed for admission of fresh air. The back and sides, in contact with the fuel, are of firebrick, but are firmly held together and connected with vertical flanges or gills by sheet-iron attached behind. The amount of air admitted through the front of the grate is regulated by a flap, which can be screwed up or down at pleasure, and which should always be kept shut until the fire has become hot. Where position, and the work to be done, will allow, two or more rooms may be warmed by one of these grates, thus effecting an economy both in coal and daily attendance.

The requirements of some of the larger London school-rooms caused an adaptation of the Boyd grate to be made to the new circumstances (woodcut No. 189). Greater power was obtained by utilizing the hearth, as well as the back and sides, as a warming surface, and by fitting flanges or gills to the whole in such a manner that the fresh air must of necessity pass over a large proportion of surface before entering the room.

The Pierce grate—an old invention—is, like the Manchester, marked by much simplicity. It differs from all the others, except one, in using fireclay, rather than iron, as a heating-surface, and claims as a result superior purity in the air. The sides and back containing the fire are of fireclay, and the air-chamber is arranged with fireclay flanges over which fresh air is admitted. In the use of these grates the warm air is usually admitted to the room at a height of six or seven feet from the floor.

The Longden grate, as the most recent invention, has had the least proof from actual experience of its merits. It differs in several important respects from any of the preceding, in particular, in aiming at the economization of fuel by the size of the fire-box and the consumption of smoke. A Section and Plan (woodcuts No. 190 and 191) are given to shew the method of construction. The bottom, made of iron in the other cases, is here of fireclay perforated with round holes smallest at the upper surface. The smoke and flame pass into a shallow flue formed between an iron frame and a fireclay back, and the former is expected to be consumed in the passage. The fireclay back is flanged, and the heating surface, here also, is free from contact with heated metal. The two arrows shewn on the plan are intended to represent the direction of the fresh air which would be admitted by trenches, communicating with the outside.

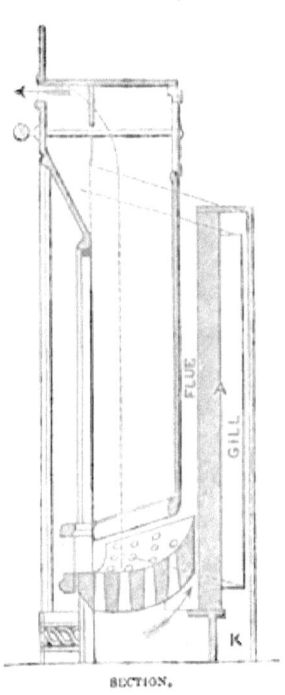

Stoves standing on the floor interfere so much with the working of a school that this reason alone is sufficient to prevent their frequent adoption in new buildings. Another reason, equally sufficient to an architect, is the difficulty of dealing with the smoke flue, as well as with the stove itself, in such a manner as to avoid unsightliness. Stoves are commonly adopted for warming easily and effectually large rooms hired for a limited period, or for such as are only likely to be occupied during the term of a short lease.

190.—THE LONGDEN GRATE.

Chief among them—

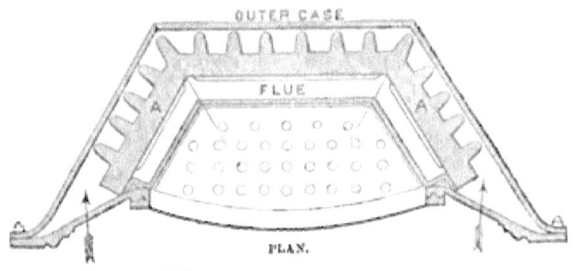

191.—THE LONGDEN GRATE.

indeed alone for school purposes—stands the gill-stove. This is of two principal kinds: one known as the Gurney, which is circular in form; another, called the Studio, which is commonly of a barrel shape. Gill-stoves possess an enormous advantage, where very large rooms are to be warmed, in that a minimum amount of heat is wasted. The common objection that they dry the air unnaturally, can be met by the provision of a trough of water at the bottom, in which the gills stand to the depth of a few inches, thus supplying by evaporation the moisture of which the air had been deprived. When placed on the floor, unless very near a wall, a *down-cast* flue becomes necessary for the smoke. This kind of chimney always requires in up-cast at least twice the length of that which had been found necessary in down-cast, without even then being quite certain of draught. Further, at the lowest point of the down-cast portion the soot must be cleared out very frequently. If it be desired to adopt the gill-stove for use in a new schoolhouse, let it be placed underground in a chamber provided for the purpose. The heating result will then be greater, the complaint of

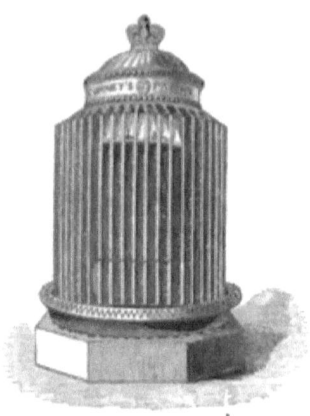

192.—THE GURNEY STOVE.

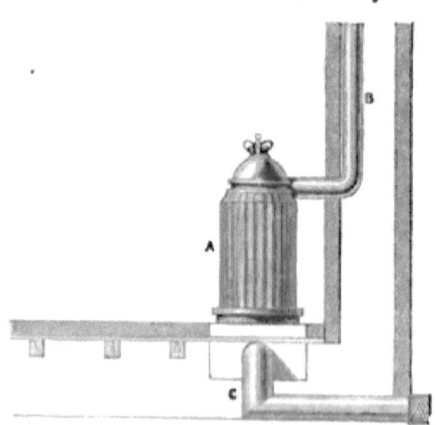

193.—SECTION OF PART OF BUILDING, SHOWING FRESH AIR IN CONNECTION WITH THE GURNEY STOVE.

Reference.
A. The stove.
B. The smoke pipe leading into brick flue.
C. The fresh-air pipe communicating with grating outside.

unsightliness will disappear, the working of the school will not be inconvenienced, and the down-cast flue will no longer be necessary. The objection remains that the air so warmed will be impure unless the water-trough in which the stove stands be continually replenished from time to time as the evaporation

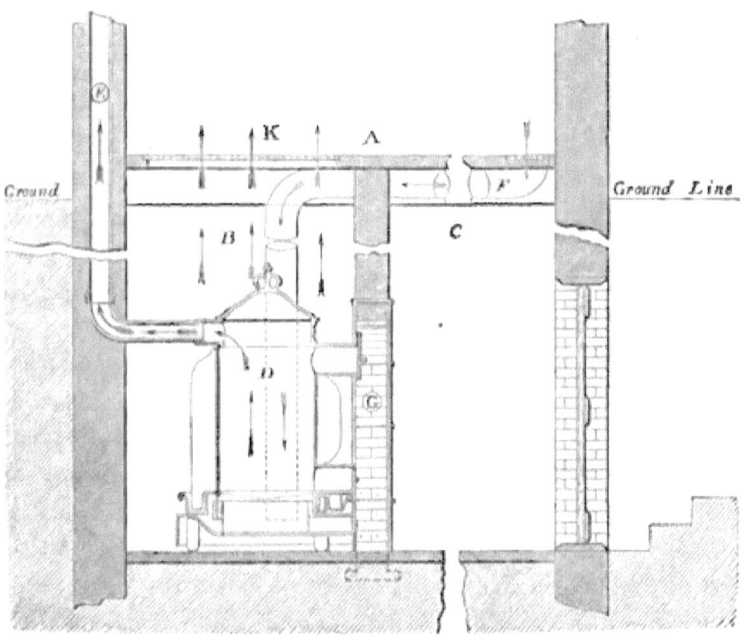

194.—THE GURNEY STOVE, PLACED BELOW-GROUND.

Reference.

A. Floor of building.
B. Apparatus room.
C. Stokery.
D. Gurney stove.
E. Smoke flue.

F. Cold air admitted from interior of building.
G. Front of stove.
K. Grating for admission of warm air to building.

causes it to disappear. No caretaker could be relied upon to do this certainly and regularly in the cellar, where any check by occasional supervision is out of the question. Nor yet could we be certain that, to save himself trouble, he would not over-fire. The gill-stove in its ultimate condition is thus reduced, however regretfully, to a place in Class No. 3, page 268, among the things to be avoided, so far as new buildings are concerned. Yet, in cases where it can be used near a wall so as to avoid the down-cast flue, without loss either of convenience or of appearance, with provision of fresh air admitted from the exterior (woodcut No. 193), and with the water-trough attended to with certainty and regularity, the gill-stove will be found a powerful, efficient, and economical means of warming.

195.—THE GILL OR "STUDIO" STOVE.

The collection, in modern times, of criminals and persons afflicted with disease of mind or body, in buildings of immense size, has rendered necessary some degree of scientific attention to the warming and ventilation of such buildings. Hence, in our various prisons, lunatic asylums, and hospitals may be found some of the best examples of different methods on a large scale. A similar necessity has now arisen in respect of large town schools in consequence of the increase in the size of the buildings. Children are collected in large numbers under one roof for purposes of training, and, in degree, the warming and ventilation assumes the importance of that applied to a hospital or gaol.

The abandonment of the open fire as the source of heat, must be invariably accompanied by increased provisions for inlet and outlet. In applying any system of artificial warming to a large schoolhouse we must be careful that the ventilation is ample, and that a condition of stagnated air is impossible. This can only be effected with certainty by gathering together all the outlet flues to one common shaft placed in a central position in the roof, and by applying the artificial extracting power of gas, hot water, or other means. Fire is the most effectual agent, but the trouble of maintaining a fire at so great a height from the ground renders it practically out of the question. Air, when expired from the lungs, flows upwards because of its greater heat and consequent lightness. Each exit-flue should therefore carry off the vitiated air at the highest point of the room, where the heat is greatest. Each inlet should supply copious volumes of warm fresh air as near the floor-line as may be. With such a system carried out thoroughly, the small amount of carbonic acid gas which will have separated itself from other impurities, become cold, and settled to the bottom of the room, may reasonably be left to itself.

The amount of fresh air to be *continuously* admitted requires some consideration, for on it depends the amount of warming-power to be provided, whether by open fires, hot water or other means. Calculating the movement through the inlets to be at the rate of 150 feet per minute, *from* 15 *to* 20 *cubic feet of air per child per minute* is required to pass into the schoolroom in a ceaseless stream.

Germany, never famous for the ventilation of its buildings, is now paying much attention to the subject in its schools all through the country. The systems are various, and generally comprise one for winter use and another for summer, as in the new Luisen Schule at Berlin. The warm-air inlet is always about 6' 6" from the floor, so as to clear the boys' heads, but the extraction is from the bottom of the room in winter and the top in summer. When there is an equalizing chamber the extraction is always from the bottom. An inspection of the heating chamber in the basement story of a new Parish School in Freidens Strasse, shewed that the apparatus was divided, on account of the size of the building to be warmed, into two portions. The principle of generating heat was not specially admirable, being that of flanged flues of iron, coming under the same category (No. 4, page 268) as the gill stove, but with liability to a still greater amount of overheating. In the class-rooms three gratings, each having a surface of 12" by 12", were visible. One at the usual height of 6' 6" was the fresh-air inlet. The two others were placed in the opposite wall, one, about six inches below the ceiling line, was said to be the extraction, the other, in the same wall about a foot from the floor, formed the opening into two flues, one leading upwards, the other downwards, and forming respectively a second inlet and outlet available for occasional use. The second inlet was said to be used chiefly to obtain heat more rapidly, or in greater volume, in very cold weather or in the early part of the day. The second outlet was doubtless devised with some vague idea in reference to the carbonic acid gas lying near the floor, for the motive power of the extraction was of the weakest kind. Too much dependence is sometimes placed on the difference in temperature between the air of the room and that outside. The hotter and lighter air will, no doubt, be forced out by the denser fresh air admitted, but the action is often of a feeble sort. If the air outside be hotter than that inside the room, the extraction flue, unassisted by any artificial power, will either cease to act or will act the reverse way. Sometimes the direction of the wind may be used as a means of forcing in air, and thus securing more rapid removal of the vitiated. In some of the recently erected Berlin schools, air is admitted by brick channels, while the polluted air is carried out at the opposite side. In Appendix A, at the end of this work, will be found some account of the methods of warming and ventilation employed in the King William Grammar School at Berlin.

Warming, by means of a hot-water system circulating in iron pipes at a low pressure, is often used by carrying the pipes themselves into the various rooms. This is unsightly, and, unless fresh air is by some means admitted directly to the surface of the pipes, so as to become warmed in its access to the interior of the room, the principle is merely that which we have already condemned under head No. 3 (page 268). Considered in reference to schools, this system under the most favourable circumstances is not the best. The various stories of a large

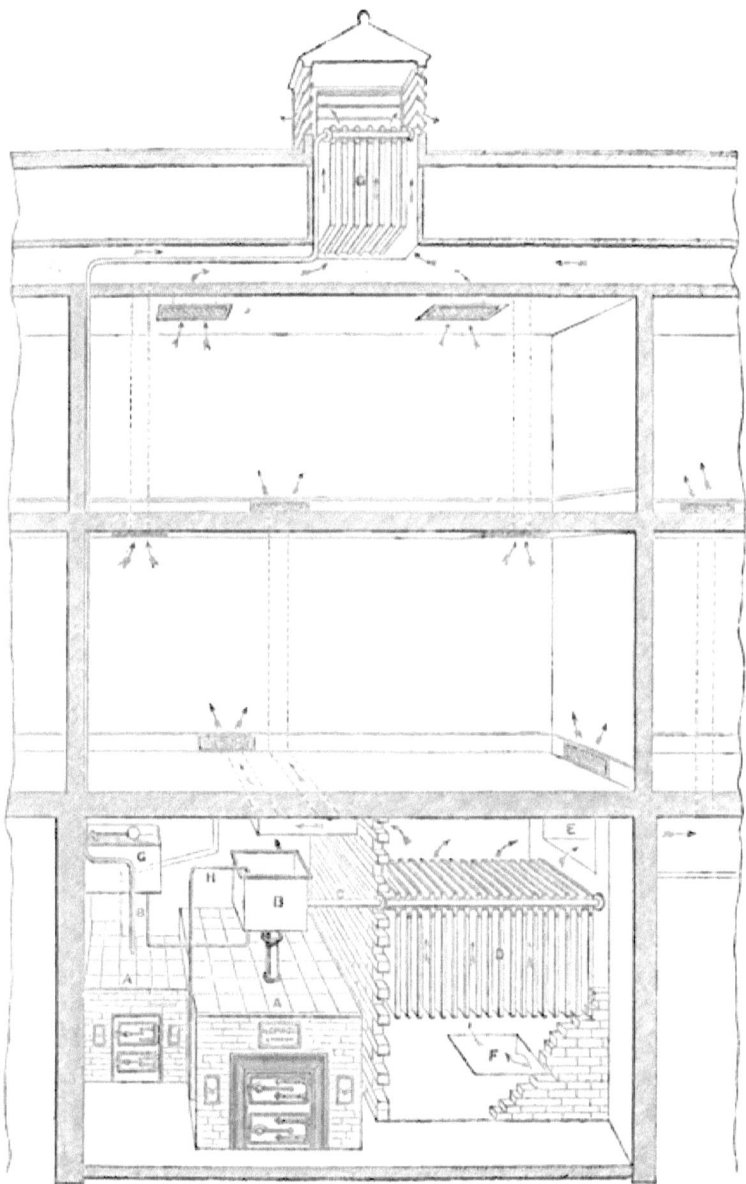

196.—SECTION OF BUILDING, SHOWING WARMING AND VENTILATING APPARATUS.

Reference to Warming Apparatus.

A. Boiler. B. Expansion box. E. Main warm-air supply.
C. Flow pipe leading into flat warming F. Fresh-air shaft from exterior of building.
    vessels. D. Warming vessels. G. Supply cistern. H. Safety pipe.

Reference to Ventilating Apparatus.

A. Boiler. B. Flow-pipe rising to warming vessels in rarefying chamber.
C. Hot-water vessels.

building cannot be warmed equally, because of the tendency of the heated water to rise at once to the highest point, unless separate pipes be taken from the boiler to each floor. And the pipes themselves (supposing fresh air to be admitted over their surface) are always more or less in the way.

The collection in one chamber in the basement of all the pipes has a disadvantage in necessitating the provision of a somewhat larger heating surface. But its advantages are very great for our purpose. The pipes are quite away from the daily working of the school, and, being all together, can be repaired at any time most easily. The warmed air can be admitted, by flues formed in the walls, at any point. And the temperature of the school can be lowered, when necessary, by fresh air admitted direct to the flues. The same principle which would dictate separate *pipes* from the boiler to each floor, in this system points to the necessity of separate *flues*, if we would avoid the heat being all sent to the highest story. This kind of warming was advocated by the late Mr. H. C. Price, who, instead of ordinary pipes, used a series of flat vessels about 3 feet square and $2\frac{1}{2}$ inches thick. His system, as carried out by the present firm in many of the larger schoolhouses in London, and combined with ventilation, is as follows (woodcut No. 194): The apparatus consists of a boiler connected by flow and return pipes with a number of the flat vessels placed vertically in a chamber so as to allow a thin stratum of air, admitted in large volume at the lowest part, to pass between and become heated. The heated air escapes from the top of the chamber (through large main flues in very large buildings) to the flues which conduct the heat to the several rooms. The apertures are placed vertically near the floor and fitted with valves controllable at pleasure. Exit flues communicate, from the highest points of the rooms, with one large central extraction shaft, heated by a small group of hot water flat vessels, similar to those already described, and fixed in the highest part of the shaft. This artificial extracting power is heated by a small separate boiler, and can therefore be worked alone in summer when no warming power for the building is necessary. In this respect the system offers an advantage over the ventilating grates, which are good for winter but provide no ventilation in summer.

Much of the success of this system depends on the freshness and purity of the air admitted at the main source before it enters the heating chamber, and on the amount of ventilation provided. By the low-pressure system used, the water in the pipes, at its hottest, never exceeds 212°, and the pipes or vessels themselves with fresh air constantly passing over their surface cannot usually maintain a temperature of more than 200°. If the air admitted be charged with particles of animal or vegetable matter, these may be gradually decomposed, and, with insufficient ventilation, would soon render the room offensive. Hence arise the two necessities pointed out, which should always be carefully attended to in work. It is important to note that the aggregate area of the inlet flues should always be rather in excess of the outlet,— the supply should be somewhat greater than the demand,—and the latter should either be of exactly the same length, or should terminate in a larger common flue of the kind already mentioned. This method, sound in principle, has been proved efficient and healthful in practice. The most economical apparatus for warming by hot water must always be that which contains a large body of circulating water, and which by reason of its volume takes a long time to cool. It is not, however, that of the smallest first cost.

In its application a few simple rules are to be noted. The inlets should never enter a room through the floor. While admitting the warm fresh air at as low a point as possible, the gratings should be placed vertically within two or three inches of the floor. Hot air rises quickly enough, without artificial assistance tending to shoot it at once up to the ceiling. Thus a horizontal direction should always be given in the first instance, as far as possible. Again, floor gratings invariably become a too convenient dust-hole, whatever pains be taken to the contrary. The position of the inlets should not be too close to the desks, but at the ends or the opposite side of the room. That of the outlets should be in the ceiling—not the wall—and as far off the inlets as possible, so as to secure a current through the entire room. It should be remembered that air in movement, unless very hot indeed, generally feels chilly, and is liable to give cold. The openings should be numerous and well distributed.

The point at which the open fire should be superseded in favour of an artificial method of warming, is not easily described, although in practice it is not difficult to determine. When we have to deal with a school for 1,000 children, placed in three stories one above the other, no doubt arises. The latter is the more economical, both in first cost and in permanent maintenance. On the other hand, the warming of an Infant school for 120, does not warrant the expense of an apparatus. It is in every way better to use a couple of little fireplaces, and to utilize their heat well. Between these two instances we may gradually approach, deciding each case according to the ease and inexpensiveness with which a system can be applied, until we reach a debateable land where it is impossible to say that one course is better than the other. In cases under this head, the doubt may be decided in favour of that system which is best for the daily work of the school and gives least trouble. Generally, experience has shown that for schools of three departments containing a total of 500 children, the open fire is yet the best. Also that an artificial system is preferable for numbers of 750 and upwards. The instances between these numbers can only be decided by the peculiarity of their plan. In the future, when more experience has rendered possible a closer comparison, the decision may be arrived at with greater certainty.

Even for a school of 750 children the *first cost* will often be least if the open grates be adopted. For the *annual* expenditure a different tale will be told, and, if we have adopted a kind of grate with a large power of consuming fuel, the yearly bill may be appalling. For establishments of this size there can be little doubt that, in spite of prejudice and the natural desire to keep down the cost of original construction, well contrived systems of artificial warming and ventilation will eventually be found the best. At present, the misfortune lies in the small number among architects and manufacturers, who will take the trouble to devise and carry out good systems in a scientific manner.

The foregoing has been written with a full knowledge of some of the various methods of warming and ventilation recommended by different medical men.

One has suggested that the floors of a schoolhouse should be all warmed by flues on the principle of the old Roman hypocaust, and that the ceiling should be entirely of perforated zinc, through which cold fresh air should be constantly admitted to the room. The Romans themselves found the hypocaust to be very expensive, and, in their time of greatest luxury, it was only to be found in the houses of the rich. The use of the chimney was not then well understood. So soon as any responsible architect has carried out the suggestion in any new building, we shall inspect it with every desire to learn. Another has written a pamphlet in which one of the points held to be essential is that none of the windows of a room should be made to open, and that water-closets should on no account be provided with windows. A third has also published a pamphlet in which the burden of the argument is directed to establish as a basis or principle that all warmed air should be admitted at the top of a room and extracted at the bottom.

Medical men seldom speak or write on the subject without displaying much scientific knowledge. The application of such knowledge in a useful, practical and simple way to the ordinary modern building is not so successful. *Primâ facie* common sense does not endorse the hypocaust. People *must* have windows— windows capable of being opened—in all water-closets and in all rooms. The theory of extraction from the bottom instead of the top may be scientifically and theoretically the best, but it is practically inapplicable to a schoolhouse. It may be perfectly true that the circulation of air in a room should be as constant as that of blood in the body. In practice it can never really be so. We go to sleep and forget all about the circulation of the blood, which continues its action without attention. If we go to sleep and forget the duties of the stoker, the fires die out and the warming power also dies a natural death. In all systems of warming and ventilation, the practical and working daily use must have a voice in the arrangements. Extraction from the bottom requires, from its great friction, so enormous a motive power as to be out of the question except in buildings of very great size, and, for school purposes, affords no advantages sufficient to compensate for reversing the order of nature.

We have endeavoured to place in a short, compact form, some of the leading principles which govern the subject both in its scientific and practical aspect. To do more than this, to lay down dogmatically that any one system is the best under all circumstances, would only be an evidence of quackery. In each case and with each fresh set of conditions, the architect should exercise his own judgment, and should invariably entrust the carrying out of the work to some engineer specially accustomed to the kind of appliances and arrangements proposed to be used.

197.—LOCALITY OF THE FIRST BOARD SCHOOL ERECTED IN LONDON.

CHAPTER XVI.

THE BOARD SCHOOLS OF LONDON.

Exceptional case of London—Board formed without option by the Act—No statistics—Proceedings taken to obtain these—Schools erected before their completion—Old Castle Street School—Harwood Road School—Its architecture—Agitation for Prussian system—Johnson Street School—Its planning —Results—New North Street School—Its plan and abandonment—Winstanley Road School—The slums in which some are placed—Eagle Court— Angler's Gardens—Style of architecture suited to London Schools—Wornington Road, Aldenham Street, Orange Street, and West Street Schools— Variation in style—Camden Street and Mansfield Place Schools—One-story School at Haverstock Hill.

THE vast and increasing amount of ignorance prevalent among the lowest class of population in the largest towns of England which no ordinary efforts seemed able to reach, induced the legislature to pass, on the 9th August, 1870, "An Act to provide for public Elementary Education in England and Wales." While enacting generally the provision of sufficient accommodation in every School District, and the election of a School Board in cases where a proved deficiency was not supplied by voluntary means, the Act treated the condition of the metropolis as beyond all doubt, directed the creation of a "School Board for London," and gave the Board positive instructions to "proceed at once to supply their district with sufficient public school accommodation."* No time was lost in proceeding to an election, and the Board thus established proceeded to obey the peremptory orders of Parliament.

At the outset of their work it became necessary first to form a correct list of efficient Elementary schools and thus to estimate the total amount of existing accommodation. Secondly, to ascertain the number of children who ought to be in regular attendance at Elementary schools. By subtracting the first number from the second, the nett accommodation required in buildings to be newly erected would be arrived at. Following the ordinary calculation that school-children formed one-sixth of the population, the gross number arrived at was 560,000, and the existing accommodation having been shown to be for 308,000, the balance amounted to no fewer than 252,000 children still to be provided for. The enormous cost involved in the erection of new schoolhouses for so large a number induced the Board to institute for themselves a searching investigation into the facts. With the aid of the last census they obtained the name of every child of school age who slept in London on the night of April 2nd, 1871, and set on foot a house-to-house inquiry as to the circumstances of each child. The results of this stupendous and protracted labour, as set forth in a large blue book, enabled the Board to arrive at a determination to build for 112,000 instead of 252,000.

During these proceedings many schools had been voluntarily transferred from their former management to that of the Board;

* Elementary Education Act, 33 & 34 Vict. s. 37 (7).

temporary premises for the more pressing districts were provided ; and, finally, at the instance of Lord Sandon, it was determined to build a first batch of twenty schools in the most destitute districts without waiting the result of the laborious statistical investigation.

198.—OLD CASTLE STREET SCHOOL.

Of these the first which we select for illustration was in point of time the first to be opened, its site and numbers having been determined on the advice of the rector of the parish, then also a member of the School Board. The woodcut (No. 197) which stands at the head of this chapter shows the kind of neighbourhood—one of the poorest in Whitechapel—in which it is located, and which is only reached by threading the way along the narrow wretched lanes leading out of Commercial Street.

The new building erected in Old Castle Street, Whitechapel, from the designs of Mr. Biven, is, in point of architectural character, (woodcut No. 198) extremely plain, as befitted the site. The main block contains in height two stories of school and one of playground. The minor projections or wings are of one story, and contain (woodcut No. 199) the two Infant schools, the class-rooms belonging to which in both cases occupy a portion of the space of the ground, or playground, floor of the Graded schools. The caretaker's apartments are similarly placed under another portion. The remaining covered space is devoted to playsheds.

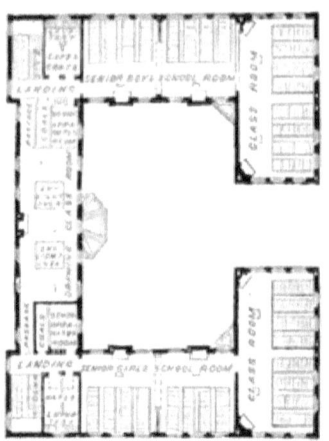

199.—PLAN OF SECOND FLOOR.

The general plan is symmetrical and intended for the separate charge (if thought desirable) of the Infant schools as well as of each division of the others. The building presents, in fact, the plan of a school of six separate departments. On the first floor (woodcut No. 200) a room in a central position commands, from a bay window, the whole of the playgrounds and playsheds. On this floor are the junior schools for boys and girls. The second floor (woodcut, No. 199) contains the two senior schools and a large room, lighted partly from the top, and intended for instruc-

200.—PLAN OF FIRST FLOOR.

tion in drawing The principle of warming adopted in this large school is that of direct radiation from open fires. The site, including a further portion not shown on the plans, occupies an

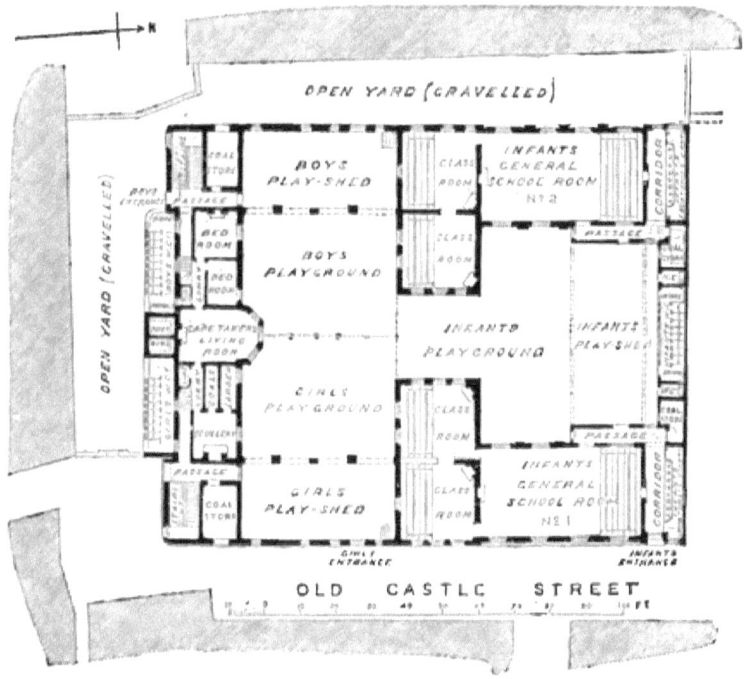

201.—OLD CASTLE STREET SCHOOL. GROUND PLAN.

area of 36,010 square feet, and cost the sum of 11,500*l*. The building contract, exclusive of furniture and fittings, amounted to 9,755*l*. The accommodation, allowing 8 square feet per head in infant schools and 9 feet in others, is as follows.—

Infant school	240
Do.	240
Junior Boys	198
Do. Girls	198
Senior Boys	192
Do. Girls	198
Total	1272

It should be noted that, in the statements of accommodation accompanying the plans of the various schools, the numbers given are not those of the children for whom seats are provided. The calculation is based on the provision of 8 feet superficial of internal floor-space for infants, and that of 9 feet superficial for boys and girls. The clear height is always 14 feet, so that each infant has an allowance of 112 cubic feet of air space, and each child in the Graded schools 126 feet. The figures properly represent the maximum, not the "average" attendance. Calculated simply on the seats provided, the Graded schools would show a superficial area of at least 10 feet to each child.

The statement of cost includes always the building, the lavatory and cloak-room fittings, the warming and ventilating apparatus, the boundary walls, and the completion of the surface of the playgrounds by tar pavement. Furniture, such as desks, tables, &c., is not included. All school apparatus is, of course, also excluded from such estimates.

In the various examples here given of the new Elementary schools of London, it may be mentioned that the building material used is always brick, generally the London "stock" brick, relieved by some intermixture of red. The stock-brick, if well burnt, is much harder and less porous than the Suffolk, or other red brick, and is therefore more impervious to damp, although the latter is infinitely preferable in point of appearance and colour. Only two or three, out of a hundred new schoolhouses, are wholly of red brick.

Among these is the subject of the next illustration (No. 202) erected,[1] from the designs of Mr. Champneys, in Harwood Road, Fulham. Close proximity to a large open green, known as Eel Brook Common, gives the red colour of the brickwork an enhanced value to the eye of the artist. The style in which the building has been thought out is a quaint and able adaptation of old English brick architecture to modern school purposes. Apart from the opinion, which may be termed that of fashion, because of its temporary nature, but which runs for the moment headlong

CHAP. XVI.] THE BOARD SCHOOLS OF LONDON. 297

after the favourite style, even when carried out in the most
tasteless and unmeaning manner, this building must be regarded
as possessing decided architectural character. The war between
rival styles has raged so long that we are in some danger of for-

202.—HARWOOD ROAD SCHOOL.

getting the existence of certain broad first principles common to
the great architecture of all times and countries, and which are
certainly never absent from the more conspicuous and representative examples. Among these first conditions of architecture
must be ranked a regard for good form, good proportion, good
grouping, and, above all, good architectural character and good

colour. A large proportion of modern buildings exhibit a total want of these essentials and sometimes of others, not mentioned, which, if based on any artistic intention, aim rather to attract the eye by startling novelties than to produce enduring impressions.

203.—PLAN OF SECOND FLOOR.

The design in question must rank as thoughtful and artistic work, whatever may be our individual preference as to style.

The plan of the Harwood Road schools (woodcuts Nos. 203, 204, and 205) is little more than a parallelogram with one projecting class-room. On the first floor a special room is provided

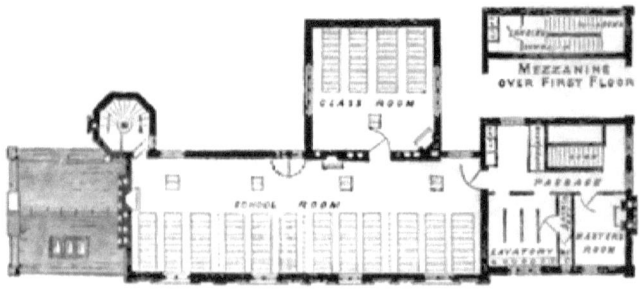

204.—PLAN OF FIRST FLOOR.

for instruction in drawing, to which is attached a small octagonal staircase for access to the boys' department situated on the floor above. A door is provided from the girls' school-room to the landing. Thus the drawing class-room can be used alternately by the boys and the girls.

In carrying out the building the arrangements have been exactly reversed, and our illustrations may be somewhat perplexing unless it is explained that the plans (Nos. 203, 204, and 205,) give the original disposition on the ground, while the view (woodcut No. 202) is drawn from the actual building.

The area of the site is half an acre. Its purchase-money was

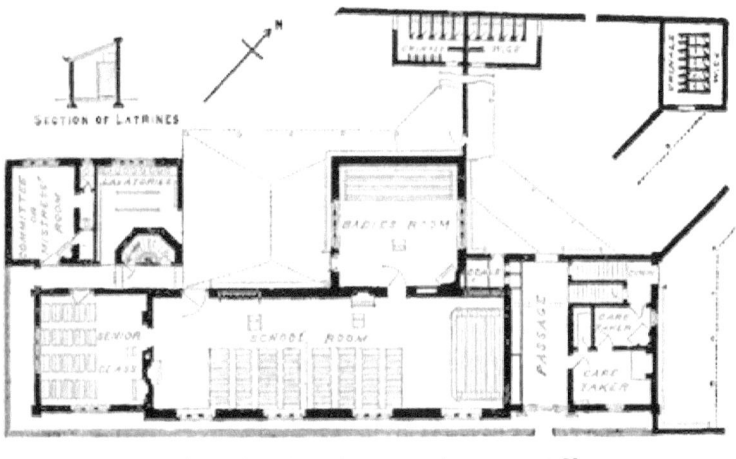

205.—HARWOOD ROAD SCHOOL. GROUND PLAN.

1,600*l*. And the contract for the building, excluding furniture, was taken at 5,716*l*.

The accommodation is as follows: viz.:—

Infants	289
Boys	219
Girls	219
Total	727

Soon after the work of supplying the deficiency in accommodation had fairly begun and several new buildings were already in progress, attention was drawn to the great difference in method of planning between that existing and proposed to be still followed in England, and that pursued on the continent. On all hands it appeared to be admitted that the English system possessed an advantage in the greater economy of teaching power. The advocates of the German system, however, urged that the best method, not the cheapest, should be adopted for the new schools of London, and claimed the superiority, as tested by results, for the foreign schools. After much consideration it was determined that the arguments were of sufficient importance to justify an experiment in the erection of a complete specimen, and that therefore a schoolhouse should be built consisting of separate rooms, each requiring the employment of a separate and fully qualified teacher. It was further held that the system could never be successful in this country without the addition for the use of the Graded schools of an assembly hall, and that therefore the school to be built should be of a large size, so as to distribute the extra cost of the hall over a large number of class-rooms. The hall, although in some sense corresponding to the Aula of Germany, is larger in proportion to the size of the school, and is really intended for the assembly of all the children of the Graded schools at one time, instead of merely serving for the occasional examination of one or two classes. The number of children,—at first fixed at 1,000—was in view of the above reason and of the great necessities of the neighbourhood increased, and it was finally decided to build for 1,500.

The Jonson Street school at Stepney, therefore, as finally carried out from the designs of Mr. T. R. Smith, is of considerable size. Its external architecture (woodcut No. 206) is of simple Pointed character constructed of grey stock brick, with a sparing intermixture of stone: such simple architectural features as are introduced being almost wholly in brick. It possesses two features different from—indeed never found in—its Prussian model. One is the presence of an Infant school, which here occupies the whole of the ground floor (woodcut No. 210) except a certain portion not required for school work and therefore devoted to an equally useful purpose as covered playground. The other is the utilisation of the hall itself for two of the daily

CHAP. XVI.] THE BOARD SCHOOLS OF LONDON. 301

classes. One of these classes may, from the position of the hall

206.—JONSON STREET SCHOOL, STEPNEY.

in reference to the class-rooms, be held at one end, and the other in a gallery provided for the purpose. The German principle of

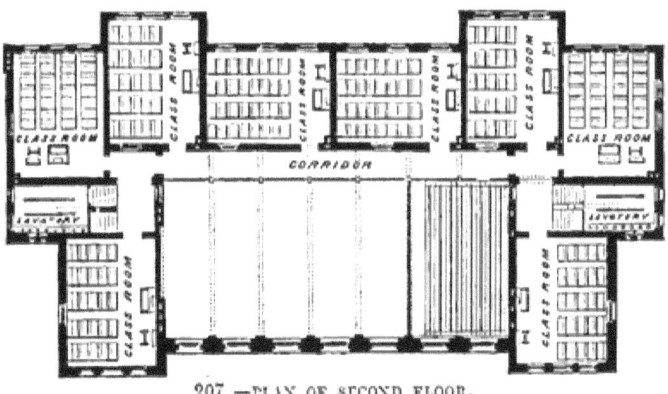

207.—PLAN OF SECOND FLOOR.

lighting from the left side of the children is strictly carried out, although other lighting is introduced in the case of some of the rooms. The class-rooms either enter from the hall direct, or from

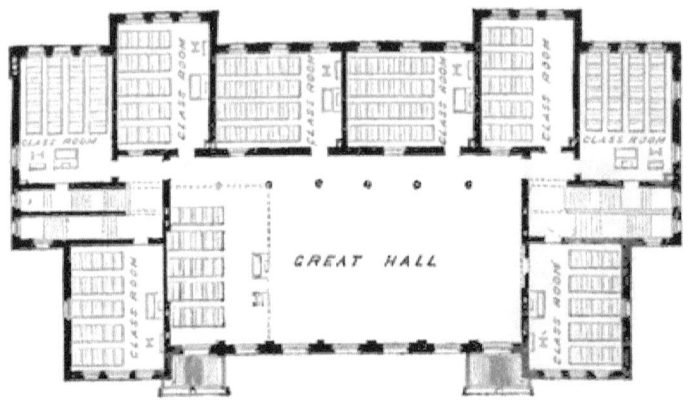

208.—PLAN OF FIRST FLOOR.

the landings in immediate contiguity thereto. The general shape of the building does not follow the regular parallelogram seen in some of the examples already alluded to in the previous chapter on German schools. The girls and infants enter from one street,

209.—PLAN OF MEZZANINE.

and the boys from another. The infant department is divided into two schools, each with its separate school-room and two class-rooms, its cloak-room and lavatory, but the covered playground is common to both. From its great size the school required two entrances to each of the Graded schools and two staircases. The arrangements are not those of a senior and junior department to each sex, as commonly understood, but approximates thereto as nearly as the class-room system admits.

The ground story, devoted to infants, is 15 ft. in clear height, and a mezzanine has been introduced into one part, as shown

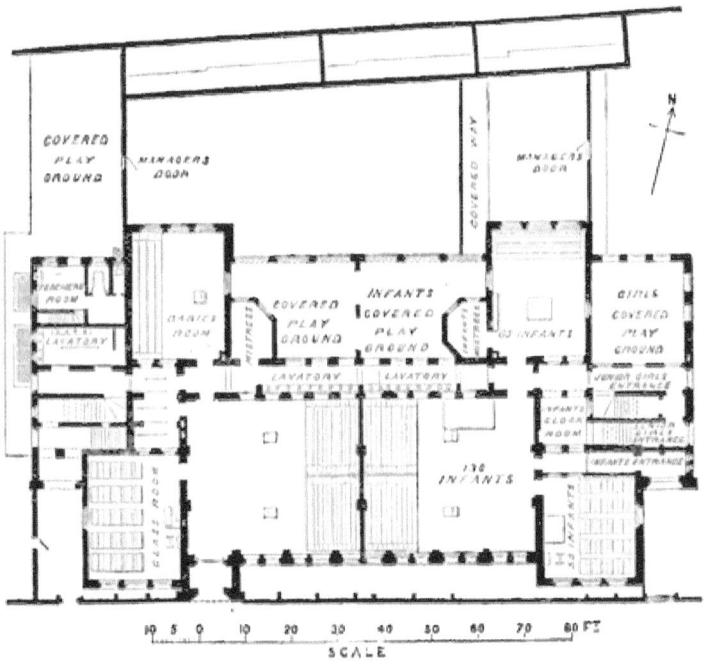

210.—JONSON STREET SCHOOL. GROUND PLAN.

by woodcut No. 209, by which a caretaker's house, teacher's room, and lavatories have been economically obtained over offices, covered playground, &c., not requiring the full height given to the general story.

Including the two classes obtained, as above described, in the hall, and again calculating the internal area at 8 ft. superficial for infants, and at 9 ft. for other departments, the total accommodation appears as follows: viz.:—

Infants	575
Girls	540
Boys	560
Total	1675

The area of the site is three-quarters of an acre, with two frontages at opposite sides, each to a separate street, the ends of the building both abutting on land soon to be covered by small dwelling houses. The land was purchased for 3,000*l*., while the cost of the building amounted to about 12,600*l*. The great size of the latter rendered any other than what we have called an "artificial" system of warming and ventilation out of the question, and it accordingly presents what is believed to be a good specimen of the method—described in a previous chapter—of warming by means of hot water vessels placed in the basement, and heated at a low pressure, and of ventilation by means of numerous outlet flues leading to one common shaft, and supplied with motive power for extraction.

The Jonson Street school cannot, when critically considered, be regarded in the light of a success which invites general imitation. One noticeable defect is the smallness of the accommodation for infants as compared with other departments. Another lies in the enormous aggregation in one building, rendering necessary the covering of too great an area of population, and the bringing of children every day from great distances, a principle condemned in its application to Elementary schools by the experience of all Europe. A third consists in the comparative uselessness of the hall. The last arises from the great annual cost of the teachers involved by the system of class-subdivision. The experiment shows that, for our purpose, the German system, if carried out on a large scale, bears its own condemnation in its size. If attempted with the ordinary numbers it would present at all points an inferiority to the English method, unless its teaching-results can yet be shown to be of a decidedly higher order.

Our next example represents the outcome, in general arrangement, of an endeavour to carry further the science of school planning by uniting or fusing together the best points of the class-subdivision system, and those more exclusively English, and also to provide for collecting on special occasions the children of several neighbouring schools. The first experiment in a new direction had shewn a considerable increase in salaries relatively to numbers taught, arising from the numerous certificated masters and mistresses required to work a system of instruction depending on the entire separation of each class in a separate class-room. The first cost of the building had not proved proportionately large owing to its great size. The London experience however, of schools built to accommodate more than 1,000 children soon showed, here as elsewhere, that the area of population which they were intended to cover involved distances too great for the younger children to walk in all kinds of weather. Smaller establishments thus asserted their superior claim to adoption and came into favour. Schoolhouses for 700 to 800 with the use of the hall also, could not be produced on the separate class system except at much greater outlay, and, as the latter continued to be regarded as indispensable to English methods of teaching, the adoption of the foreign plan could only be considered desirable in the solitary school in each division of the metropolis where special arrangements might justify a special scale of expenditure.

The building proposed for New North Street, Shoreditch, belonged to this latter class. In that neighbourhood it was resolved to adopt the separate class system combined with the advantages of a hall, and to provide means for throwing hall and class-rooms together into one great space where great assemblages for children might at times take place for special purposes. Owing to contemplated changes in the neighbourhood involving the demolition of a large number of dwelling-houses of the poorer kind for the construction of a new street, it was found that great disturbance to, and partial removal of, the very population for whom the school was intended would ensue. The erection of the large building was therefore abandoned in favour of a more modest scheme. Its provisions, however, contained the only example of this particular idea, and its plans may be here given, if only as a record of the kind of thought and intention then prevalent in London as to the requirements of some kinds of schoolhouse.

The external architecture (woodcuts Nos. 211 and 216) was intended to be studiously simple, because of the shut-in nature of

211.—NEW NORTH STREET SCHOOL. NORTH ELEVATION.

the site which would prevent any but the upper portions from being seen. Its style is an adaptation of the old brick treatment formerly common in the metropolis depending almost entirely on good form and proportion supplemented by good colour in brickwork. In London this is obtained by the use of stock brick as the staple, with red brick used somewhat in the position of stone dressings.

The ground floor (woodcut No. 214) was to have been occupied by two similar and separate departments of an Infant school, all the arrangements of which may be seen on reference to the plan, except that the desks in the school-room are not shown in their places in a line between the larger and smaller galleries.

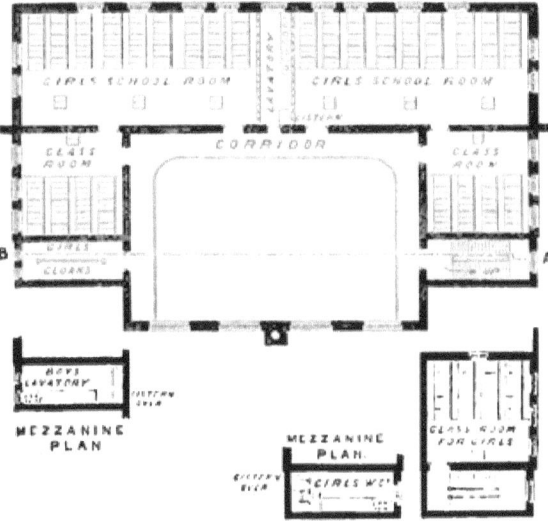

212.—PLAN OF SECOND FLOOR.

The first floor (woodcut No. 213) would have formed the floor

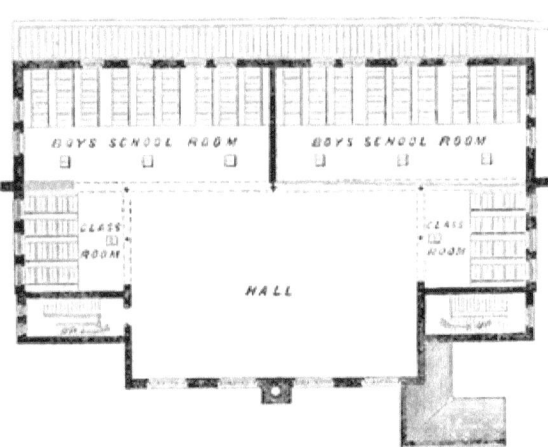

213.—PLAN OF FIRST FLOOR.

of the hall, as in the case of the Johnson Street school, and the

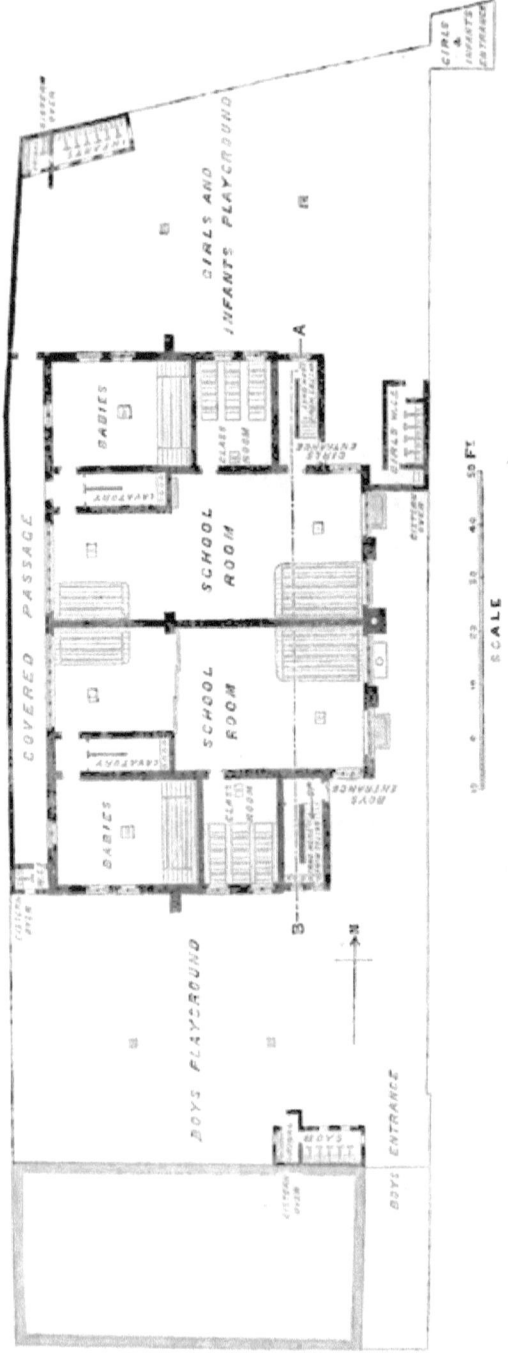

214.—GROUND PLAN OF NEW NORTH STREET SCHOOL.

CHAP. XVI.] THE BOARD SCHOOLS OF LONDON. 309

arrangements proposed were such, that a speaker standing at one side and looking towards the school-rooms, could command almost the entire area when the sliding partitions were drawn back.

The second floor (woodcut No. 212) contained the Girl's school,

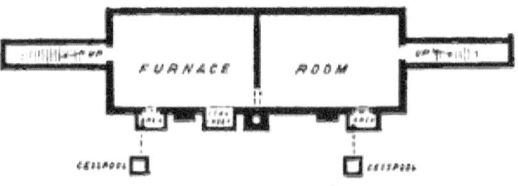

215.—NEW NORTH STREET SCHOOL, BASEMENT PLAN.

approached from a gallery running round three sides of the hall. This last would have been available for a further number of spectators or audience in connection with any meeting taking place on the floor of the hall below. One of the class-rooms for girls,

216.—NEW NORTH STREET SCHOOL, WEST ELEVATION.

corresponding to one class of boys which could be held on the floor of the hall, was arranged in a portion of the building intended to be carried to a greater height. The Girls' and Boys' departments had thus an equal provision of two school-rooms and three class-rooms.

The area of the site is somewhat less than 20,000 feet, and its cost was 9,550*l.* The cost of the building, as contracted for, was 9,930*l.* The accommodation provided by the plans was as follows viz. :—

Infant School		287
Do.	do.	287
Boys'	do.	300
Girls'	do.	300
	Total	1174

The plans for warming and ventilating the building were identical in principle with those adopted at Johnson Street, Stepney.

The school in Winstanley Road (woodcut No. 217) in a newly built neighbourhood, close to the Clapham Junction Station of the South-Western Railway, possesses a more castellated character than perhaps any other among the Board schools. Placed on a difficult site, too small in reality for the number of children to be accommodated, it demanded planning which should utilise every available inch of land in the best manner. To the east and south are streets 40 feet wide, while the other boundaries abut on land more or less built upon. The *main* lighting of the Graded schools is therefore secured from the north, and the principal building is only kept back sufficiently far from the northern boundary as to prevent future interference therewith. It would have been possible to build one block of three storeys high and to place the Infant department on the ground floor. This plan is more economical, but is accompanied by an incurable disadvantage in the fact that the width of room which is found to most useful for Graded schools is too narrow for the proper management of an Infant school, and, when placed one above the other, all must be of the same width. For this reason it is believed that the plan of which Winstanley Road school is an instance, and which deals with the Infant school as a separate building of one storey while the two storeys of the other departments are raised on a tier of

Fig. 8.—SOUTH VIEW OF WINSTANLEY ROAD SCHOOL.

low piers and arches so as to obtain covered playgrounds about 8 feet to 9 feet high, is decidedly to be preferred. The whole building is not so high, by 5 feet, as that involved by the other plan; covered playgrounds are obtained, and greater convenience is secured in some respects by the power of passing under the building, when, if the Infant school intervened the circuit of the school-house would have to be made to reach the back.

The main staircase is, of necessity, placed towards the front, and is treated as an octagonal turret, terminating in a belfry containing the school-bell. It should be stated that, so far as practical use is con-

CHAP. XVI.] THE BOARD SCHOOLS OF LONDON. 313

cerned, this staircase is not quite perfect. It is what we have called a "following" staircase (page 208, ante), and by the porches which flank the base of the turret, the boys obtain access on one side and the girls on the other. As the two staircases thus placed in the same position on plan proved to be insufficiently separated for the two departments of the school, the well-hole

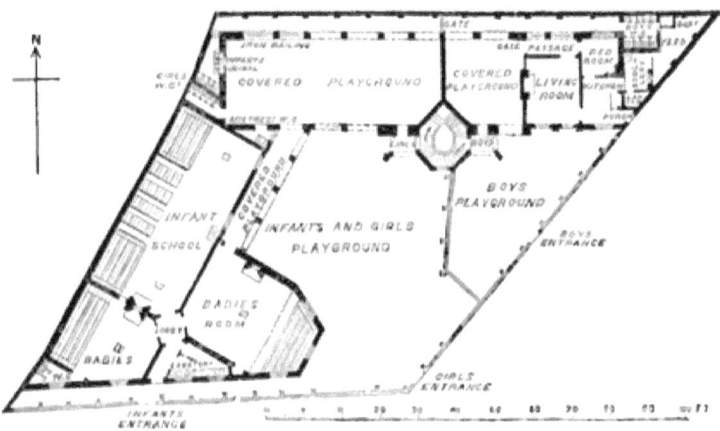

219.—WINSTANLEY ROAD SCHOOL, GROUND PLAN.

has already been provided with a wooden cylinder which reduces the arrangement to that of one of the larger staircases leading to a mediæval belfry. Another point not to be imitated is the *size* of the staircase. The clear width of the turret is 11ft. 6in., and that of the stair 3ft. It is better, as a rule, not to adopt a circular staircase for the use of children, but where its use is exceptionally desirable, as in this case, the size of the turret should not be less than 13ft. 6in, and that of the stair 3ft. 6in., so as to provide a sufficient width of step on the inner side nearest the handrail.

The ground floor (woodcut No. 219) shows the arrangements of the Infant school, in which the lighting is all placed with satisfactory results at very high points in the walls or in the roof. An arcaded covered way connects this with the main building, both as to practical use and as to architectural unity of design. At the eastern end is provided a caretaker's house and a room for the use of teachers and managers.

The first floor (woodcut No. 220) contains the Boys' School,

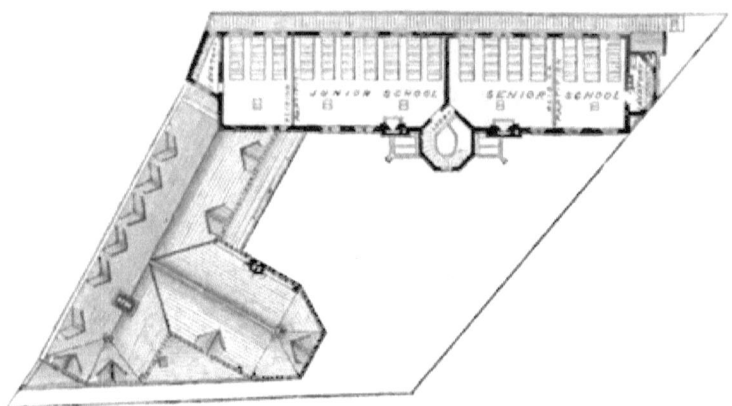

220.—PLAN OF FIRST FLOOR.

and is almost identical with the floor above, which accommodates the girls. The whole of the warming is by Boyd's stoves.

The area of the site is only a quarter of an acre, and was purchased for 1,000*l*.

The accommodation is as follows, viz. :—

Infant School	....	316
Boys'	do.	248
Girls'	do.	248
	Total	812

A sculptured panel towards the eastern end of the southern façade representing Knowledge strangling Ignorance has been executed from a model prepared by Mr. Spencer Stanhope. The same model has been applied to several other schools.

The locality in which the Old Castle Street school stands has been illustrated by the woodcut which heads the present chapter. A large proportion of the buildings erected by the School Board

CHAP. XVI.] THE BOARD SCHOOLS OF LONDON. 315

have been of necessity placed in situations little better, while some occur among the vilest slums to be found in the whole metropolis. Eagle Court, a narrow alley turning out of St. John's Lane, Clerkenwell, (woodcut No. 221) ranked among these.

221.—EAGLE COURT, CLERKENWELL.

So lawless were some of its inhabitants, that on the first commencement of the new schoolhouse which now occupies its northern side, and is shown on the left side of the drawing, it became necessary to protect the workmen from violence by a guard of police. In such a neighbourhood, shut in from all possibility of being seen, the plainest of plain structure could alone be suitable.

The necessity of avoiding any infringement of surrounding windows coming within the legal category of "ancient lights,"

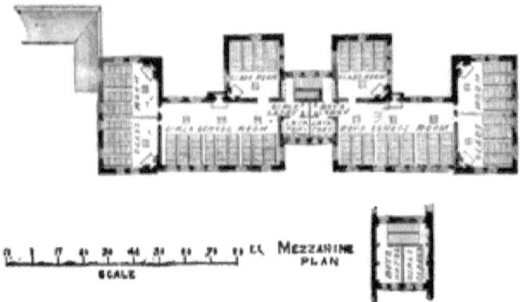

222.—PLAN OF FIRST FLOOR.

dictated the maximum height of the new building, and limited it to two storeys. The Boys' and Girls' schools are therefore both on the first floor (woodcut No. 222) approached by two separate staircases having opposite entrances, and planned so that both only occupy the space of one. A recess is formed in the centre of the front, in order to obtain end-lighting to the school-rooms, and therefore side-lighting to the children. Without this, or

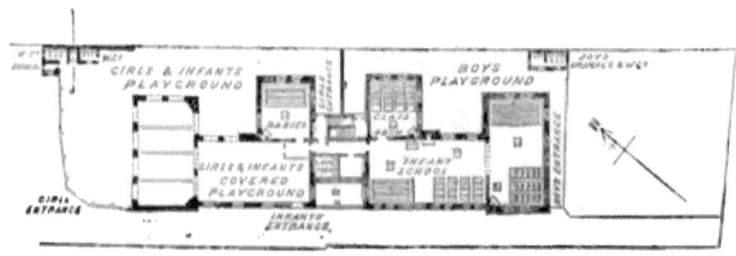

223.—EAGLE COURT SCHOOL. GROUND PLAN.

some similar provision, sufficient light could not have been obtained in this narrow court without recourse to lighting through the roof by means of dormers or skylights.

The space under the two Graded schools is occupied on the ground floor (woodcut No. 223) by the Infant school and by the covered playground devoted to the use of the girls and infants.

The site contains an area of 12,068 superficial feet, and, being at the same time of purchase covered by small tenements, cost 6,180*l*.

The expense of building amounted to 5,030*l*.

The heating and ventilation is by open grates with warm air chambers, and by extraction flues.

The accommodation is

Infant School	....	326
Boys' do.	....	246
Girls' do.	....	246
	Total	818

The Schoolhouse in Angler's Gardens, Islington, exhibits a plain specimen of external architecture. The site has one narrow frontage to a street, and provides a large quantity of land in the rear, with no other frontages except to the two narrow alleys or footways, respectively on the north and south. These alleys are inaccessible to carts and carriages, and have narrow entrances, leading out of the street. This school, therefore,—like some others—will only be seen in its entirety by those curious enough to penetrate to the back of the rows of street houses.

On reference to the ground plan (woodcut No. 227) it will be seen that the Infant school is provided in a separate building placed towards the main street, while the Graded schools are placed across the middle of the back land, and are raised on piers and arches, as in the case of the Winstanley Road school, to secure good covered playgrounds. In this, as in all other examples, the desks in the Graded schools are placed with their backs to the north, as the first essential of the plan.

The Infant school has its senior section placed on the first floor (woodcut No. 225), where also is a large room, with an easterly

CHAP. XVI.] THE BOARD SCHOOLS OF LONDON. 319

aspect towards the street, available for use either as a Committee Room, Teacher's room or for a Drawing-class.

The first floor plan of the Graded schools (woodcut No. 226) is

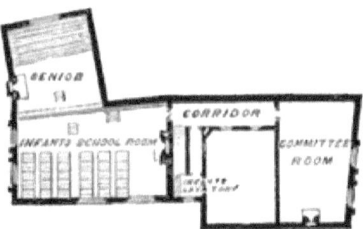

225.—PLAN OF SECOND INFANT SCHOOL ON FIRST FLOOR.

devoted to boys, and forms the first example given of what may, in the main, be regarded as the model arrangement for a school

226.—PLAN OF GRADED SCHOOL OVER COVERED PLAYGROUND.

of the kind. The points attained may be tabulated as follows : viz :—

 1. No space wasted in corridors.
 2. Double exits, so that one part can be cleared without disturbance to the remainder.
 3. The same number of class-rooms as classes in the school-rooms.
 4. Side-lighting in every case.
 5. Class-rooms arranged in pairs for working by a certificated master and pupil-teacher in each case.

The plan also shows how the lavatories and cap and cloak-rooms are obtained in mezzanine floors, accessible from the respective landings.

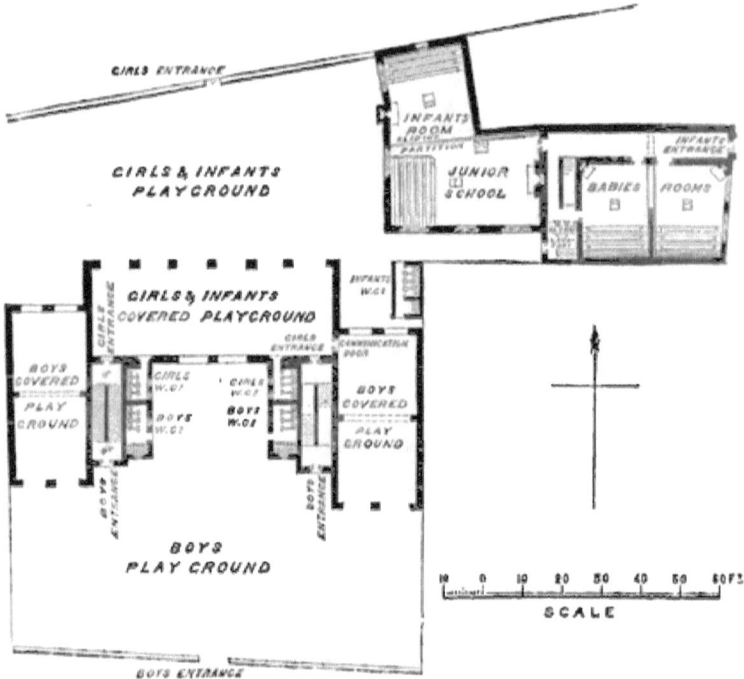

227.—ANGLER'S GARDENS SCHOOL. GROUND PLAN.

The area of the site is 19,575 superficial feet, and its cost amounted to 4,522*l*.

The cost of the building was 8,700*l*.

The accommodation is as follows : viz. :—

Infant School	....	433
Boys'	do.	325
Girls'	do.	332
	Total...............	1090

Among so large a number of new schoolhouses, some are fortunate in being placed in positions where they can be easily seen, and it becomes of some importance to consider what general architectural treatment should be bestowed, and, in short, what style is the most suitable. To enter into the abstract merits of different styles is out of the question here. Yet none will deny that fit expression of the purpose to be served by the building must guide us at the outset. If a church should at once be recognised as a church by the character of its architecture, and a prison as a prison, so should a schoolhouse be immediately known as a home of education. It is clear also, that a building in which the teaching of dogma is strictly forbidden, can have no pretence for using with any point or meaning that symbolism which is so interwoven with every feature of church architecture as to be naturally regarded as its very life and soul. In its aim and object it should strive to express *civil* rather than ecclesiastical character. A continuation of the semi-ecclesiastical style which has hitherto been almost exclusively followed in England for National schools would appear to be inappropriate and lacking in anything to mark the great change which is coming over the education of the country.

In endeavouring to give fitting architectural character and expression to the Board schools of London it is not necessary, because of the non-dogmatic character of their instruction, to abandon all indigenous architecture and to seek something wholly new or "original." History shews that, in all previous cases, new wants have been met by new developments of the then existing or prevalent manner of building. Our difficulty lies in the fact that, considered as a vernacular or universally-practised art, architecture has not had a being for many years, and there is, consequently, no prevalent architecture of good type from which to develope. The common buildings have not displayed either architectural thought or artistic instinct. The days of the guilds of trade have died out so far as they can be said to affect the buildings of the country. The time when a man's ambition to found a house for his descendants was fitly seconded by the workmen's pride in rendering it a worthy monument of their

Y

skill in handicraft, has given place to the day of the speculative builder whose soul is occupied by the scale of rents which he can exact for his scantily-built mansions, glorious in wealth of cheap stucco and ill-designed ornamentation without and within. The ill-treatment of the builder's workman, or the want of sympathy between himself and his employer, has in course of years too often erased from his very nature the old instincts of good workmanship, which in other branches of art and manufacture yet cause the name of "English" to be synonymous with superiority in the eyes of the foreigner. The easiest way instead of the best has so long been practised, that the old English methods of construction and durable finish have nearly died out, and can no longer be found to exist among workmen, except in the heart of distant rural districts. The decadence of the old spirit of unity among workmen has corresponded with the decline of architectural art generally. From having a wholesome rivalry in the same direction architects came to wage a war of different and conflicting styles, wherein the presence of a particular shibboleth was exalted to the neglect of the vital first principles of art, and which, with occasional periods of repose, has not since ceased to rage. At the present day each enthusiast has his favourite kind of architecture, which he often prefers to the most scholarly and admirable productions belonging to another order of taste. Some would urge a recurrence to the classic times of Greece, others to the meridian of Gothic art in the fourteenth century, while a third class of advisers hold that originality is the chief thing, and that no necessity exists for drawing on the accumulated wealth of idea in the past. Maintaining, as we do, the necessity of developing our new wants from some already existing or past style as a nucleus or source, the difficulty lies chiefly in the selection of the most suitable. Preferences may exist as to this or that kind of architecture. Confining the question to English ground, the fragments remaining to us here and there among the wreck of the past, so far as they embody the results of genuine artistic thought, are the more valuable the nearer they approach our own time. Why should we take a great leap over the intervening centuries, and neglect the works therein produced, in order to reach an ideal thirteenth-century style belonging to a time of widely different popular habits? Specimens of good and thoughtful brickwork in sufficient number still remain scattered among the old architecture of the city and its suburbs, to form the basis of a good style suited to modern requirements. Hackney and Putney, Chelsea and Deptford, all furnish old examples. In London, the plainer and less expensive buildings, forming by far the most numerous class, must always be constructed of brick.

It is, therefore, of some interest to consider from what source we may chiefly derive our brick school-architecture. Brick was extensively used by the Romans in the buildings erected by them in this country; but, after the Norman conquest, the stone-built building came to be preferred, and the use of brick to any architectural purpose seems to have died out among us till the time of the Tudors, although extensively used in mediæval times on the continent of Europe. After its re-introduction, some time elapsed before brickwork came to be employed easily and naturally apart from the copyism of stone architecture.

The only really simple brick style available as a foundation is that of the time of the Jameses, Queen Anne, and the early Georges, whatever some enthusiasts may think of its value in point of art. The buildings then approach more nearly the spirit of our own time, and are invariably true in point of construction and workmanlike feeling. Varying much in architectural merit, they form the nucleus of a good modern style. In looking to the art of this period as a basis, a servile copyism need not be attempted, for it may not be impossible to accept its spirit and yet to clothe our own rendering with new form and a higher sense of architectural being.

The superstition which implies that a building must be architecturally good because evincing on the part of its designer a preference for the pointed arch, has long ago vanished before the results of modern experience. Good artistic work is not common to any one style alone.

The four new schoolhouses which we next proceed to describe, are located in different quarters of the metropolis, and are all based in *motif* of architectural design on the idea that, in London, good architecture of simple type may be produced on the model of the old brick architecture of London.

The first of these occurs in one of the newly-formed districts in the N.W. of London, known as Portobello, where the schoolhouse has been built (woodcut No. 228) with one of its frontages to Wornington Road, the other (if front it can be termed) being towards the main line of the Great Western Railway, which passes in a cutting at a level of some feet below. As in the cases of Winstanley Road and Eagle Court, already described, the general plan of Wornington Road School has been greatly affected by the conditions imposed by the nature of the land to be built upon. Where the height of the building exceeds one story, nothing can be worse than back to back planning. Summer ventilation is rendered almost impossible, the master cannot have the same facilities for supervision as when the class-rooms are placed across the ends of the school-room, the intervening doors cannot be made with panels of clear glass without producing one of the worst results of the wide school-room, viz., that of the children sitting facing each other, and the master must continually be compelled to turn his back on his class for the purposes of an inspection which, by another plan, can be made without leaving his place or interrupting his class.

In the present instance, as shewn by the ground plan (woodcut No. 230), it has been considered best to occupy the portion of ground nearest to Wornington Road, by two Infant schools of one story placed back to back. Both buildings can thus only have windows on one side, but each has also ample and highly placed light from dormers in the roof. The lavatories and cloak-rooms are near the entrances. The babies' rooms can be entered or left without disturbance with the

228.—WORNINGTON ROAD SCHOOL.

principal work of the schools, and W.C.'s are provided at a short distance. By this plan the youngest children have not the necessity of traversing a long distance between the street and their school-room, and the elder girls easily proceed to their

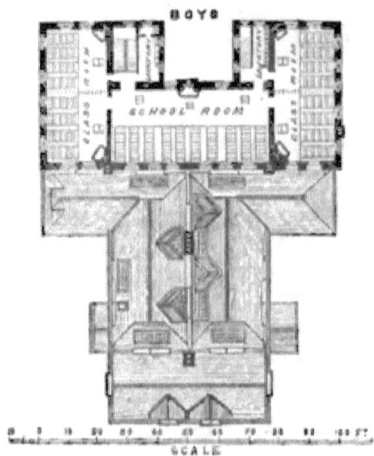

220.—PLAN OF FIRST FLOOR.

own department after leaving their younger brothers and sisters in the Infant school. At the N.E. side both schools have access to a covered playground obtained under the Graded departments. This direct access from the school to the covered playground is always desirable for infants, if it can be obtained without damage to light or other greater necessity of the building. In the present instance, owing to the different level of the ground at this point, a few steps have become necessary which, in the abstract, are not desirable. Under one portion of the Graded schools a room for teachers has been provided, and also a caretaker's house, the latter so placed as to command the playgrounds from windows on both sides.

The boys' entrance is obtained, apart from those for girls and infants, from a narrow lane which bounds the S.E. side of the site.

CHAP. XVI.] THE BOARD SCHOOLS OF LONDON. 327

The plan of the Graded schools is alike on both floors. The woodcut (No. 229) shews the arrangements of the first floor used by the boys, which are similar in principle to those in the corresponding position at Angler's Gardens, but differ in the circumstance of the children in the school-room having no side lighting.

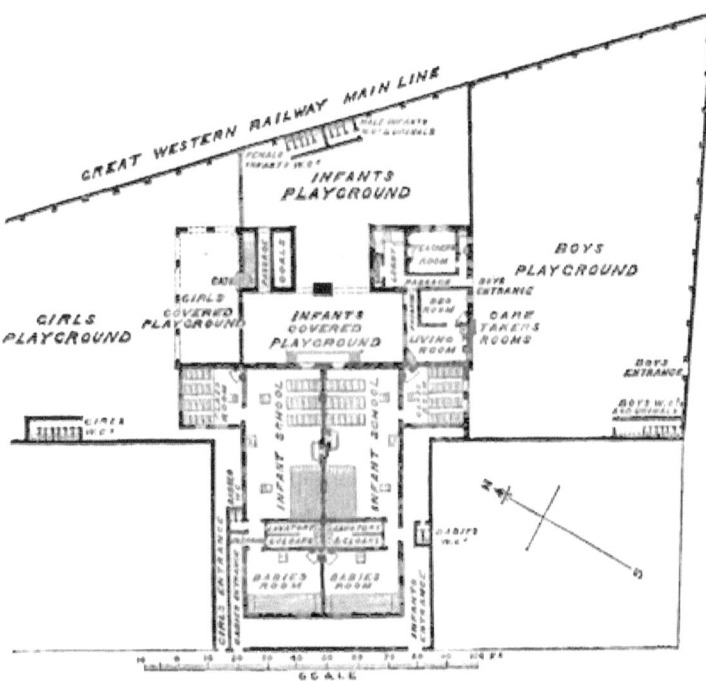

230.—WORNINGTON ROAD SCHOOL. GROUND PLAN.

The area of the site amounts to 31,000 square feet and its purchase was effected at the sum of 1870*l.*

The building has cost in process of erection the sum of 7845*l.*, and accommodates the following number of children, viz. :—

    Infants 246
    Do. 246
    Boys 306
    Girls 318
            Total............... 1116

The warming and ventilation is by open ventilating grates, &c., as in Winstanley Road.

Another example with the same kind of architectural character, but tending more towards a classic spirit, is that erected (woodcut No. 231) in Aldenham Street, immediately to the west of the St. Pancras Railway station. The site has two frontages, one to the street from which the building takes its name, and another to Hampden Street. The former being the better street and possessing the greatest width, the front of the new schoolhouse would naturally be placed thereto if possible. Such a position secures also the northern aspect, against which the backs of the line of desks in the Graded schools can be placed, and provides sunny aspects for the playgrounds thus naturally placed towards the south. The boys' entrance in the boundary wall towards Hampden Street, is so placed as to form a vista termination towards Middlesex Street, which runs southward from the former, and almost at right angles. The view is taken from the Aldenham Street side. It is studied chiefly in reference to symmetry and simplicity. It will be noticed that dwelling-houses flank the new building on both sides.

A reference to the plan of the second floor (woodcut No. 232) shews how the dimensions of the site have necessitated a modification of the usual plan in which a pair of class-rooms is simply placed across the ends of the school-room. Right angles and square corners are always preferable for the interior working of a school. They look best and are more useful for the disposal of cupboards or apparatus. In the present instance, the length of frontage proved to be insufficient for that of the required school-room when added to that of the class-rooms. It is true that one pair of class-rooms might have been placed in the usual way, and the other projected wholly into the playground. The plan actually carried out possesses, however, the advantage of affording side-lighting which could not have been obtained by the other. The girls are here placed on the first floor, and the boys on the second, a point often in practice decided by the position

231.—ALDENHAM STREET SCHOOL.

of the respective entrances in relation to the landings. The plan of the girls' school is, of course, exactly similar to that of the boys'.

One feature, not usually to be found in the Board schools, is here seen, in the provision on each floor of a corridor connect-

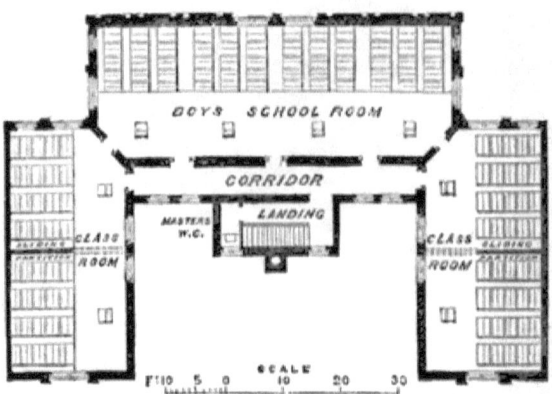

232.—PLAN OF SECOND FLOOR.

ing the class-rooms, and available for cap and cloak-room use As a rule, this corridor has the grave objection of preventing summer ventilation. In the present case, the thorough current so desirable for the stagnant atmosphere which sometimes exists in hot weather is obtainable by another disposition of the windows and by the system of ventilation adopted.

In this, as in most other cases, the plan of the lowest floor is determined by the necessities of the upper. Regard both for space and economy pointed to the desirability of placing the Infant department under the Graded portion. The ground plan (woodcut No. 233), therefore, does not present quite so compact and perfect a school as might be wished. In spite of difficulty, the main lighting of every gallery is from the side. The boys are admitted only from Hampden Street, while the girls and infants may enter by either of the two frontages according to convenience of distance from their homes.

The area or superficies of the Aldenham Street site amounts to only 13,950 feet, and was obtained at a cost of 8310*l*.

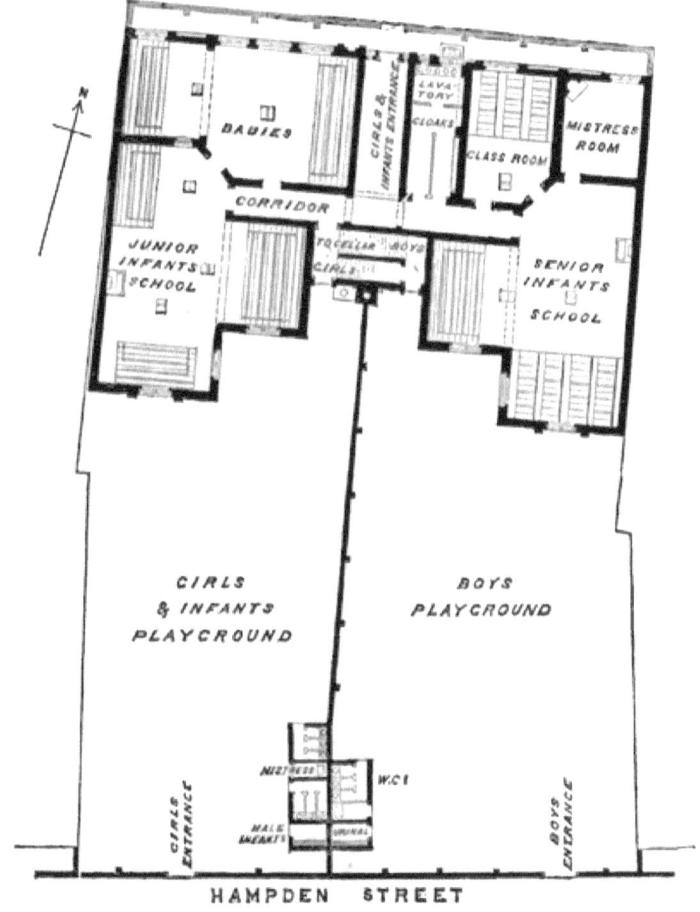

233.—ALDENHAM STREET SCHOOL. GROUND PLAN.

The building has been erected at a cost of 6780*l*. The accommodation is as follows, viz. :—

Infants	420
Boys	313
Girls	359
Total	1122

The warming and ventilation is by an artificial system of the kind already described, but the method of extraction employed is somewhat different. When the warming apparatus is in work,— that is to say, during the winter months,—the boiler fire forms the motive power, the main extraction-flues being placed so as to surround the iron smoke-flue, which is, of necessity, very hot, and promotes an upward current. When no warming is necessary, and the apparatus is consequently not at work, the ventilation is maintained by a small boiler then used to warm the flue. This portion of the work has, in the present case, been carried out by Mr. Phipson.

Another example of a schoolhouse treated in the manner of old London brickwork has been built (woodcut No. 234) in Orange Street, which is situate in one of the densest quarters of Southwark. The confined nature of the neighbourhood, the narrowness of the streets, the desirability of allowing the rays of sun to enter the playground, all suggested that the building should be set back from the street to the furthest extremity of the land. As it happens, this arrangement also enables the passer-by to see the building better than if it had been brought up to the street-line. The back of the desks in the Graded schools is placed to the N. as one of the first conditions to be attended to.

The ground plan (woodcut No. 236) shews how the entrances are placed in communication with covered ways of sufficient size to serve also as covered playgrounds, and leading direct to the school doors. The site is limited in size, and every foot of space has been considered of value. The space at the back, only left from the necessity of securing the lighting of the rooms above, has been roofed with glass and fitted as a cloak-room. In principle, the Infant department presents nothing differing from other examples, though in each case the application of the same principle in a different way may cause the plan to appear at first sight as different. The gallery in the principal room can be shut off on the occasion of a lesson by means of a wooden partition sliding horizontally. Towards the street and also towards the narrow

234.—ORANGE STREET SCHOOL.

passage on the east side, the open iron railing originally intended has been substituted by a high brick wall as a species of boundary affording better protection in so rough a neighbourhood.

For the two floors of the Graded schools one woodcut (No. 235) suffices. In both cases the plan is the same, except that the boys'

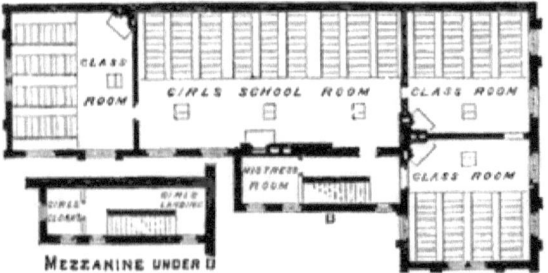

235.—PLAN OF SECOND FLOOR.

school-room, which is on the first floor, has a row of windows behind the line of desks, while the corresponding room on the second floor, devoted to girls, is lighted on the same side by dormers in the roof. It will be noticed that, here, no two class-rooms are connected by sliding partitions. Teachers are not wanting who prefer the connection to be made by an ordinary door, and this method has been carried into effect in the Orange Street school. Three classes are taught in the school-room, and three class-rooms are also provided. The size of the school hardly justified the expense of a second staircase, but the single staircase provided is placed so as to allow separate access to the majority of class-rooms without entering the school-room. One advantage gained by placing the Girls' department on the highest floor is, that thereby greater space can be obtained in the mezzanine floors for the greater quantity of feminine clothing in cloaks and shawls as compared with that of boys. Few of the latter, attending Elementary schools belonging to the Board, are found to possess an overcoat.

The site contains 12,500 superficial feet of land, obtained at a cost of 3175*l*.

The building has been erected for the sum of 5840*l*.
The accommodation is as follows, viz. :—

Infants	297
Boys	250
Girls	262
Total	809

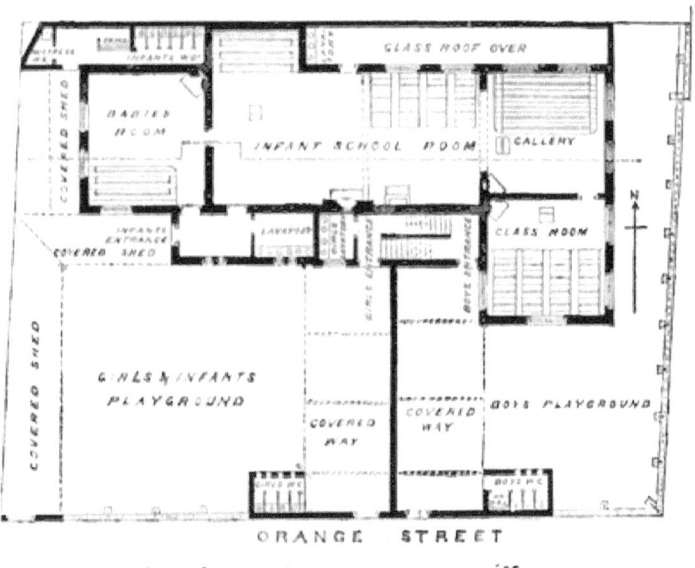

236.—ORANGE STREET SCHOOL. GROUND PLAN.

In the densely populated neighbourhood lying on the western side of London Fields, an open public recreation ground in the School Division of Hackney, much difficulty was experienced in obtaining a site for the necessary Elementary school at a sufficient distance from those of a voluntary nature already in operation. In all neighbourhoods where the Church of England has been vigorously prosecuting the work of education it is found that schools are planted precisely where mostly required. A conspicuous instance occurred in Lambeth, where it was found impossible to obtain a site within easy walking-distance of the children's homes except in close proximity to an already existing school, and where the reluctantly selected spot received, when hotly challenged, the cordial support not only of the Education Department, but of the House of Lords, to whom an appeal to the contrary had been made. The invariable rule of the School Board has been to supplement, rather than to supplant, existing educational organisations when the latter were proved to be of an efficient kind. A stern determination has existed not to countenance buildings or standards of teaching unworthy of the nation and which the Education Department had refused to recognise, and to perform thoroughly the work committed to the Board by the provisions of the Education Act and by the choice of the people in the persons of its representatives. An equally strong determination has been shewn—as could be proved by numerous instances if it were any part of the duty of this record to do so —to offer no unnecessary or unavoidable rivalry to those schools which had borne "the burden and heat of the day" before Government took up the task of dealing with the notorious deficiency of the metropolis in respect of education. A necessary reform had to be carried out; and it had to be carried out apart equally from the spirit of fanaticism on the one hand and that of destructiveness on the other. In the present case, a site with two frontages was, after conference with the promoters and managers of some of the local schools, at length obtained at the S.E. corner of the London Fields.

The schoolhouse in West Street, Hackney, as represented in the Frontispiece, thus came to be located on a site whereon three good playgrounds could not be obtained from lack of space. As one of its fronts looked directly on to the "fields," it was decided to regard that public ground as available for the boys' playground and to devote any space unoccupied by buildings to the joint use of the girls and infants. The plan of other schools, adopted with the object of providing covered playgrounds, here became manifestly unsuitable, and the building is accordingly of three stories, so disposed as to leave the greatest amount of open space. The access required to the upper stories, occupied by the respective sexes, rendered necessary two staircases in different positions,—one, for the boys, accessible from the London Fields side, the other, for the girls, entered from the playground.

From the ground plan (woodcut No. 238) it will be seen that the general form of the building assumes a resemblance to an

237.—WEST STREET SCHOOL. PLAN OF FIRST FLOOR.

irregular L. The Infant school is arranged in two departments, to the junior of which is attached the babies' room, and to the senior a separate class-room. In the former the use of galleries preponderates, in the latter benches and desks. This floor, generally, is not so satisfactory as could be wished, owing to the manner in which desks and galleries unavoidably face each other. The boys' latrines are placed on the ground level, close to their street entrance.

The two upper floors may be considered to be represented by the plan of the first floor (woodcut No. 237) occupied by girls. There are three class-rooms in each case, and the school-rooms

z

are divided into two sections. Without this latter arrangement, the number of class-rooms would scarcely have been found sufficient for so large a school.

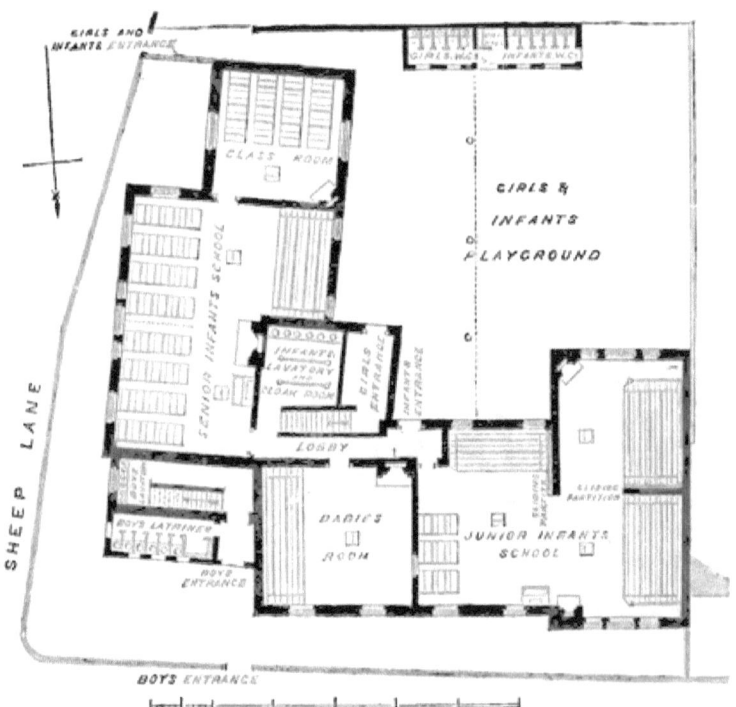

238.—WEST STREET SCHOOL. GROUND PLAN.

In the planning of this building the subject of lighting from the side has received some degree of attention.

The site contains a superficial area of only 10,296 feet, and cost the sum of 1452*l*.

The expense of the building amounted to 7234*l*.

The accommodation is as follows, viz. :—

Infants	420
Boys	343
Girls	359
Total	1122

An architectural treatment not based on that of the old London brickwork has been adopted in some of the new buildings. In these a pointed character of civil type has been aimed at. Cusps, crockets, finials, and tracery are usually excluded, and any features possessing marked ecclesiastical association avoided as far as possible. Whether we like it or not, the education of the people is now governed by the lawyer rather than the clergyman, and the forms of mediæval architecture run the risk, when applied to buildings intended to be used as Elementary schools, of possessing no artistic meaning or of being simple anachronisms. Two examples of this different treatment may be given.

The first is that erected in Camden Street, Camden Town, which is treated externally, with angle buttresses and terminals. The schoolhouse (woodcut No. 239) stands on a piece of vacant land which had for some years been used for the manufacture of mushroom spawn. One of the conditions attached to the purchase, at the sum of 2840*l.* of the 28,436 superficial feet of land forming the site, was that the children should have no entrance from the principal street. The covenant was imposed with the idea that the intermittent presence of a thousand children suddenly let loose, might tend to the depreciation of the dwelling-houses in Camden Street. It so happens that the plan of the building is better on the whole than it could have been if arranged with front entrances. The general arrangement is that of one-story Infant schools on the ground level, and of two-story Graded schools raised on piers and arches, as at Winstanley Road. The babies' rooms are, however, obtained under two of the class-rooms of the upper school, a plan not as a rule desirable, because of the comfortless and costly amount of height thereby obliged to be given to the covered playgrounds, if we would have the same level maintained in each floor of the upper building and the rooms underneath of the regulation height. In the present instance, a difference in the levels pointed to the exceptional arrangement.

The ground plan (woodcut No. 241) shews the general arrangement. The boys approach the building by a separate

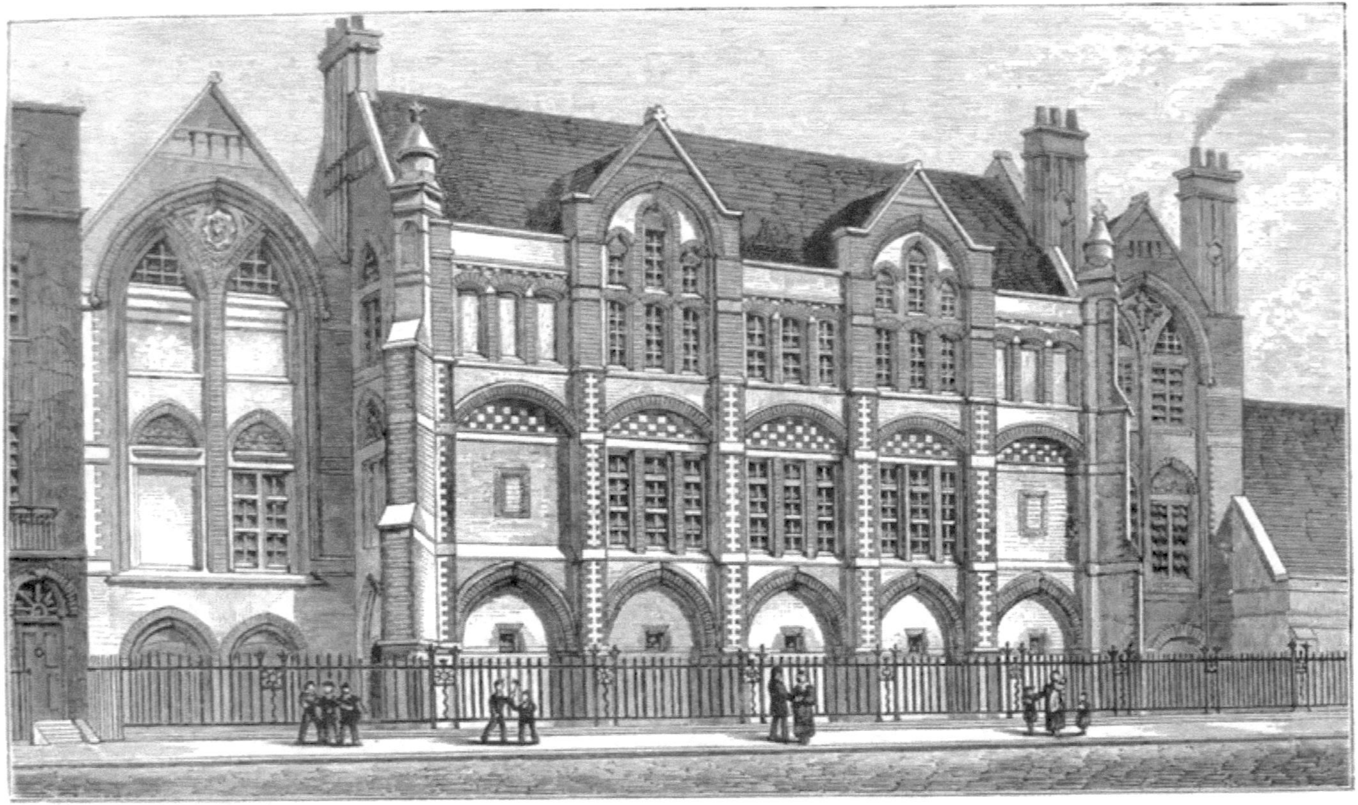

street from that used for the entrance of girls and infants. And there is also a direct communication between the Infant school-room and the covered playground, which the drawing does not clearly show. The babies' room, the senior class-room and the school-room itself are all capable of sub-division by means of sliding partitions.

On the upper floors (woodcut No. 240) the fittings in the school-rooms may be grouped either for three classes of 40 each,

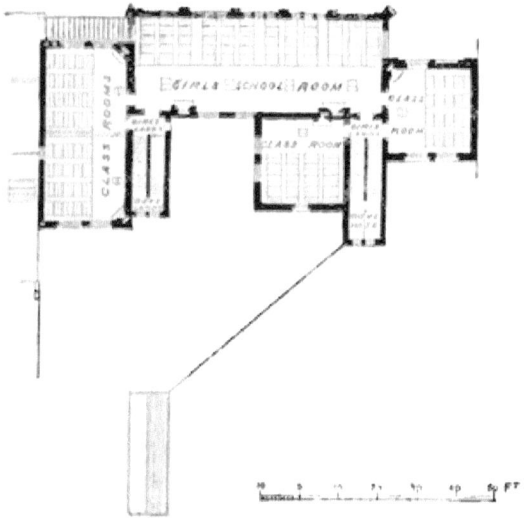

240.—PLAN OF FIRST FLOOR.

or for four classes of 30 each, with the same numerical result. On each of these floors there are four class-rooms, two being arranged as a pair in one case, but the arrangement being impracticable in the other, because of the necessity of building sufficiently far away from neighbouring window lights of sufficient age to rank technically as "ancient." There are also two staircases arranged as in the Angler's Gardens' instance.

This schoolhouse forms another of the earlier cases where open fires were adopted as the medium for warming large numbers. Were it to be erected now, it would undoubtedly be the subject of a complete and economical artificial system of warming and ventilation.

The cost amounted to 8750*l*.

The accommodation is as follows, viz.:—

    Infants 428
    Boys 337
    Girls 327
        Total 1092

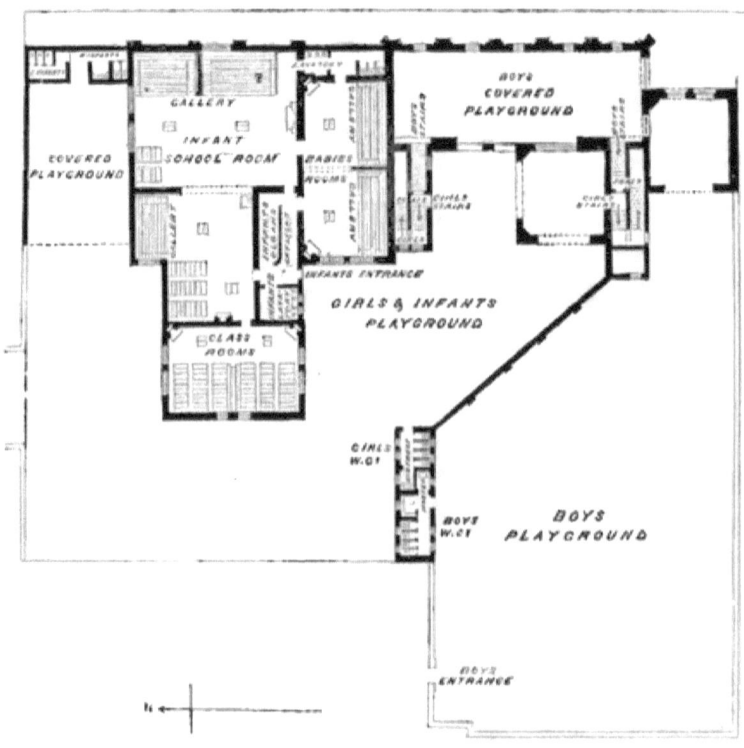

241.—CAMDEN STREET SCHOOL, GROUND PLAN.

The second example which we give, as following in the Pointed style, is the large building (woodcut No .242) erected in Mansfield Place, Kentish Town.

Whether in completeness of plan, or in character of architec-

242.—MANSFIELD PLACE SCHOOL.

ture, this schoolhouse must be ranked among the most conspicuous of those designed and contracted for during the first three years of the Board's existence. Its fault lies in being somewhat beyond the mark of an Elementary, and suggesting in appearance rather the uses of a Secondary or Grammar school. Too great an amount of appearance has not, however, been the usual characteristic of the new buildings, and, among many which have been architecturally starved, there can be little harm in

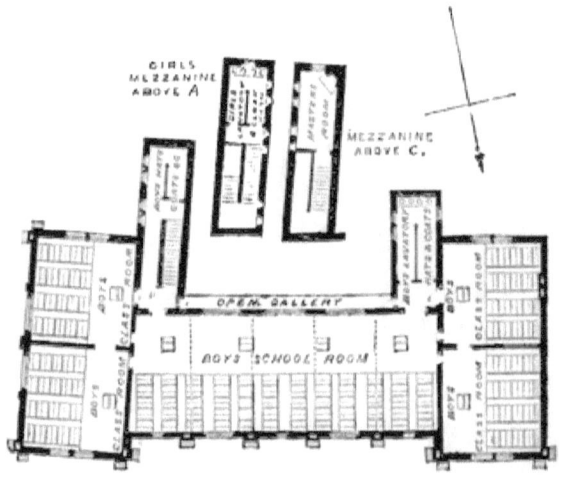

243.—PLAN OF SECOND FLOOR.

providing an exception. The general treatment differs from that of the Camden Street School, in which the buttresses of stock-brick are *edged* only with red, and angle terminals rise some feet above the eaves-line. Here the *whole* of each buttress is built of red brick, and the stock-brick facing of the general wall is relieved by patterns of red. Great simplicity is preserved, yet moulded stone is used to the pointed arches and other features, and even the window-heads are of that material. The architectural effect is chiefly produced by general forms and a balance of colour. The latter would have been increased in value had the roof been covered with red Staffordshire tiles instead of blue Welsh slates.

The general notion of arrangement is that for which we have, in other cases, already expressed a preference. The Infant schools are on the ground level. The Graded schools, of two stories, are lifted on piers and arches in order to obtain for all three departments ample playgrounds protected from the weather. The main frontage is towards Mansfield Place, but there is also another of very narrow dimensions in the adjoining street, through which a portion of the infants may obtain access.

The ground plan (woodcut No. 244) shews little more than

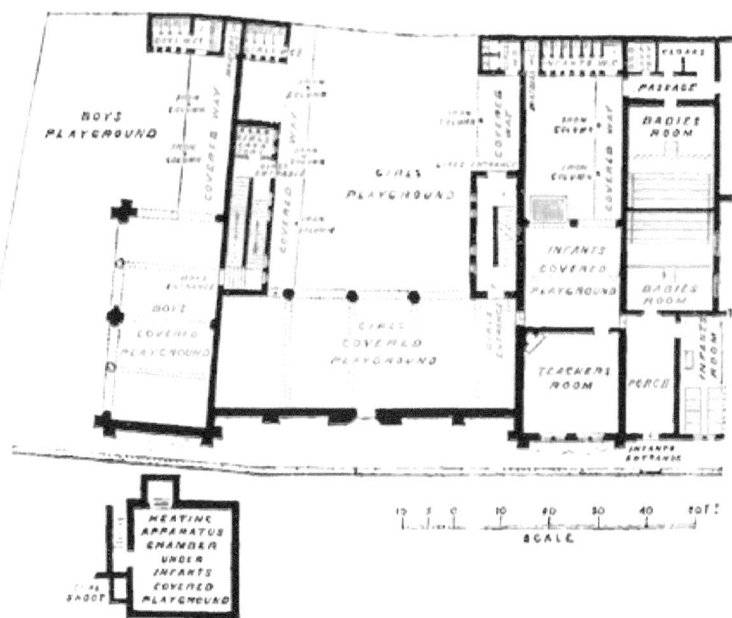

244.—MANSFIELD PLACE SCHOOL. GROUND PLAN.

the playground arrangements and the commencement of the Infant school, which lies to the right. Two staircases are here provided to the Girls' school and one to the Boys', one of the former being of the double kind, two in the space of one, separated by a wall. Each of the girls' staircases enter from the playground, and are placed in reference to the building so as to clear one part of the school-room, or two of the class-rooms, without disturbing the remainder. The only reason against giving the boys the same provision, is, that access could not be obtained to the second staircase except by passing through the girls' playground. The same, or nearly the same, result has been obtained by connecting the staircase with the further end of the school-room and with the two remote class-rooms by means of an external corridor (woodcut No. 243) supported on corbels. The boys in these portions can thus reach the staircase without passing through the school-room, and the arrangement does not interfere with the light or ventilation of the latter.

The building is warmed and ventilated throughout in a most complete manner by Messrs. Price & Co., in the manner described at page 286, and illustrated by woodcut No. 194.

The area of the site measures 17,945 superficial feet, and was purchased at the sum of 2750*l*.

The building has cost the sum of 9600*l*.

The accommodation is as follows, viz.:—

Infants	462
Boys	340
Girls	327
Total	1129

Among the sites purchased on reasonable terms is one at Battersea, where the fee simple in possession of an acre and three-quarters of land was obtained for 660*l*. In this case the schoolhouse is erected of two stories in height, and a splendid playground remains.

Another at Haverstock Hill is so singular in shape as to require a block plan (woodcut No. 245) for explanation. It contains a superficial area of 45,600 feet; but, owing to awkwardness of shape, its really useful size may be taken as much less. In reference to its utilisation, the erection of a schoolhouse of one story only was decided upon, being the only complete instance among the new Board schools. It will be noticed that the land has three points of access, one from Haverstock Hill, another from

Craddock Street, and a third from Truro Street, the first and last being of most value for use. It then became necessary so to dispose the buildings as to use the ground to the best advantage, having regard to the position of the entrances and to the retention of an already existing house, available for use as a master's residence. The Infant school is accordingly planned in

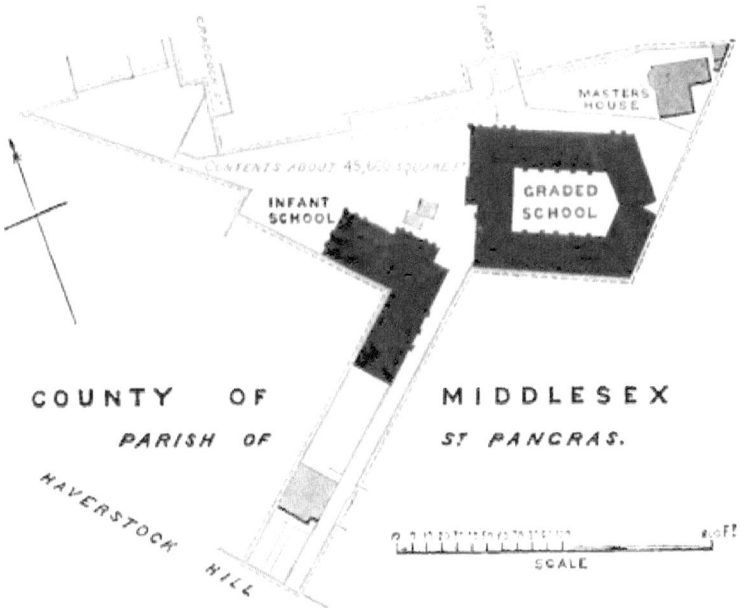

215.—HAVERSTOCK HILL SCHOOL. BLOCK-PLAN SHEWING SHAPE OF SITE.

the form of an L and packed against an external angle of the S.W. boundary. The Graded schools are placed against an internal corner, and the master's house looks—as it was built to look—down the garden. We have not considered it necessary to give a detailed plan of the Infant school, for the simple reason that the only remarkable feature of this group is the arrangement of the Boys' and Girls' departments.

The detailed plan (woodcut No. 247) shews these arranged almost symmetrically and opposite to each other, while the

space thus enclosed is roofed in and floored. At one end the plan follows the boundary line at an obtuse angle (which is reversed on the other side), with the object of avoiding waste of land. The central court or hall is to be used generally as a girls'

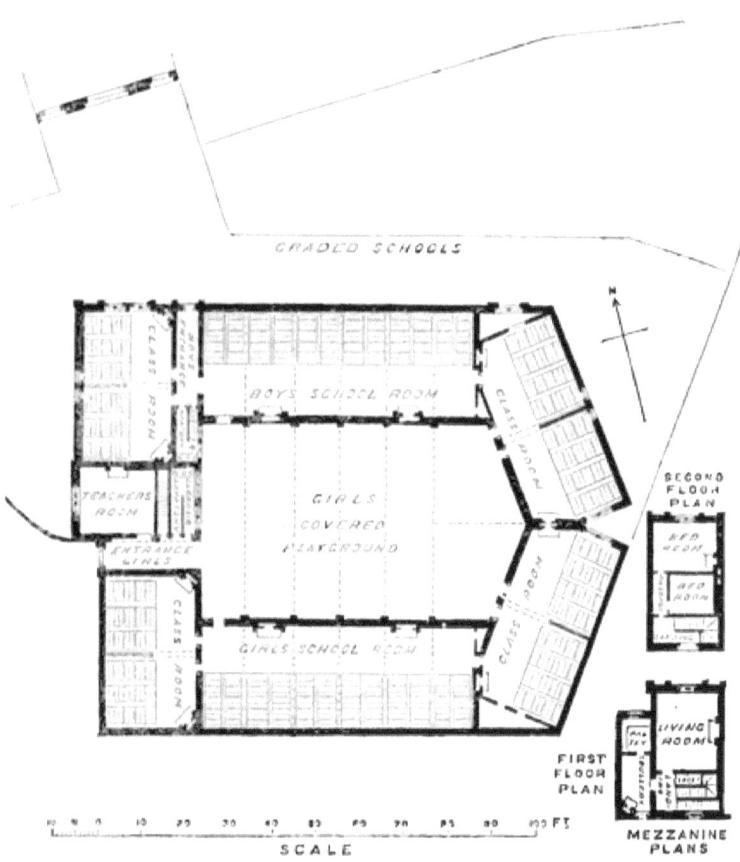

247.—HAVERSTOCK HILL SCHOOL.

play-room, but it is also designed for use as a hall of assembly, for addresses, lectures, the distribution of prizes, and other purposes. With this object arrangements are made for passing, when desired, warmed fresh air from the backs of four of the fireplaces.

The site has cost the sum of 2270*l*., and the buildings 9270*l*.

The accommodation is as follows, viz. :—

Infants	447
Boys	306
Girls	306
Total	1059

The cost of warming by any kind of apparatus would be considerable for a building straggling over so much ground. On the whole, the old method was considered to be preferable, and the schoolhouse is accordingly warmed by open fires of the ventilating kind.

The total cost of the sites and buildings cannot, at the time of writing, be precisely ascertained. Among the 86 schoolhouses erected during the three years ending November, 1873, when the life of the first School Board for London terminated, the average cost has probably been about 7l. 10s. per head, and that of the sites about 3l. 10s. per head. The total cost would thus be about 11l. per head.

The following total shews at a glance the number of schoolhouses in each school division of the metropolis, either completed, in course of erection, or contracted for up to November, 1873 :—

Division.	Schools Completed.	In course of Erection.	Contracted for.	Total.	Accommodation.
Chelsea	3	1	...	4	3,306
Finsbury	4	3	4	11	9,793
Greenwich	6	...	2	8	6,036
Hackney	3	6	4	13	14,216
Lambeth	4	6	3	13	11,424
Marylebone	...	4	5	9	7,957
Southwark	4	4	2	10	10,081
Tower Hamlets	12	4	2	18	16,812
Totals	36	28	22	86	79,625

CHAPTER XVII.

INDUSTRIAL SCHOOLS.

The Prison—The Reformatory—The Industrial School—Special object of the latter—Powers of the Industrial Schools Act which can be employed by School Boards—Programme of the kind of building to be erected—Playground and lavatory arrangements—Barnes' Home and Industrial School.

No work, professing to deal with School Architecture in its bearing on Public Elementary education, can be regarded as complete without some reference, however short, to the subject of Industrial schools. In country districts, and in the smaller cities, no question may arise, but in the great towns of the land, members of School Boards cannot long exercise their functions, without discovering the necessity of understanding precisely the powers and responsibilities attached to their office in respect of this class of establishment, and the most desirable course of action to be pursued as cases arise from time to time in which children ought neither to be sent to a prison, nor to a reformatory, and are yet beyond the reach of ordinary school influence.

The Prison is well known to be penal, and, generally speaking, deals with the advanced in crime. The Reformatory, also penal, deals with those who are just entering on a criminal career. The former is the receptacle for those condemned to undergo certain specified punishments. The latter, as its name implies, is the place of half punishment, half discipline, for those of early years still deemed capable of reformation.

The Industrial school stands on the border land between vice and virtue; and is intended to prevent as many as possible from entering, at an early age, the lands beyond from which return is difficult. Its inmates are not criminal, and are not punished in any manner pointed out by law, but consist of the youth of both sexes, who, by reason of their unfortunate parentage, their equally unfortunate companionship, or their want of submission to proper control, may be expected soon to become positively criminal, unless immediate steps be taken to arrest their course, and lead them in a better direction.

In carrying out the compulsory clauses of the Education Act, well-organised School Boards employing proper officers to enforce the attendance of children, cannot fail to find instances which show the provision of some other place than the Public Elementary school—attended by the children of the working people of the locality—to be manifestly required. At a very early period in their history, the School Board for London determined to put in force the powers conferred on them, and to avail themselves of the Industrial schools already in existence. Five special officers are devoted to the work of seeking out suitable cases, and their labour is divided among the most populous parts of the town. Up to November, 1873, more than 2100 cases had been investigated, and 999 cases had been sent to certified Industrial schools. The vacancies in the schools of the immediate neighbourhood being soon exhausted, the Board had to enter into agreements with sixteen country, as well as with eighteen metropolitan schools. They also made grants, amounting to 3700*l*., for the enlargement of five schools, in order to secure permanently 400 places in those schools for cases sent at their instance. The total annual cost to the Board of each child was from 7*l*. to 8*l*., the remainder being defrayed by the Treasury, by private subscriptions contributed to the funds of the Industrial schools, and by the earnings of the children.

It may be noted, therefore, that by the Industrial Schools Act, 1866 (29 & 30 Vict. c. 118), the classes of children, not being paupers, who may be detained in certified Industrial schools under the authority of a magistrate, are set forth in ss. 14, 15, & 16, as follows, viz. :—

"14. Any Person may bring before Two Justices or a Magistrate, any Child apparently under the Age of Fourteen Years, that comes within any of the following Descriptions, namely:

"That is found begging or receiving Alms (whether actually or under the Pretext of selling or offering for Sale any Thing), or being in any Street or public Place for the Purpose of so begging or receiving Alms;

"That is found wandering and not having any Home or settled Place of Abode, or proper Guardianship, or visible Means of Subsistence;

"That is found destitute, either being an Orphan, or having a surviving Parent who is undergoing Penal Servitude or Imprisonment;

"That frequents the Company of reputed Thieves.

"The Justices or Magistrate before whom a Child is brought, as coming within One of those Descriptions, if satisfied on Inquiry of that Fact, and that it is expedient to deal with him under this Act, may order him to be sent to a Certified Industrial School.

"15. Where a Child apparently under the Age of Twelve Years is Charged before Two Justices or a Magistrate with an Offence punishable by Imprisonment or a less Punishment, but has not been in *England* convicted of Felony, or in *Scotland* of Theft, and the Child ought, in the opinion of the Justices or Magistrate (regard being had to his Age and to the Circumstances of the Case), to be dealt with under this Act, the Justices or Magistrate may order him to be sent to a Certified Industrial School.

"16. Where the Parent, or Step-parent, or Guardian of a Child apparently under the Age of Fourteen Years, represents to Two Justices or a Magistrate that he is unable to control the Child, and that he desires that the Child be sent to an Industrial School under this Act, the Justices or Magistrate, if satisfied on Inquiry that it is expedient to deal with the Child under this Act, may order him to be sent to a Certified Industrial School."

The powers of School Boards to deal with Industrial schools are derived from Sections 27 and 28 of the Education Act, which confer the same right of contributing money, whether towards the building, alteration, enlargement or management, the purchase of land, or the support of the inmates, as is enjoyed by a prison authority under Section 12 of the Industrial Schools Act.

The whole wording of the Act tends to indicate that the Industrial school is *a school for the neglected*. To such an extent is this the case that, under Section 15 (already quoted), any child who has once been the inmate of a prison, is excluded from the benefit derivable from admission to an Industrial school.

The wording of Section 16 (already quoted) is very peculiar in reference to the representation of the *parent* that he is unable to control the child. It may be and often is the case, that the fault lies rather with the parent than with the child. The scope of the Act is, therefore, great, and its working, in connection with the enforcement of any compulsory by-laws, should be carefully watched by those who have the responsibility of this particular action of a School Board.

The amount of juvenile refractoriness, profligacy, or delinquency which in its actions keeps just beyond the reach of the ordinary criminal law, forms, among the population of large towns, so large a percentage of the children of school age, that in many instances the question must arise, whether it is better to contribute towards existing Certified Industrial schools, or whether it may not be necessary to erect special buildings to be worked in connection with the Board Schools of the district.

It is clear that children of this class must be removed from contact or association with other children into buildings where they may receive proper care. The Boards must decide how, in their own districts, this shall be done. In cases where the erection of new buildings has been determined under the provisions of the Act, some consideration should first be given to the future treatment of the inmates. The plan of the building will depend upon the uses to which it is to be put.

Dense ignorance is one of the first sources of evil to be removed, —hence, a school-room will be necessary. For positive good, the useful employment of the hands by the exercise of some honest trade is desirable,—therefore, workshops will be required. The children will all sleep on the premises, thereby requiring dormitories. They will be fed there, thus rendering necessary a dining-room and kitchen. The Industrial establishment is neither a workhouse, a school, nor a boarding-house, and yet combines some of the features of all the three. The inmates must be boarded, lodged, educated, and trained to manual labour. The manner in which this simple yet comprehensive programme is proposed to be carried out, will control the shape of the building to be erected.

In any new building intended for use as an Industrial school, accommodation should not be provided for more than 100 inmates, this number having been proved by experience to be as great as one Superintendent should in fairness be expected to have at one time under careful supervision.

The daily life would probably be arranged so that one half of the entire number would be at work in suitable shops, while the other half would receive instruction in properly arranged school premises. The best plan for the latter has received sufficient discussion in previous pages. That of the former would depend to some extent on the kind of work to be done. In view of the fact that, at certain times, it would be desirable to assemble the whole of the children, the school-room and work-rooms might be connected by folding doors, so as to be thrown together when desired.

The boarding arrangements would suggest a dining-room sufficiently large to hold the whole of the inmates at one time, although it would not be absolutely impracticable to issue rations to two batches in succession. Having regard to facility of management, a room for the entire number is preferable, and it should be used exclusively for meals. Complete kitchen arrangements and offices would necessarily follow, considered in relation to the number and daily wants of the inmates and necessary officials.

Lodgings, in the form of dormitories for the whole number, must be provided. These should be planned to accommodate about 10 boys in one room, each boy (or girl) having a separate bed. In connection therewith a teacher's room should be provided on each floor between two dormitories, and with a small opening or window in each of the walls between. This arrangement secures ample inspection, and admits of the doors being locked for additional security.

Other features, inseparable from an establishment of the kind, also follow, such as Waiting-rooms for visitors, Porter's room, Master's or Teacher's rooms, Servant's rooms, Governor's rooms, &c., &c. A portion sufficiently isolated from the rest of the building will be devoted to the use of the sick, or those suffering from infectious disorders, for the hospital ward will be as indispensable here as in a workhouse. A distinct building in the form of a separate cottage is better than any portion of the main establishment.

Whatever the plan of the building or its internal arrangements, the class of children to be dealt with renders necessary the exercise of unusual care in the provision of ample and well-appointed playgrounds. Size is the first necessary element, protection from weather the second, and carefully-considered gymnastic appliances the third. If real educational influence is to be exerted, the children must first be made happy by the training influences of the "uncovered school-room." A portion of the covered playground should be built off and set apart as a bath-room. Where circumstances admit, it should be sufficiently large to allow of instruction in swimming.

The lavatory arrangements should, in principle, be precisely the opposite to those of the Elementary school. In the latter the child should seldom require to wash; the basins are consequently limited in number, and the water applied separately to each. In an Industrial school the process is indispensable,

CHAP. XVII.] INDUSTRIAL SCHOOLS. 357

and the whole number should wash in four divisions of 25 each, the water being turned on to the whole of the basins at once by

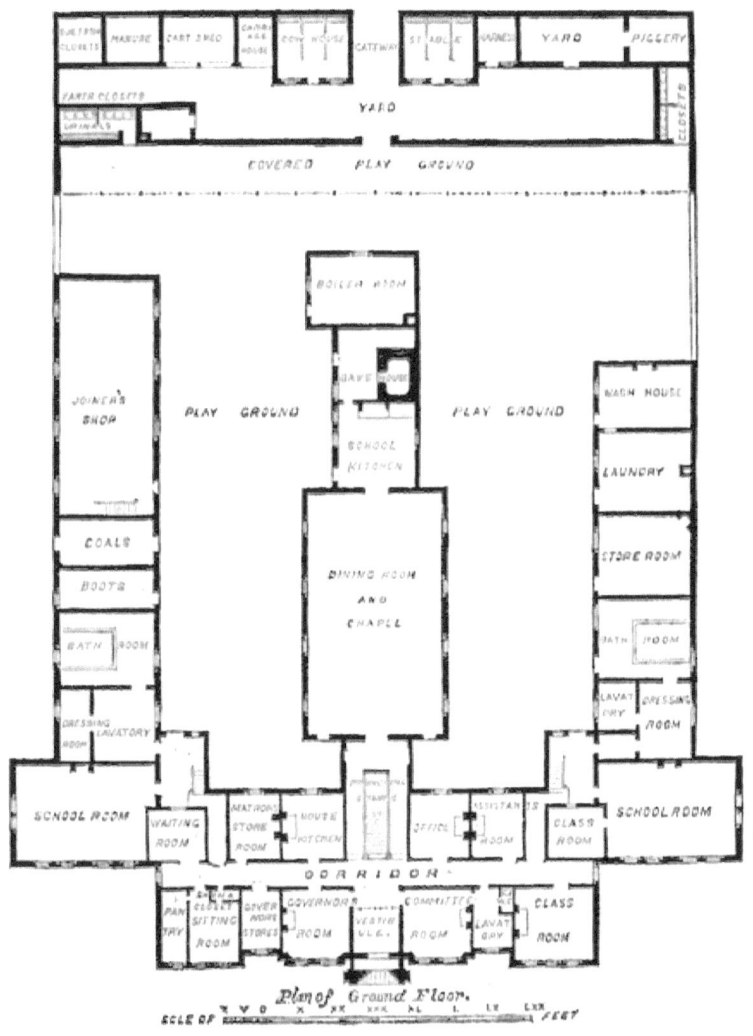

248.—INDUSTRIAL SCHOOL AT ARDWICK. GROUND PLAN.

the attendant. A long room like a corridor—measuring perhaps 50 feet by 7 feet—is required. Along the wall not occupied by basins, should run a wooden rail, with hooks for towel, brush, and comb, in a space of six inches to each boy. One looking glass should be provided for each four boys. This kind of arrangement is in existence at the east end of London, and is considered, from its working, to be exactly suitable.

Among existing certified Industrial schools in or near the metropolis, which may be visited by those desiring to study further the subject, may be cited those respectively at Feltham, Surrey, and at Burdett Road, Tower Hamlets.

As an illustration of a building of this class actually erected, but without setting it forth as a perfect model, we give (woodcut No. 248) the plan* of Barnes's Home and Industrial School at Ardwick, near Manchester. In this instance the accommodation is for 200 children, and the building has been placed in an agricultural district, so that sufficient land could be obtained for instruction in agricultural, gardening, and other outdoor pursuits, in addition to the usual indoor industrial occupations. The dining-room is intended for use also as a chapel. In connection with an establishment of this size, the provision of a separate building for purposes of worship would be warranted, and would certainly be more decorous. In cases of smaller size, the use of the school-room or work-room, or both combined, is preferable to the dining-room. The first floor contains Governor's, Matron's, Master's, Assistant's and Servant's bed-rooms, also bath-room and dormitories. The sick ward is on the second floor, isolated as much as possible. The cost of land, buildings, and fittings amounted to about 12,000*l.*

* From the *Builder* of Sept. 24, 1870.

249.—A GALLERY LESSON.

CHAPTER XVIII.

(BY JOHN F. MOSS, ESQ.)

SCHOOL FURNITURE AND APPARATUS.

Advantages of good furniture—School desks, old and new—Dimensions suggested by Education Department—Convertible desks—Specimens in other countries —German opinions—German dimensions—The dual arrangement—Dutch and American desks—English designs—Swedish desks—Graduation of desks in class—Drill for dual desks—Drawing—Teachers' desks—School cupboards —Fenders and fire irons—The cooking stove—Minor details—Easels—Lessonstands—Blackboards—Maps—Diagrams—Models—Abaci—The French "compendium"—Kinder-garten apparatus—Infants' hammock.

SUITABLE appliances are to the teacher very much what proper tools are to the handicraftsman. The furniture and apparatus should be carefully adapted to the kind of instruction to be given, and the most approved methods of imparting it. All ulterior considerations must be held subservient.

The furniture of the school-room should be graceful in form

and good in quality and finish. Children are particularly susceptible of surrounding influences, and their daily familiarisation with beauty of form or colour in the simplest and most ordinary objects, cannot fail to assist in fostering the seeds of taste, just as daily discipline tends to promote habits of order. Furniture finished like good cabinet work is more likely to be respected even by the mischievous school-boy than that of an unsightly or rough character.

Whether this be so or not, it cannot be denied that particular combinations of form and colour exercise an important influence on the minds of the young and the ignorant. The stained glass, pictures, and images used by the monks, formed a powerful—if not the principal—means of instruction in the middle ages. In our time it is desirable to extend the process of education, at least during school hours, by the adoption of good and tasteful designs as well as of superior workmanship for the necessary mechanical aids. The insensible influence thus exerted will not be without due fruit in future years, and, in the present, will assist in promoting a love for the school. Good taste and good workmanship do not really involve increased cost of manufacture.

One of the results of the want of education in the present day is that of misapplied labour, sometimes producing costly furniture of bad design.

During a large proportion of the time spent in school the scholar must be seated. In providing for this condition, his comfort as well as the convenience of the teacher should be considered. If compelled to assume an uneasy or restrained posture, the pupil will of necessity have his attention more or less distracted from work, and the teacher will suffer in consequence. Hygienic principles should determine the height of seat and desk and the relation of each to the sitter, in reference to the support to be given to the various members of the body.

The rude furniture upon which the school-boys of a generation ago did penance is happily becoming obsolete, but in the clever contrivances which modern ingenuity has produced for its improvement, we have some which are almost as ill suited to the legitimate requirements of the school-room.

The desks used in the earlier British schools, though somewhat Spartan in their conception, were better than the quaint writing boards hung against the walls of the first schools established under the auspices of the "National Society," whilst these in turn were considered preferable to the double desks at which scholars sat face to face, and which are still used in some old-fashioned Proprietary and other schools.

The "Rules for Planning and Fitting-up Schools" issued by the Committee of Council on Education led, in Government-aided schools, to the general adoption of parallel groups of desks, graduated according to the ages of the scholars. The heights therein suggested have for many years formed the standard upon which a variety of other desks have been constructed.

For the younger children the height of a desk having slightly inclined top was fixed at 2ft. 2in., and the height of the seat at 1ft. 2in. The medium heights were 2ft. 3½in. to the top of desk and 1ft. 3in. top of seat; whilst for the elder children the height of desk was 2ft. 5in., at the outer edge, and the distance from floor or sleeper to seat top was 1ft. 4in. Desks with flat tops were directed to be 1 inch lower at the front. The width of the top was to be 1 foot and that of seat and book-shelf 7 inches in each case. Between the desk and seat a space of 3 inches was allowed, and, where the teacher required gangways behind the desks, (imperatively necessary for convenient working,) the minimum distance between the rows was to be 12 inches. An allowance of 18 inches on the desk and bench for each child was made for junior classes; but in senior classes each scholar was allowed 22 inches. Plain iron standards were recommended as supports.

The common desire of school managers to provide special accommodation for occasional tea-meetings or other social gatherings, has not always tended towards real improvement in school desks. The reverse has too often been the case. Since the first issue of the " Rules," many attempts have been made to produce desks capable of transformation into a variety of forms, some of which were entirely incompatible with the one object of paramount importance. Mistakes so engendered are unfortunately too liable to be repeated and perpetuated in schools where public interests and educational advantages should alone be considered. We sometimes "follow the leader" without due reflection.

The English "convertible" or "reversible" desk has no counterpart in the public schools of Germany, and is equally unknown in the educational establishments of Switzerland or Holland. The direction of endeavour has been to improve the desk in reference to school use alone. Among German educationists the subject has given rise to much discussion. Eminent men of different professions—physicians, architects, and school professors—have united in giving careful attention to the various points involved, having for their object the comfort and healthfulness of the scholar, and the facilitation of the work of the teacher.

Dr. Wiese gives a synopsis of various writings bearing on the best methods of uniting the seat with the desk, and on its general formation, a matter which he declares to be as important as the proper lighting of the school. Among the works alluded to are those of Zwey, Dr. Falk, Dr. Frey, Dr. Cohn, Dr. Kleiber, and Dr. Virchow. The official commissioners have also made many suggestions. The weight of opinion is to the effect that the height of the seat should correspond with the length of the scholar's leg, from the knee to the sole of the foot. There must be no stretching of muscles: therefore the sole of the foot must rest on the floor or upon some flat surface. If the seat be too high, the swinging of the foot in the air causes a compression of the blood-vessels and nerves of the hinder part of the leg and knee: if it be too low, the thighs of the scholar are pressed against his stomach to the disadvantage of health. Speaking of Secondary schools, it is set forth that the height of the seat should be, according to the age of the scholar, from 15 to 18 inches, and, where the seats are higher than the lower part of the leg of the scholar, a footboard, about 2 inches high, should be used. The breadth of the seat must be at least a good half of the length of the upper part of the leg, from the seat to the knee-cap, that is, one-fifth the length of the body. In order to prevent the scholar slipping forward, the seat should be slightly declined backwards. The height of the desks should be so arranged that the under part of the arm may rest comfortably on the desk top, and that the powers of vision may not be strained, or, in other words, that the normal distance of vision be preserved. Desks which are too low, cause, by the bending of the scholar, a pressure on the chest and lower parts of the body: while those which are too high cause the right shoulder to be so lifted as to remove the upper part of the arm so far from the body, that the lower arm cannot be laid flat on the table, thereby causing the arm to be unsteady and easily tired. Accordingly the height (provided the top of the desk be inclined) should not be less than 28 inches. The breadth of the desk should be 12 to 14 inches, according to the size of the scholar, with a sloping surface of 10 or 12 inches, to which is joined a narrow horizontal surface at the top for the reception of writing materials. For the sloping part, an inclination of 1 to 2 inches is sufficient. Writing can be seen better, or be better overlooked, on a somewhat sloping surface.

Notwithstanding the length to which this discussion has been carried in Germany, no exactly uniform system for settling the dimensions of desks and seats seems to have been hit upon which could be universally accepted. The consequence is that even in newly erected schools widely different standards are adopted.

Herr Weyer, Stadt-baurath of Cologne, has, in the desks introduced a short time ago (woodcuts Nos. 250, 251, 252, and 253), adopted the heights and sizes given in the subjoined table, which is here quoted by way of illustration, though at variance in some details with what is recommended after full enquiry for adoption in England.

It will be noticed that there are seven different proportions of desks, each of which is used throughout one class-room.

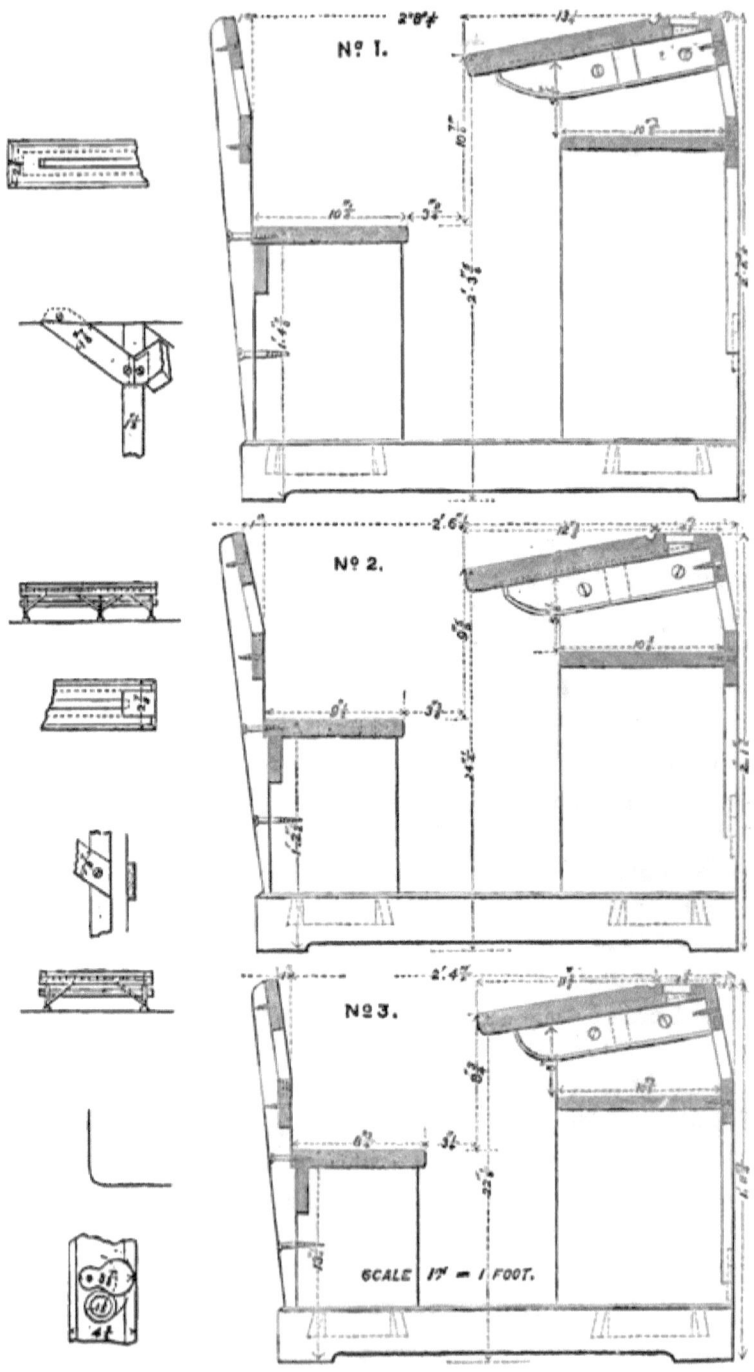

250, 251, 252.—THE COLOGNE SCHOOL DESK.

CHAP. XVIII.] SCHOOL FURNITURE AND APPARATUS. 365

The difference between Rheinlandish and English inches being very small, viz.: 36″ Rhein. = 37″ Eng., the dimensions in this schedule could be given in English inches in some cases only, the same as in the original, the omitted fraction say of less than $\tfrac{1}{8}''$ being considered of no practical value:—

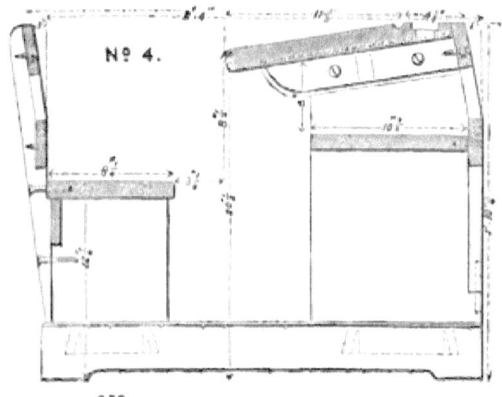

253.—THE COLOGNE SCHOOL DESK.

	No. of Class (Division).	I.	II.	III.	IV.	V.	VI.	VII.
	Age of scholars in years	5–6	7–8	9–10	11–12	13–14	15–16	17–18
a	Height of same in inches	41–44	45½–48½	49½–52½	53½–56¾	57¾–61	62–65	66–69
	Average heights ,,	43	47½	51½	55½	59¾	64	68
b	Height of seatboard ,,	12¾	13¼	14⅜	15¾	16¾	18	19
c	Width of seatboard ,,	8¼	8¾	9¼	9¾	10¼	10¾	11¾
d	Height from top of seat to front edge of desk	8¼	9	9½	10¼	11	11⅜	12⅜
e	Height from floor to front edge of desk ...	20¾	22⅜	24¼	26	27⅞	29⅜	31½
f	Height from floor to back edge of desk ...	22¾	24½	26	27⅞	29¾	31⅜	33⅜
g	Distance between edge of desk and seat (measured in projection)	3	3¼	3¾	3⅜	3¾	4	4¼
h	Width of top of desk, horizontal part	4¼	4¼	4¼	4¼	4¼	4¼	4¼
i	Do. inclined part	11¾	11¾	12¾	12¾	13¾	13¾	13¼
k	Depth of book-shelf	10¼	10¼	10¼	10¼	10¼	10¼	10¼
l	Distance of the same from top of desk ...	5⅛	5⅛	5⅜	5⅜	5¼	5⅜	5⅜
m	Sloping of back of seat ...	1	1	1	1	1	1	1
n	Depth of seat	27¼	28½	30¼	30½	31⅜	33¾	34¼
o	Width of seat	17¼	18½	19¾	20¾	21¾	22½	23½
p	Length of desk with 4 seats	5·1	6·2¼	6·6¼	6·10¾	7·2½	7·6¾	7·10¾
q	Do. with 5 seats	7·3½	7·8⅜	8·1¾	8·6¼			

The mistaken notion, that desks arranged for seating children in pairs entail a waste of floor space, appears to have prevailed in some high class schools. Many most experienced teachers have, however, expressed a decided preference for the arrangement. In the case of desks made in lengths sufficient to seat four or five children (without folding top), as largely used in German schools, it is impossible to have proper facilities for entry and exit, and at the same time to keep the inner edge of the desk sufficiently near to the seat to give proper accommodation for writing.

Dr. Wiese says the opinion is gaining ground that comfortable sitting is, for school work, of more importance than comfortable standing. The nearer the desk be placed in reference to the seat the more favourable will it prove for the position of the body and for the comfort of the eyes, which, in this case, are not strained by being brought too near to the writing. The scholar who sits too far from the desk, either bends too much, and thereby hurts his chest and eyes, or he glides too far forward on his seat and so gets an unsteady position. "It is therefore desirable," says the writer, "that the inner edge of the desk should be distant from the front of the seat only about one inch." It is recommended that the vertical distance from the desk to the seat-top should be the length of the fore-arm, or one-sixth the size of the body of the scholar. Too great a distance encourages crooked growth, for the scholar while writing has his body weighing on one arm, instead of having the arm naturally resting on his body. If the difference in height between desk and seat be too slight, then the chest sinks and the back is bent out so as to encourage stooping.

Considerable difference of opinion exists as to the best shape and position for the back. The plan adopted in most German and Dutch schools is that of making the front of one desk act as the back of another (woodcuts Nos. 254 and 255). But, according to Dr. Wiese, some German physicians recommend a low back which will support the lower part of the body, leaving the upper part free, *i.e.*, a back 3 inches broad, and 5 to 6 inches in height above the seat, and slightly inclined. Dr. Frey advises the

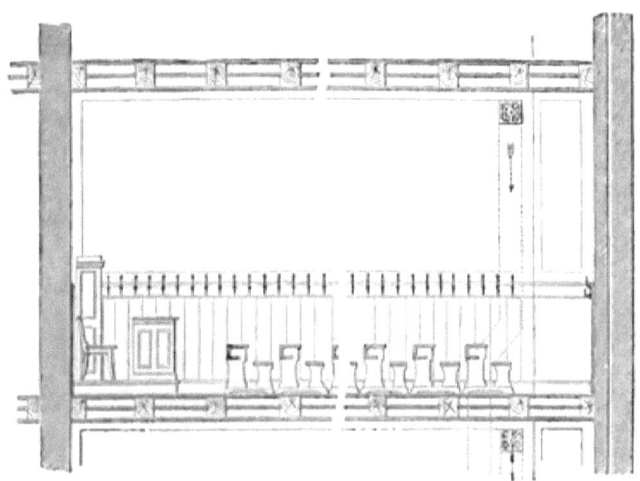

254.—DESKS AND SEATS AS ARRANGED IN A GEMEINDE SCHULHAUS, BERLIN.

provision of an upright independent back (fixed upon the seat), which besides possessing sanitary advantages keeps the scholar better in his place. This plan has been adopted in the

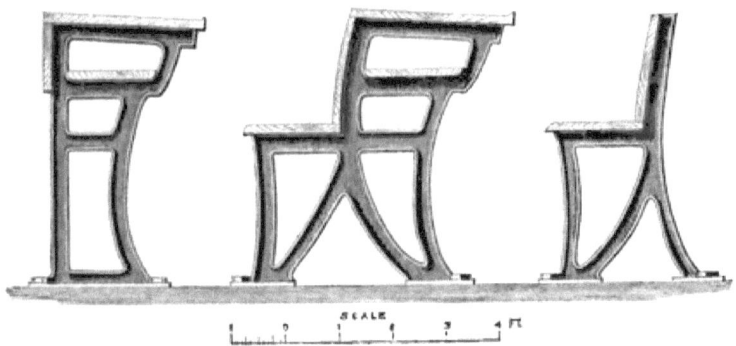

255.—SECTION OF DUTCH DOUBLE DESKS WITH SEATS ATTACHED.

double desks used in some American schools, but a method more commonly adopted in the United States, is to provide detached seats or chairs even with the double desks, as shown in woodcut No. 256.

So important is it considered in Germany to give proper support

256.—AMERICAN DOUBLE DESK WITH SINGLE SEATS.

to the scholar while sitting, that, by an instruction of the Royal Government at Treves, in the year 1865, it was decreed that *all* school benches in that district should be provided with backs. An " Instruction " issued at Stutgardt also prescribes that there shall be backs firmly connected with the seats of all school desks.

It is sometimes fancied that the subject of the school desk is of little importance. It is hoped that in the preceding remarks we have established the contrary. In truth there is much more than mere sentiment involved. Anything which contributes to proper physical training and to the development of a robust, healthy, and vigorous people cannot be deemed a minor or unimportant matter.

The advantages of giving ready access to the teacher need not here be discussed, nor is it necessary to repeat the arguments in favour of seating the scholars in pairs, which actual working has abundantly proved to be both economical as to space and particularly advantageous to the efficient working of the class.

In the article of this work devoted to the subject of desks as affecting the planning of buildings, the desk best adapted for

CHAP. XVIII.] SCHOOL FURNITURE AND APPARATUS. 369

general use in English schools has been sufficiently described. Its proportions have been settled mainly on the basis of the most

257.—MOSS'S PATENT SCHOOL-BOARD DESKS (REGISTERED DESIGN FOR STANDARDS).

approved foreign models, with some modifications suggested by the most recent experience of their use.

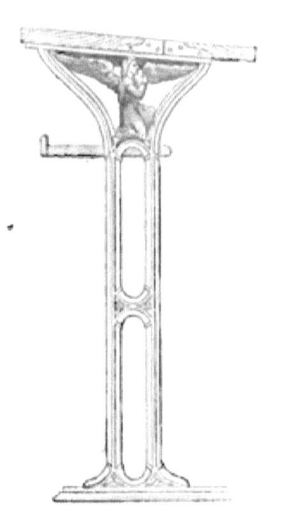

258.—THE "ANGEL" STANDARD.

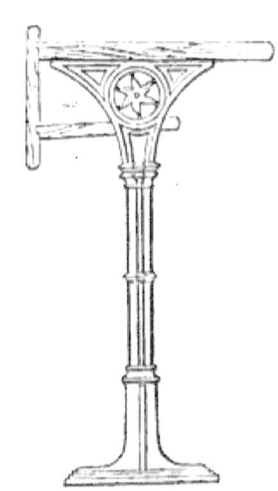

259.—ALTERNATIVE DESIGN FOR DESK STANDARD.

The subjoined illustrations (woodcuts Nos. 257, 258, and 259) will serve to indicate how the shape of the standard or iron support may be varied: the more tasteful the design the better: but extravagant elaboration should be avoided, and care should be taken to guard against affording awkward projections either for the lodgment of dust, or for interference with the children's clothing.

The desk shewn in woodcut No. 260 is one which was introduced in some of the first schools built by the Sheffield and other

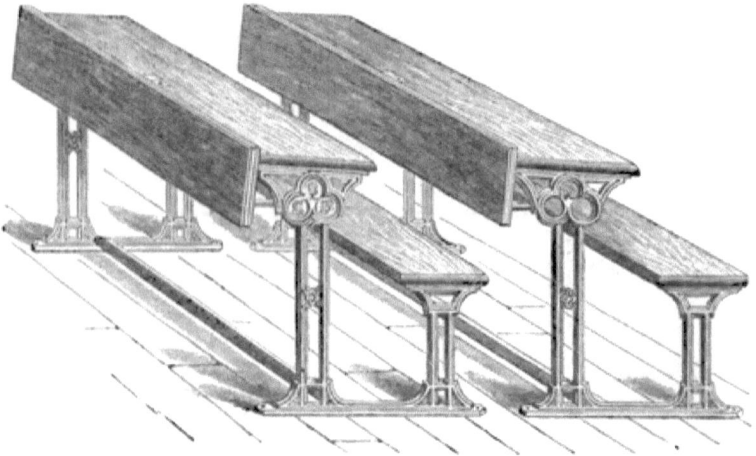

260.—THE SHEFFIELD DESK.

School Boards soon after the Elementary Education Act of 1870 came into operation. It has been much approved of by many experienced educationists, but does not possess all the advantages of the improved double desk already described. The initial letters in the trefoil at the head of the iron support may be varied to suit the town in which the desks are adopted, or the spaces may be left blank.

Of the desks hitherto used in the large public schools of Germany, the older kinds are cumbrous and even unsightly, but in most cases there is provision for the comfortable disposition of the sitter while attending to oral lessons.

Among the more recent German desks are those in use at the

King William Grammar School at Berlin, constructed to seat from four to five scholars, of which we are able to give a drawing (woodcut No. 261).

261.—SCHOOL DESKS FROM THE KING WILLIAM GYMNASIUM, BERLIN.

As in the case of the Dutch desk, the front of one forms the back for another. In order to render this available, they must be placed close behind each other, and some of the scholars are therefore not easily accessible to the teacher, a difficulty which would have been obviated by the dual arrangement to which the Germans are now themselves tending.

In the Victoria School for Girls at Berlin—perhaps one of the best furnished schools in Continental Europe—desks for seating the children in pairs are adopted. Similar desks are also being introduced in some of the best new schools in various parts of the Empire. Save in the matter of length, they seem to be modelled very much after the older desks, which are arranged for the seating of four children together. The spaces in front of the bookshelves (usually left open in the ordinary German school desk) are filled with plaited cane. Both here and at the Pro-Polytechnic school in Chemnitz, the space allowed per scholar under the dual arrangement is greater than can well be afforded, or would be necessary, under similar arrangements in English public elementary schools.

In Sweden, and also in many public schools of America, the single arrangement applied both to seat and desk, is in use for each scholar. In both countries lockers have been provided to the desks, while in the former the most approved desk is fitted with a lifting seat made self-acting by means of a lever and weight (woodcut No. 262). The single desks used in America are usually independent of the seats, the latter being in the shape of chairs fixed on iron supports. The finish displayed in the wood work employed is frequently above all praise. The designs used in the iron work would not generally excite admiration.

The single desk entails too great an expenditure of space, and is on this account deprecated by some of the highest American authorities, who now generally recommend the use of desks for seating the children in pairs.

262.—SWEDISH SINGLE DESK WITH LIFTING SEAT.

263.—AMERICAN SINGLE DESKS AND SEATS AS USED IN PRIMARY SCHOOLS.

Many different patterns of the double desk are used in America, and, in the construction of some, much ingenuity is displayed, though not always with useful result. Some are fitted with movable seats, very much after the fashion of the Swedish single desks.

264.—AMERICAN DOUBLE DESKS WITH SEATS ATTACHED.

In a number of the schools in England which have been fitted with desks on the dual principle, the whole have been fixed on sleepers (or scantling). The principal reason urged in favour of the plan, is that it gives greater facilities for the removal of dirt in sweeping the floors, and where it is adopted the depth (or thickness) of the sleeper can easily be varied between the first three and the last two rows.

In grouping the class the object is to enable each child to see the teacher, and the teacher to command the face of each child. The use of different sizes of desks in the same class—the higher ones being placed in the rows in the rear—is one step towards the attainment of this object. With many teachers it is found sufficient, while others prefer that the two last rows should be brought into further prominence by being raised on sleepers as shewn in the section (woodcut No. 265). This again can be attained in two separate ways. Either the two last rows may remain on the same level of 4 or 6 inches above the floor line, while the desks themselves differ in size, or the arrangement may be carried still further by placing the fourth row on a four-inch sleeper, and the last row on an eight-inch sleeper. The most generally useful plan, and therefore that which is now recommended, is that of fitting the last row with desks of the size next above that of the four front rows, and that the two rear rows should be raised together on sleepers of one height—say of 4 or 6 inches. Woodcut No. 265 shews this arrangement, and it is

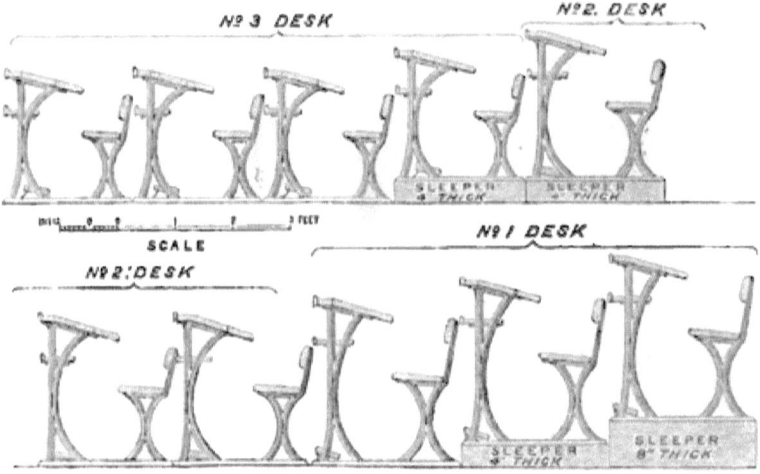

265, 266.—SECTION SHEWING ALTERNATIVE METHOD OF GRADUATING DESKS IN CLASS.

believed that it will be found simplest and in every way most satisfactory. Woodcut No. 266 shews the other plan, which not only contemplates a different size of desk for the two last rows, but the raising of each on steps of four or six inches one above the other. When we recollect that a large proportion of teachers are quite content with the whole of the desks and seats being placed on the level floor with only the fourth and fifth or sometimes only the last row of a greater height than those in front, it will be seen that objections may be entertained to this plan. Everything which tends to increase the number of steps or even to introduce them in any way, should be guarded against as far as possible. The comfort of the children and the convenience of drill are better attained if the floor be maintained perfectly flat; and only with the object of giving better command to the teacher, and a better view of the teacher to the scholars, should the use of steps be entertained. Some teachers object to the use of sleepers on account of the necessity for the desks being screwed down, whereas if the desks be placed on the floor they can be more easily removed, and the grouping altered, the weight being sufficient to keep them steady.

Those who are much interested in Infant training, or who, like the Wesleyans, hold that the gallery should be largely used in teaching, may probably advocate the graduation of the entire class. The old practice of the Home and Colonial School Society involved this, as shewn in our illustration (wood-

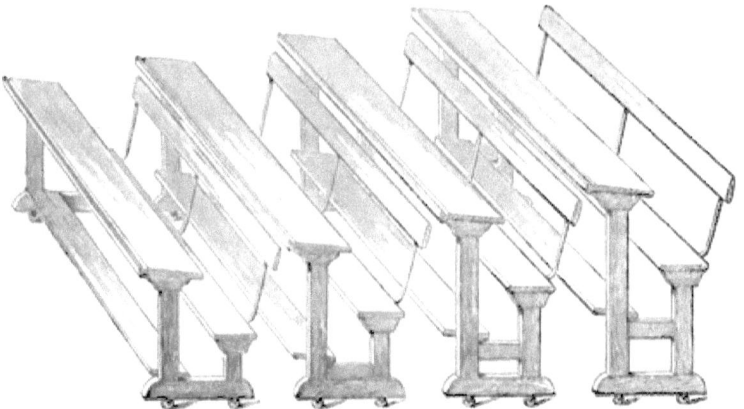

267.—HOME AND COLONIAL SCHOOL SOCIETY'S DESKS (MOUNTED).

cut No. 267). The desks in use were of the long kind—seat and desk being formed in one frame working by means of the construction at the bottom on sleepers. Generally, no backs were used to the seats, and each length of seat and desk was placed on castors, so as to be moved about at pleasure. The graduation of this seat must be considered excessive, and only necessary when much greater depth of class is used than that of five rows, for those in the rear have to climb to their places somewhat awkwardly.

An experienced teacher will have no difficulty in arranging the drill to be employed in connection with the dual system. The scholars on their part will soon learn to move with greater smartness and precision than is readily attainable under the old system. The importance of doing things in a regular and orderly manner cannot be too forcibly impressed. A definite system of drill is therefore essentially necessary. Some teachers prefer to use numbers only, while others incline to a code in which the orders are given by words.

The following suggestion for a code of drill illustrates our meaning. The teacher may give his orders either by words or numbers :—

(1)—"*Return.*"
At the word "Return," the hands should be raised to grasp the slate.

(2)—"*Slates.*"
At the word "Slates," the slate should be smartly lifted and placed in the groove in front of the desk without noise. The hands should then be lowered.

269.—"RETURN."

270.—"SLATES."

If books have been used, as in the case of an arithmetic lesson, the additional command may follow; (1) "*Return*" (2) "*Books.*" At the second word "Books," the books should be placed on the shelf under the desk, and the hands brought back to their original position.

378　　　　　SCHOOL ARCHITECTURE.　　　[CHAP. XVIII.

(3)—"*Lift*" (or "*Raise*").

At the word "Lift," the edge of the flap should be grasped.

(4)—"*Desks.*"

At the word "Desks," the flap should be raised quickly but without noise, and the hands dropped.

271.—"LIFT" (OR "RAISE").　　　272.—"DESKS."

(5)—"*Stand.*"

At the word "Stand," the scholars should rise smartly with arms straight by their sides.

(6)—"*Out.*"

At the word "Out," the scholar at the *right* end of the desk takes one step to the right and a short step to the *front*. At the same moment the scholar at the *left* end of the desk takes a step to the left and a short pace to the *rear*.

273.—"STAND."　　　274.—"OUT."

The sixth movement leaves the children standing in Indian file down the respective gangways. The command may then be given, "*Quick*"—"*March*," the word "Quick" being simply used as a caution. At the word "March" all the files move off with the *left* foot, taking care that regular paces be maintained. When the last scholar in each file has reached the front of the desks, the word "Halt" or "Right turn" may be given. In returning to their places, the scholars should be marched in files, halted and fronted, then afterwards seated at the given words:—

(6)—"*Sit.*"
At the word "Sit," the scholar *opposite the right end of the desk* takes a short step to the *rear* and another to the *left*; at the same moment the scholar *opposite the left end of the seat* takes a short pace to the front and another to the *right*.

(5)—"*Down.*"
At the word "Down," each scholar takes his seat smartly.

(4)—"*Lower*" (or "*Close*").
At the word "Lower," the top edge of the flap should be grasped with both hands.

(3)—"*Desks.*"
At the word "Desks," the flaps should be quietly let down.

The subject or kind of lesson should here be announced.

(2)—"*Take out.*"
The hands should be raised to slates or books as required.

(1)—"*Slates*" (or "*Books*").
The slates should be lifted from the grooves, and placed gently on the desks, the scholars "Dressing by the right" or taking time from the teacher who signals with the hand.

For drawing lessons desks of special form need only be provided in schools where the instruction is of an advanced kind. In the drawing classes of an Elementary school the ordinary desks will be found sufficient. Woodcut No. 275 is an illustration of a neatly contrived and convenient drawing-easel used in the higher class for the study of art at the Victoria School for Girls in Berlin.

The proper preservation of books and apparatus is a matter of great importance, particularly in large schools, where the damage arising from a great many small carelessnesses may lead to a total result of serious waste. The maxim, "A place for everything, and everything in its place," was appropriately printed in

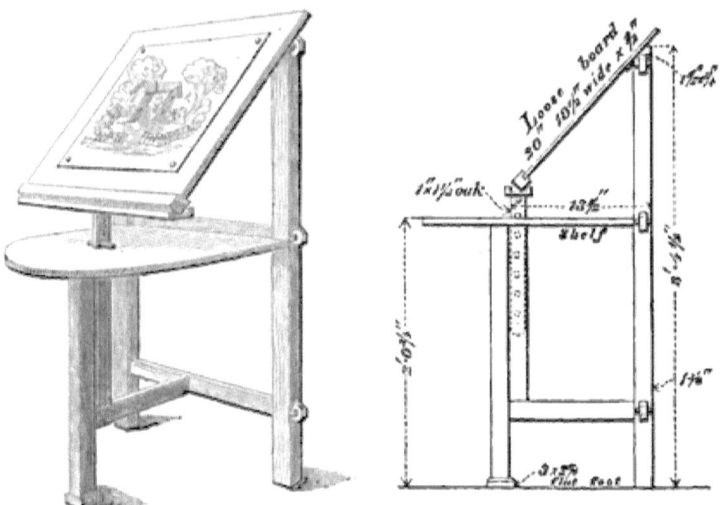

275.—DRAWING EASEL IN VICTORIA SCHOOL, BERLIN.

large type some time ago by one of the educational societies, for the purpose of being hung in the school-room. The same moral should be exemplified in all the various appointments of the school. If proper receptacles be not provided for books and apparatus,

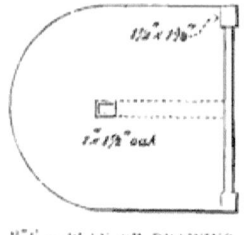

276.—PLAN OF DRAWING EASEL.

disorder and destruction will quickly follow. If habits of neatness and carefulness are to be inculcated, a good example must be set in the neatness with which the school properties are kept. As an illustration of the care taken in every little detail, it may be mentioned that in many American schools pieces of washleather or bottles containing acid are constantly kept on the teachers' desks, for the instant removal of the least inkstain accidentally caused on the surface of the beautifully polished furniture.

A moderately large and substantially made pedestal desk should

CHAP. XVIII.] SCHOOL FURNITURE AND APPARATUS. 381

be placed in the principal room for the use of the head teacher, in such a position as to afford him the best opportunity of watch-

277.—HEAD-MASTER'S DESK.

ing what is going on. In addition to the provision of drawers in this desk, there may be a cupboard in front for the reception of the surplus stock, stores, and maps. In a well-managed school the surplus stock will be kept as small as possible. Woodcut No. 277 represents a master's desk designed for a school having 250 or 300 children in attendance. With a proper number of pupil-teachers' desks, this will probably afford all the storage room ordinarily required.

The height should be sufficient for standing at, and the platform on which the desk stands should be six or eight inches from the floor. A tall stool with back should be made to match. In a few American schools the front portion of the head-teacher's desk is made to hold the school lending-library, and is consequently fitted with glass doors.

278.—MASTER'S CHAIR.

As a rule, libraries will be more conveniently kept in separate cupboards or cases.

In the secondary schools of Germany, teachers are often provided with desks or cupboards to which they can only obtain access by turning their backs upon the children. Under all

279.—CLASS-ROOM OF A GEMEINDE SCHOOL, BERLIN.

circumstances, however admirable the discipline of the school, the teacher should as far as practicable be in a position to observe constantly what is going on. It is not uncommon to find in the same school a rostrum 12 or 18 inches high, with boarded front, against which the desk is placed. The black-board is fixed on the wall behind, an arrangement which cannot be commended. Where the board is a fixture, it cannot be reached with sufficient ease. Woodcut No. 280 serves to illustrate the kind of teacher's desk platform which may be

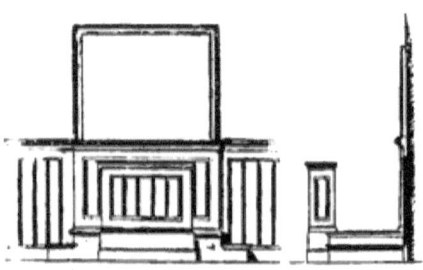

280.—TEACHER'S DESK AND BLACK-BOARD FROM THE KING WILLIAM GYMNASIUM, BERLIN.

found in many German schools. In Girls' and Infants' schools, the head-teacher's desk need not be so high as that described for the use of the master in Boys' schools.

251.—HEAD-MISTRESS'S DESK WITH FLAT TOP.

Where no desks of special kind are provided for sewing-classes, the tops should be flat and sufficiently low to be reached by the teacher while seated (woodcut No. 281). The front edge of

282.—MISTRESS'S WORK-TABLE WITH FOLDING TOP FOR SEWING CLASSES.

the top should be marked with inches. Some mistresses prefer for use as work-tables a desk with folding cover, under which needlework and sewing materials may be stored without disarrangement (woodcut No. 282). In this, as in all other kinds of mistress's desk, there should be at least one drawer fitted with small partitions or compartments.

283.—PUPIL-TEACHER'S DESK.

In the pupil-teacher's or assistant-master's desk, placed in front of each class, it will be well to devote some space in which books, slates, pens, and ink may be kept under the charge of one of the assistants. Individual responsibility is thus fixed. The desk here represented (woodcut No. 283) is fitted with perforated trays, so that after a writing lesson, the ink-wells may be collected and placed beyond the reach of dust or accident. A locker is provided in which the class register and the teachers' books may be kept.

In some establishments, as in the model school at Belfast, a "press" is inserted in the wall opposite to each class. The pupil-teacher's desk, arranged as above described, is considered to be preferable.

High stools or chairs should be provided for all teachers, and it should be part of the head-teacher's duty to see that these are used. An experienced inspector of schools, whose attention has been specially directed to this point, states that he has known many instances of deplorable results arising from young people, particularly girls, being allowed to remain for many hours upon their feet, day after day, while engaged in teaching.

If the use of cupboards be preferred to a full supply of teachers' desks, as already described, there should be in Boys' schools one of adequate size to every 120 children, and in Infants' schools about one to every 150 children. For a Boys' school, they should be four to five feet wide, seven feet high, and twelve inches in clear depth. For girls they should be made double, with the lower part projecting. Their position should be

284.—CUPBOARD FOR BOYS' SCHOOL.

somewhere opposite the desks of the children for whose use they are designed. (Woodcuts Nos. 284, 285, 286, and 287.)

285.—CUPBOARD FOR BOYS' SCHOOL.   286.—MISTRESS'S CUPBOARD.

Where cupboards are only used for the storage of extra stock, or saleable stationery, they should be placed either near the head-teacher's desk, or in some position easily accessible without distraction or disturbance to the children during lessons.

In schools where no mistress's room is provided, there should be for the mistress a special cupboard, about three feet wide, with a single door about 2 ft. 4 in. This should be placed, if possible, in a recess fitted with open shelves above. When there is a mistress's room a few brass pegs and a housekeeper's dwarf cupboard would alone be necessary. In other cases, of course, the above remarks as to cupboards for mistresses apply also to those for masters.

287.—CUPBOARD FOR GIRLS' SCHOOL.

The fender used in an Elementary school should comprise arrangements for protecting the children from falling into the fire, and for preventing hot cinders from rolling on to the wooden floor.

It should therefore be designed as an open guard, so as not to obstruct the heat, and should have a plate of metal a few inches high round its lower part to serve as fender. It should be hooked at its highest point firmly to the wall on each side to prevent its being pulled over by a child. (See woodcut No. 288.)

The only fire-irons necessary in a school are the poker, the shovel, and the cinder-sifter.

The coal-scuttle should be strongly made, fitted with a handle at the top, and lined with zinc. It should be about 18in. high at the highest part, and one should be supplied to every fireplace.

In Girls' schools it will be well to have one class-room fitted with a common cooking stove, at which lessons in domestic economy may be practically illustrated, as occasion may require. It need be no more costly than an ordinary register stove; but even in the case of higher schools it will be found of great service, if only teachers can be found willing and able to devote themselves with sufficient zeal and enthusiasm to this important branch of woman's education. It will also be found useful for those teachers and children who may live at some distance, and are therefore obliged to take meals at the school.

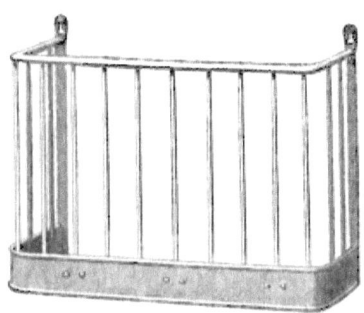

288.—SCHOOL FIRE-GUARD.

If sufficient space be not afforded in the pupil-teacher's desks for the reception of all the inkwells used by the class, loose trays should be provided, and care taken that none of the wells are left in the desks after the writing-lessons for the day are over. Without this precaution the ink will be spoiled with dust, and items of waste, apparently trifling, will entail considerable expense in the course of the year.

Porcelain inkwells are generally most approved, and are produced in a variety of shapes at small cost.

Attempts have been made to dispense with the daily removal of inkwells from the desk and *vice versâ*, by the introduction of sliding or hinged covers. No contrivance in this direction has, as yet, been found to act so well or to be so free from disorder as to warrant the expense and trouble their use occasions in Elementary schools.

Blinds should be neither white nor too opaque. Perhaps the best material is the blue linen of which the workmen's blouses in France are made. Many of the Board schools in London are being fitted with this material. It possesses the advantage that

c c 2

"it will wash," and is also of a beautiful colour. The difficulty of obtaining it in sufficient width for blinds without the introduction of vertical seams is, however, as great as that of inducing English makers to produce good colours. School blinds, if required at all, should be of sufficient width to prevent the admission of sunlight between its edge and the window jamb.

A board on which may be placed all notices relating to school matters is necessary, and should be fixed near the door in each school-room. It should contain space on which may be chalked the number of scholars in attendance, morning and afternoon, and also the average for the preceding week. In some cases it may be deemed advisable to exhibit on the notice board the names of all late comers and absentees, with the view of checking irregularity.

Each school-room should be furnished with a glazed frame for holding the time-table, having its back made removable by means of hasps, or other contrivance, so that the sheet may be changed as required from time to time.

Schools should be well supplied with dusters, towels for the lavatories, and other little requisites, the absence of which, however seemingly unimportant, may occasion much more annoyance and discomfort than might be imagined without experience. The children should be encouraged to provide themselves with penwipers. The separate lockers provided in "single" American school-desks are fitted with india-rubber to prevent noise from the shutting of the lid. To prevent scratching, it is a common practice to bind slates with either india-rubber or lint.

The children's lavatories should be fitted with rollers of hard wood or brackets. The rollers should be two feet in length, and conveniently placed at the side for their use. The material of the towels should be provided by the managers, and the necessary seaming, &c., executed by the children in the Girls' school.

The pegs for caps, bonnets, cloaks, &c., should be made of wrought iron, never of cast. The kind shewn in the woodcut

CHAP. XVIII.] SCHOOL FURNITURE AND APPARATUS. 359

(No. 289) is recommended as most adapted to withstand the
daily wear and tear of school use.

Mats should be of very thick cocoa-nut fibre.
Those manufactured at the Institutions for the
Blind are usually found to be very good and

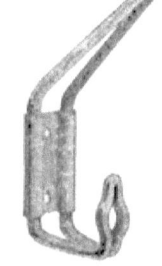

289.—HOOK FOR
CAPS, BONNETS,
&c.

should be properly sup-
...re may

...om on
central
... of the
...eping,
should,
...ith the
...ill be liable to result in

...here should be a thermo-
...in regulating the temper-
...ly be advisable to have
although this is advisable
one hand and insufficient
...orms but a small criterion
of the efficiency of ventilation.

The fittings of teachers' rooms should comprise an umbrella
stand, a table, a few chairs, linoleum or carpet to the floors, and
a portable lavatory of iron or tin, which can be supplied at a cost
of about 30s.

Easels should be supplied for each class, and proper arrangements should be made for their safe custody when not in use.
Perhaps a strong bracket fixed in the wall opposite the group of
desks will be found the most handy method. The most strongly-
made easels are liable to breakage if not handled and stored with
due care. Their material should be some kind of hard wood, such
as birch, beech, oak, or ash. For displaying maps and large

diagrams it is desirable to provide framed easels with T slides in centre, capable of being adjusted to any height. In the Royal Gymnasium at Chemnitz, Saxony, an attempt has been made to provide easels possessing a certain amount of artistic character. The example given (woodcuts Nos. 291, 292) is worthy of emulation, both in regard to easels and other minor matters. In Austria easels made of wrought iron and fitted with sliding tubes and other complicated arrangements have been introduced. As a rule, the simpler the construction the more serviceable the articles will prove in use. The junior

290.—FRAMED EASEL WITH T SLIDE.

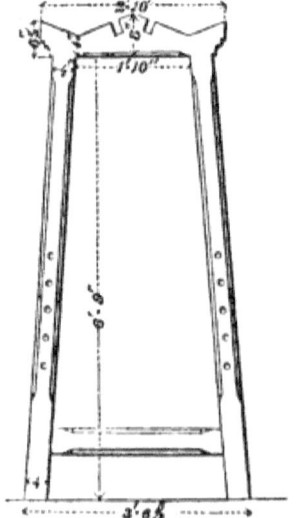
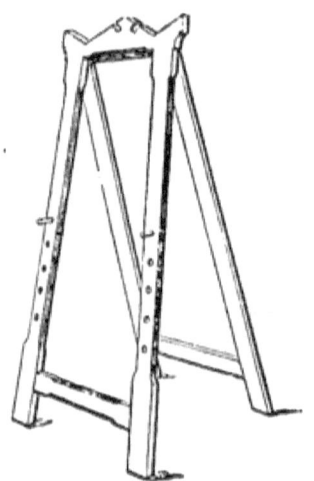

291, 292.—EASELS FROM THE ROYAL GYMNASIUM, CHEMNITZ.

classes of Primary schools and all classes of Infant schools should be provided with lesson-stands. For lessons in form, size, and colour, handy little stands, with small trays attached on which to place small models, &c., should be supplied.

When giving object-lessons, the teacher will find a small portable table of service. It should be strongly made, but not unnecessarily heavy, and contrived with reference to being readily stowed away in a small compass. It is always inconvenient to occupy the floor space with much or unnecessary furniture. Woodcut No. 293 represents a table of this kind used in some American schools, which, with some modification, might be with advantage introduced into English Elementary schools.

293.—PORTABLE TABLE, OR "OBJECT" LESSON STAND.

Blackboards or large framed slates are better made movable than fixed on walls. They may be procured from any school furniture manufacturer, either arranged to swing in portable frames, or loose for use with easels. Blackboards should be strongly framed or "iron-tongued" to prevent warping, but should not be too thick and heavy. If large framed blackboards be used, they should either be made adjustable to any height, or swing on pivots, in order to give the teacher facilities for adjustment to a slope for writing.

294.—BLACKBOARDS.

Bad writing even on a blackboard, must be carefully guarded against. One or two of the blackboards in the senior department should be ruled with white lines, so as to be useful for music lessons. The swing boards or slates of the form illustrated (woodcut No. 295) will be found convenient. In France, where the metric system is universal, the blackboards commonly used in Infant schools are made of the size of one metre each way, the object of the adoption of this uniform dimension being to accustom the eye of the child to judge of the common system of measure.

295.—BLACKBOARD SWUNG ON PIVOTS.

Great latitude is allowed in England, even to managers of schools receiving Government aid, as to the provision of educational appliances. Beyond seeing that there is adequate desk accommodation and the provision of a couple of sets of reading books and a few minor matters, Her Majesty's Inspectors do not interfere. The result in cases where funds were raised with difficulty has often been that the supply of apparatus has proved sadly deficient both in quality and quantity. Under the new order of things no such conditions need be apprehended. Nothing short of thorough efficiency in regard both to furniture and apparatus should be accepted as sufficient or satisfactory.

In Germany the state authorities exercise a much more rigid supervision over the arrangements and appointments of the schoolroom. By a decree, promulgated in the state of Wurtemberg during the year 1873, even the means of punishment were prescribed, and thenceforth teachers were directed to administer flogging only with "a thin switch of the length of half a metre." Equally definite are the instructions with regard to the gymnastic apparatus and other matters, especially those affecting the cleanliness and healthfulness of the scholars, a point in the past too much neglected in German schools.

It is not our purpose to detail all the requisites for every kind of well-appointed schools, but simply to select for remark a few of the more prominent items. A school furnisher's catalogue will probably supply all the additional information required by those of experience in the work of education.

A set of bold clearly lettered maps is indispensable. If a globe be used for illustrating the shape of the earth, a map of the world on Mercator's (or Gall's) projection should be included in the set of maps. It is desirable for the higher standards that there should be a good map of the county in which the school is located. Blackboard outline maps, with movable indicators for enabling the scholars to mark the names of towns, &c., when called upon in succession by the teacher, are used in Swedish and other continental schools. Though interesting as a means of combining amusement with instruction, they are by no means indispensable. Ordinary blank outline maps should, however, always be provided.

Relief maps or charts are extensively used in the schools of Prussia, Saxony, Austria, and other countries of Continental Europe. They are found to be of great service in demonstrating various points in Physical Geography. There are no English maps of this kind, published as yet, which seem so admirable as those produced in Germany. Undoubtedly, if well done, they would be a great boon to the teacher. Of course only a small tract of country should be portrayed, and that on a large scale, so that the mountains and valleys, lakes and rivers, may be clearly brought out.

296.—MAP HOOK.

There should be a strongly mounted terrestrial globe, with a compass showing the cardinal points, also a chart illustrative of Geographical terms. For advanced classes a celestial globe, and apparatus for illustrating the motions of the heavenly bodies, &c.,

may with advantage be added. These can now be had at a very reasonable price.

Without seeking to rival in English Elementary schools, the lavish liberality exhibited in many continental examples, an attempt should be made, where funds permit, to provide collections of specimens and such apparatus as may be attainable at a moderate outlay, for use in various departments of scientific study, especially in those which may have an important bearing upon the trades of the district in which the school is established. For these collections, neat cases, fitted with glazed fronts, should be provided, and all specimens placed therein carefully classified and labelled.

A small cabinet of objects or "products" can be obtained at a

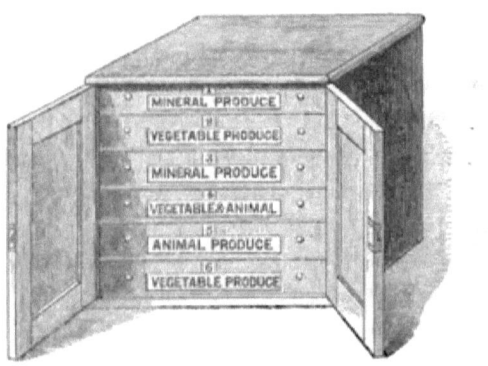

297.—CABINET OF OBJECTS.

very trifling outlay in the form shown in woodcut No. 297, and properly stocked for ordinary object lessons.

Diagrams displayed upon the walls of the school-room serve a most useful purpose. The scholar will intuitively gain much valuable information on a variety of subjects in which, by this simple means, he may be interested. Diagrams are also of great utility in direct teaching. Those displayed on the walls should be periodically changed. Some difficulty may perhaps be experienced in choosing a moderate stock from among the immense number of specimens published, and which are by no means

CHAP. XVIII.] SCHOOL FURNITURE AND APPARATUS. 395

uniformly good. The selection should include a few for illustrating the various processes of manufacture, details of machinery, diagrams illustrative of the great principles of science, such as the Laws of Matter and Motion, Pneumatics, Hydrostatics, Hydraulics, &c., &c., also Physiological, Botanical, and Ethnological representations. No school should be without a good set

res.

ers are best illustrated by a set of models lent No. 298). For cheapness, however,

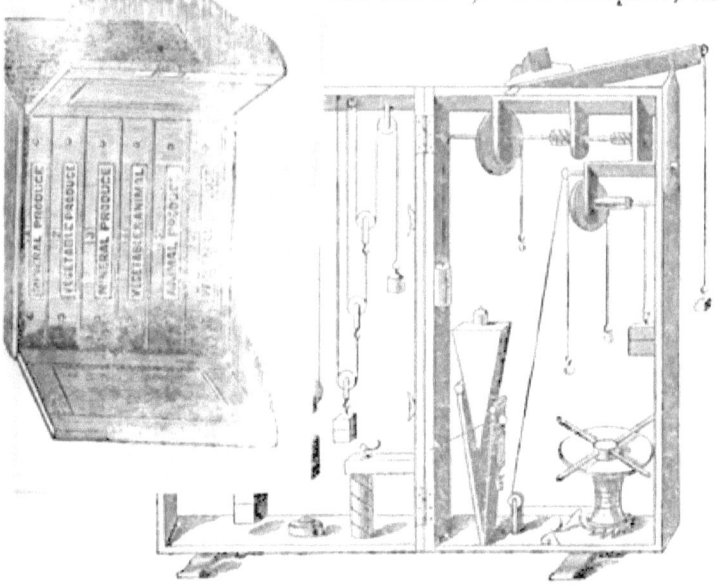

298.—MODELS ILLUSTRATIVE OF MECHANICAL POWERS.

some will prefer a diagram, which will answer the purpose tolerably well.

Where model drawing is taught, a large-sized set of wooden and terra cotta models—and, if possible, a few other examples as approved by the Department of Science and Art—should be obtained. Towards the cost, grants are made to public Elementary schools, or properly organised art classes, under conditions concerning which the Secretary of the Department will supply all needful information on application by letter addressed to South Kensington. Without some provision of this kind no school can be considered well furnished. It is difficult, however, to draw the line so as to decide at what point the multiplication of models and other appliances of this nature should be checked in particular kinds of schools in the interests of economy. The Gewerbeschulen of Prussia, and other Trade schools of the Continent, possess not only fine collections of examples for drawing from the cast, but those for modelling in clay. In some cases the models include machinery and other appliances not essential in an ordinary school.

A set of full-size weights and measures will be found of great service in assisting the formation of correct notions on size and proportion. Woodcut No. 299 is an example of a cabinet furnished for this purpose, and used in French schools. There must be a chart of the metric system displayed in every public Elementary school, and a set of the new weights and measures would be a valuable addition to the stock.

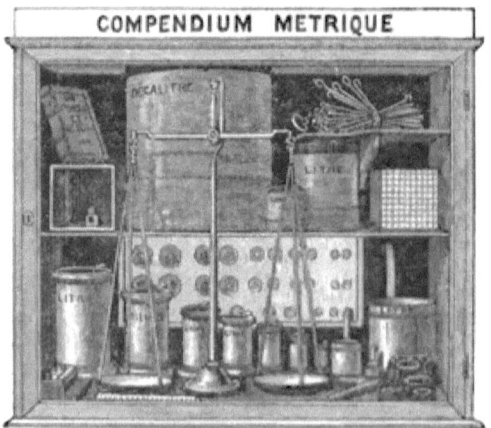

299.—CABINET FOR WEIGHTS AND MEASURES.

No one can examine an exhibition of school furniture and apparatus on the Continent without being struck with the great amount of care, ingenuity, and taste expended on comparatively small matters. Attempted changes have not always led to improvements, but the productions of so many fertile brains are sufficient to show the importance attached to the attainment of completeness in all the appointments of the school-room. An example of this is furnished in the immense variety of arithmetical or counting frames, and alphabet or spelling boxes, which have been produced; some of which, although very interesting as specimens of ingenuity and manipulative skill, are perhaps unnecessary for the ordinary school-room.

Every Infant school should have a good abacus of the ordinary kind, or a simple counting frame, like those used in some of the lower schools of Germany, which have coloured circles on strips of card placed in a case behind movable slides. Simple apparatus of this kind will be found useful occasionally in the lower classes of the Elementary school, as will also be good alphabet boxes or reading stands with movable letters.

For the very small children, a strong single wire with a few coloured balls should be provided.

No Infant school should be without a good-sized alphabet box, with loose letters to be arranged in grooves by the scholars in turn as called upon by the teacher. Children are naturally much interested in doing anything of a mechanical nature of themselves, and the above is perhaps one of the best methods for enabling children to acquire quickly a knowledge of the alphabet, and the use of letters in the formation of words. The alphabet boxes may be procured mounted on stands with a blackboard behind. In Germany a more favourite plan is to place the blackboard under the grooves for letters, so that the teacher may chalk in bold characters the words to be formed.

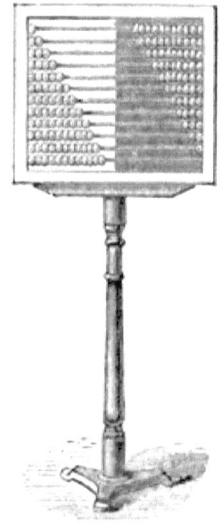

300.—ABACUS ON STAND.

Much more elaborate is the French *compendium*, a piece of furniture uniting in itself all the particulars of abacus, *syllabaire*, and blackboard. A good deal of ingenuity is displayed in the mechanism of this rather complicated article, to the disadvantage, it is feared, of

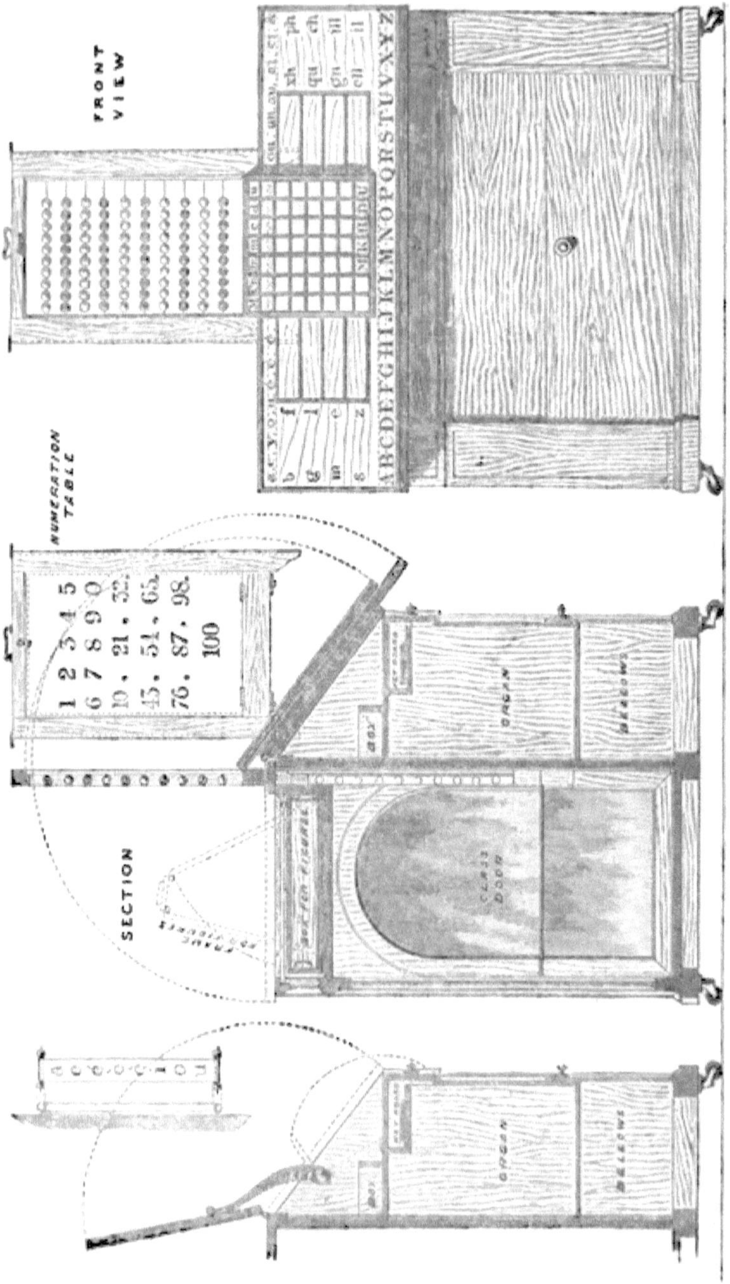

201.—FRENCH "COMPENDIUM."

practical utility for ordinary purposes. At the school connected with the Ursuline Convent in Paris, the compendium in use comprises not only an abacus, a syllable-box, and a board with place for chalk and sponge, but also an organ! (Woodcut No. 301.)

Those who have made the training of little children their study, will readily appreciate the importance of providing in Infant schools a variety of appliances, by means of which instruction may be imperceptibly imparted by being associated with amusement. Such appliances need not be costly; but in the hands of a skilful teacher may be made to do good service. The infantile mind is easily interested in simple things, and, if judiciously trained, will early acquire a thirst for knowledge; while any attempt to cram it with dry matter-of-fact learning unsuited to its age would be worse than useless. In no country, perhaps, has more attention been given to the early training of infants than in Holland, and nowhere is its value more generally acknowledged, more particularly in the advantages it gives in fitting the child for its after school career. "We try to impart knowledge to the children without letting them know how or when they acquire it," said the teacher of a Dutch Bewaar-schule, and with the aid of a few simple appliances she succeeded admirably, not only in thoroughly interesting her class, but in giving, at the same time, much sound and useful instruction. The children were entertained: their attention was fixed, and they gave ample evidence by their ready answers to skilfully put questions, how fully they entered into the spirit of the lessons. Much—in point of fact nearly everything—depends upon the tact of the teacher; but without proper aid in the shape of apparatus her work will be exceedingly difficult.

The "Kindergarten" system originated by Fræbel supplies an excellent medium for accomplishing good results. The appliances hitherto used are perhaps not all that might be desired for large schools; but as the plan becomes more generally adopted, experience will from time to time suggest further improvement. It has been aptly described by the Rev. M. Mitchell,

M.A. (one of Her Majesty's Inspectors of Schools), as a system which, " though intellectual, is truly infantile; it treats the child as a child; encourages it to think for itself; teaches it by childish toys and methods, gradually to develop, in action or hieroglyphic writing, its own idea, to tell its own story, and to listen to that of others. The grand feature of the system is occupation. The child is taught little: it simply produces for itself. It has toys given it of the simplest sort: straight bits of stick or peas soaked in water. It is shewn how to use them, and becomes an architect and an inventor. Churches, towers, houses, and mechanical adaptations swarm from the newly-acquired power. Again, with cubes of wood the ideas of the child take a more solid form: it learns the weight, number, and size of the articles, adapts them to their places and fits them together; weaves with strips of coloured paper webs of varied beauty and certain significances of forms; pricks out patterns with a needle, and even cuts clay and models it."

The first "Gift"—as the classified collections of Kindergarten Toys are called—consists of half-a-dozen coloured worsted balls, &c., &c.; in the second box are two cubes, a cylinder, and a ball in wood with strings attached; the third contains a cube in wood divided once in every direction, the fourth a cube divided into planes cut lengthways; the fifth is an extension of the third "Gift," containing a cube in wood divided into twenty-one cubes, six half cubes, and twelve quarter cubes; and the sixth is a similar extension of the fourth "Gift." Small sticks are supplied for fastening together in various shapes with peas, also sticks for being laid in different forms and for plaiting, patterns and cards for perforating, coloured strips of paper for plaiting, miniature models and clay or wax for modelling, hardwood pieces for making up outline forms of familiar objects, and other contrivances for interesting and instructing the child.

In many of the Infant schools on the Continent, long low oblong tables, with surface painted in one inch squares, are used for Kindergarten and other exercises: in others, the ordinary school-desks with flat instead of inclined tops are made to serve the purpose. The latter will generally be found most convenient, especially as, with the system of seating the children in pairs, the teacher will the more readily gain access to every child in the class; whilst the children themselves are comfortably seated, with proper support given to their easily tired backs, besides being more readily moved on the frequent changes so necessary in the working of an Infant school.

Pictures of animals and other illustrations should be exhibited around the school-room. For lessons on common objects, specimens or models are preferable to the most cleverly executed diagrams, and a fair collection suitable for Infant school teaching may be obtained at no great expense.

Of the boxes of appliances for illustrating lessons in form and colour usually supplied, the larger size should be selected: the smaller ones, though a trifle less costly, are of comparatively little use in dealing with a large class.

Models of clock faces with moveable fingers should not be omitted from the list of requisites: some are made with the fingers connected so as to imitate very perfectly the dual motion on an ordinary clock.

During the intervals of play, moderate-sized wooden bricks and a few other simple durable toys may with great advantage be introduced in the recreation ground. Children will find in their use much amusement under the care of the teacher who knows how to keep the little fingers properly employed, and the playground may thus become well nigh as important a place of instruction as the class-room.

Cots are provided in some Continental Infant Schools, so that the very young children, when tired and sleepy, may be put comfortably to rest until school duties are over, and they can be taken charge of by elder sisters or parents. Woodcut No. 302 is an illustration of the hammock used in the babies' rooms of French schools, and is an admirable contrivance in its way. The *crèche* connected with a few English Infants' Schools has similar accommodation; but all the arrangements necessary for very young children do not generally obtain, and School Boards have no legal power to expend money for children under the age of three.

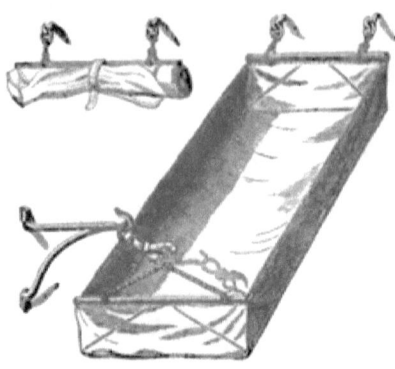

302.—HAMMOCK FOR INFANTS, USED IN THE FRENCH ASILES COMMUNAUX.

While it will be readily understood that many of the articles mentioned in these pages are desirable adjuncts to a well-appointed school, it may also be inferred that there are other matters, important in their bearing upon the advancement of a higher system of education, which are not touched upon. For instance, it may be mentioned, as an indication of the manner in which aids to practical teaching may be extended, that in connection with the Higher Burgher School in Utrecht there is a garden in which are planted by the pupils botanical specimens arranged and classified in their proper order. In the Kingdom of Sweden, at the end of the year 1871, there were 2078 schools having their own ground for planting; and during the year 1868 there where 25,612 children who had received instruction in gardening. Teachers of elementary schools in that country are now obliged to learn gardening before leaving the normal schools. Belonging to many of the higher schools of these and other States may be seen collections of specimens, scientific instruments, and models which would not discredit the public museum of a large town.

What we have attempted has been to give an outline, shewing the amount of furniture and apparatus fairly within the reach of promoters of public schools in England, and to touch also upon some points deserving of special consideration which might otherwise escape attention.

APPENDIX A.
DESCRIPTION OF THE KÖNIG WILHELM GYMNASIUM AT BERLIN.
PRINCIPALLY DERIVED FROM THE OFFICIAL ACCOUNT BY GERSTENBERG.

The general description of this building, terminated at page 94, is now supplemented by the further and more technical description there promised.

The principal front measures 211ft. 3in., and has a central projecting entrance. There are two wings, each connected at the rear with the main building, their depth being 80ft. 2in. That of the main building is 85ft. 6in., and the height from ground-line to top of roof 69ft. 3in. Two courts are formed between the main building and the wings, each 38ft. 5in. long, and 23ft. 4in. wide. These are bounded towards the adjoining property by wooden sheds 13ft. 8in. inside, and 14ft. 4in. high in the centre. Being used only for the storage of agricultural apparatus, they are shut off from the corridors with doors to keep out the boys.

The construction is of thick brick walls, with arched vaults below. In addition to the basement there is a ground floor, and first and second and attic floors. The frontages of main building and wings are stuccoed, the plinth being smooth, and resting on a granite curb 14 inches high. The ground and first floors are jointed in parallel courses, while the second and attic floors have pilasters supporting a rich Corinthian cornice.

The roofs are nearly all on the lean-to principle, inclining towards the courtyards, and are covered with zinc. The rainwater is not utilised, but is conducted through pipes into the drains, and from thence into the sewer in Bellevue-street.

Round the building, level with the pavement, runs a margin of asphalte, 2ft. 7in. wide, and next to that, on the same level, a pavement of mosaic stones, 12ft. 4in. wide. Adjoining is the roadway on a little lower level, 18ft. 6in. wide, paved with stones in small squares.

The building has three entrances: that in centre principal frontage is distinguished by a portico of two columns, surmounted by a cornice of artificial stone, and finished at the level of the first floor with an open balcony; the two secondary entrances are placed at each side of building.

In addition to the rooms devoted to educational purposes, the building contains cellar spaces in basement floor, apartments for the caretaker, an office for the director, a masters' and committee-rooms, and twelve rooms on the attic-floor not at present appropriated to any use.

The class-rooms, with the exception of those in reserve, are placed on the ground and first floors. In the right (*i.e.*, south) wing are the elementary and lower collegiate classes; in the left (*i.e.*, north) wing, the middle and higher collegiate classes. The rooms are so arranged in each wing that the junior classes are nearest to the exits and latrines. On the second floor are, in addition to the reserve-classes, the rooms not in daily use (*i.e.*, the great hall, drawing-class-room, singing-class-room, and library).

The ground floor contains the spacious vestibule, situated in the centre of the building, 40ft. 10in. long, 35ft. 5in. wide, and 16ft. 4in. high, leading directly into a main corridor of the same height, and 9ft. 11in. wide, in a line with which are the two corridors, 14ft. 2in. high, and 9ft. 2in. wide, of the wings. In each wing a stone staircase, 8ft. 8in. wide, leads to the attic-floor. These are also approached from the side entrances by passages 9ft. 10in. wide and 11ft. 4in. high, in direct communication with the latrines.

The caretaker's apartments are situated in the main building at the back of the corridor, and 4ft. 8in. above the general level of the ground-floor. The apartments consist of three rooms and a kitchen on the ground-floor, and two small spaces in the basement. The different class-rooms are all entered directly from the corridors.

In addition to the physical class-room, which is 37 ft. long, and 18ft. 6in. wide, with seats for fifty-six pupils, the north wing contains three collegiate classes, of which the two first are 27ft. long, and 18ft. 6in. and 20ft. 6in. wide respectively, while the third is 29ft. 7in. long, and 18ft. wide. Adjoining the physical class-room is a store-room, 21ft. 6in. long, by 18ft. 6in. wide, for the apparatus connected therewith.

The south wing contains four collegiate classes of similar dimensions to those in the north, also an inspector's room.

Each class-room is lighted by two windows, measuring 5ft. 2in. wide, and in the middle 10ft. 3in. high. The height of the class-room is 14ft. 4in.

The first floor contains in the north and south wings each five class-rooms of similar dimensions to those below, except that one collegiate class, and one elementary class, have a width of 18ft. 6in., and a length of 37ft. Each class-room is entered from a corridor 9ft. 2in. wide in direct communication with the stair-cases. The height of the class-rooms is 14ft. 4in., as in the case of the ground-floor. The main corridor communicates immediately with the director's office in the principal front, and the adjoining masters' room is used as a committee-room. Both these rooms are of the same size. In the triangular corners at the back are lavatories. The ante-space in front leads to the balcony above the principal entrance. The two rooms in the main building at the back of the main corridor are appropriated to the purposes of natural history collections.

The second floor contains in the left wing the drawing-class-room, situated at the front, 18ft. 6in. wide, and 76ft. 2in. long, with 48 seats, with windows facing towards the north wall. Also two reserve classes and an ante-room 18ft. 3in. wide by 43ft. 3in. long, and 19ft. 4in. high. The class-rooms are 14ft. 6in. high, and the drawing-class-room 16ft. In the south wing and facing south are two rooms for the masters' library, and adjoining it the pupils' library, and a reserve class. The singing-class-room also communicates with the corridor. It is 18ft. 3in. by 43ft. 3in. and 19ft. 4in. high, and has seats for 68 pupils.

The singing-class-room, together with the ante-room in the wing, are used as ante-rooms to the great hall. This latter occupies the whole of the central portion of the building on this floor, and measures 77ft. 3in. by 42ft., by 30ft. 9in. high, and contains seats for 300 people. It was intended to place in one angle at the rear an organ, the other being occupied by a staircase, forming the only approach to an attic above the hall. In the recess formed by these inclosures stands the platform, raised six steps above the level of the hall, and in front a podium raised 12 inches above same, forming a stand for singers and pupils at the public examinations, which form the only use of the hall.

The attic-floor consists of twelve rooms and corridors, lighted by skylights in the roof. The rooms average 18ft. 6in. square, and 11ft. 4in. high, that above the drawing-class-room being 10ft. 3in. high. These rooms are not at present in use. At the back portion towards the courtyard, one in each wing, are the rooms, already mentioned, for the solitary confinement of pupils who have transgressed the rules of the College. Those who see the mention of these rooms with horror may be comforted by knowing that they are not without windows. The remainder of the space on this floor is principally used for the cisterns for the water-heating. The wings are separated by the Great Hall in the centre. Above the great Hall in the roof-space are the ventilation cylinders for the sun-burners, and a clock and chimes.

The cellar space in basement averages 9ft. 3in. high, and contains, under the physical apparatus-rooms, a laboratory which may be approached direct from same by a circular staircase.

In each wing are boilers for the water-heating. For the air-heating there are four furnace-rooms in the middle of the building, behind which is a passage connecting the two courts by a staircase, leading to the caretaker's apartments. Under the latrines are spaces for the receptacles of soil therefrom. The remainder of the spaces are used for the reception of fuel, for which there are two shoots provided in the principal front. The heights of the stories are, from floor to floor :—Basement, 10ft. 9in. Ground-floor, in vestibule, 18ft.; in class-rooms, 15ft. 8in. First-floor, 15ft. 8in. Second-floor, in class-rooms and libraries, 15ft. 9in., in drawing-class-room, 16ft. 11in., and in the anterooms to the Great Hall, 20ft. 6in.

The construction now claims some attention. When building,

a good bottom was found at a depth of 5ft. 6in. below the surface. This was further excavated 6in. for foundations. The paving of cellars is at least 1ft. 4in. above the highest point of any water in the ground. The walls are protected from the rising damp by horizontal courses and vertical isolated walls of clinkers set in cement. The horizontal course is formed by three layers of clinkers over the whole surface of main walls at the level of pavement of cellars. The vertical isolated walls, running round the whole of the main walls, stand on the horizontal course, and reach to the level of ground. Their thickness is 5in., and they are separated by a space of 2in. from the main walls. At the level of the ground the vertical isolated wall, and the 2in. space, is covered by a course of bricks laid lengthwise, and this again is covered with a layer of asphalte, reaching 1in. beneath the granite curb.

The ceilings of the vestibule, passages, staircases, and corridors are strongly arched on iron bearing-bars; the corridors on attic-floor are ceiled in plaster in imitation of arches. The beams forming the ceiling of the great hall are carried by trusses 13ft. 11in. apart, and the ceiling is divided into panels. The other rooms have flat ceilings between the beams and walls. In the classr-ooms the beams run parallel to the frontage, and are placed 3ft. 4in. from centre to centre, and average 18ft. 6in. long 11in. by 9in. The bearings are strengthened by iron bars running crosswise to beams—one in rooms 18ft. long, and two in rooms 20ft. long. The beams forming the floor of great hall are carried above the director's and committee-room by an iron framework 16in. high, which rests on the front wall and wall of corridor and partition-wall between the director's and committee-rooms. The beams above the natural history collection-rooms rests on a girder, supported by two cast-iron columns in the partition-wall between the rooms. The staircases have a tread of 13in., and a rise of $6\frac{1}{2}$in. to each step. The cellar-floor is of brick, and that of the passages, corridor, and vestibule is of brick, covered with a layer of asphalte $\frac{1}{2}$in. thick. All the other rooms have deal floors of 10in. boards $1\frac{1}{4}$in. thick. The roofs are all covered with zinc. The window-sills of the class-rooms are 2ft. 9in. to 3ft. 1in. high from the floor, and in the corridors towards the courts 4ft. 10in. The inner entrance-doors in the passages, and the glass doors which shut off the corridors from the staircases and vestibule, are made to swing by means of cords and weights. All the rooms used by the pupils are panelled up to a certain height with boards 1in. thick and 8in. wide, in moulded frames 1¼in. thick.

The latrines are shown on the basement and ground-floor plans. The water-supply thereto is from two reservoirs placed immediately above. They contain 100 cubic feet of water each, and are filled from the mains, a water-gauge acting the part of a ball-tap, and keeping the water always at one level. Lead supply-pipes ¾in. diam. convey the water from the reservoirs to the funnels or pans under the seat, every two funnels terminating in a cast-iron vent-pipe, with simple stench-trap, from which a pipe, carried through the cellar arches, discharges into a tub placed beneath. In each tub is an upright 5in. brass pipe, with slit-like openings, through which the soil, after dissolution in the tub, reaches the ordinary 8in. cast-iron soil-pipe in the cellar, then the earthenware pipe-drain in the court, and finally the main sewer. For the purpose of removing and cleaning the tubs (an operation rarely necessary) the discharge-pipes are connected with the tubs by a short length of indiarubber tube fastened above to the pipe by twisted brass wire, and beneath to an upright socket fixed to the lid of the tub. In each corner of the chamber beneath the latrines is a cast-iron box with stench-trap, receiving the discharge from the urinals and the surface water from the courts, and connected with the iron soil-pipe passing through the cellar. The urinals are flushed direct from the mains.

In order to protect the reservoirs from the severe frosts, a brass gas-stove, placed in the chamber beneath the latrines, is connected with them.

The account of these sanitary arrangements, taken carefully from an authorised description, makes clear this part of the subject, but it is difficult to understand why, with a good water-supply, such a complicated system became necessary. The constant flushing of the closets and urinals, as originally intended, was found to cause an undue expenditure of water, and the process was accordingly limited to three or four times a day, lasting from one and a half to two hours altogether.

In considering the proportions of class-rooms, it was thought that by adopting lengths of 27ft. and 29ft. 6in. respectively, and admitting a sufficient amount of light, the black-board would be visible to the pupils as easily as in shorter rooms. Experience, since the completion of the building, has shown that this is so to a certain extent. The two end seats in each row, those furthest from the windows, have not quite so good a light as the others because of reflected light falling on the black-board, which is a fixture. The use of the deeper grouping of class, as in the case of Prima, appears here to have led the teachers to prefer the wider and shallower shape. But the deeper plan 18ft. 6in. to 20ft. 6in. has the advantage of more economical construction.

The Desks have been graduated to suit the ages of the pupils and the progress of their studies. Experience of their suitability shews them to have completely answered their purpose, except that the projections $1\frac{1}{2}$in. above the level of the top of the desk (which were executed for the purpose of raising the back of the seat and protecting the pupils in front from being splashed with ink by those behind when writing), have the disadvantage that the pupils are obstructed in moving their books. In the back rows the pupils can also conceal articles from the teachers behind these projections. In both the grammar and elementary classes the desks are placed 1ft. 6in. from the window-walls, and 3ft. 1in. from the door-walls, and at least 2ft. from the master's desk, which is 4ft. wide. Opposite the master's desk is a central gangway 1ft. 6in. wide. In the elementary classes the master can approach each pupil by means of a longitudinal gangway 1ft. wide behind each second row of desks. The teachers desire a similar gangway in "Sexta" and "Quinta," on account of the writing-lesson. Further, there is a desire for a gangway 1ft. wide between the last seat and the wall in all the classes, for the purpose of controlling the pupils. The desks are screwed to the floor with iron knees $1\frac{1}{2}$in. long. The seat is supported by a rail underneath. The teacher's seat has a fixed desk. The blackboard consists of two slates, each $\frac{3}{4}$in. thick, fixed in a wooden frame with a close joint. In the window-wall in each class-room is a cupboard for books, &c. The caps and cloaks are hung in the class-room on hooks screwed to panels on the door-wall and black-board-wall at a distance of 6 to 7in. apart, and numbered, so that each pupil has his own hook.

In the physical class-room the desks are similar to those in "Secunda," and each two rows are elevated 1ft. above those immediately in front. The isolated table for apparatus has a slate top, and is lighted by two gas-burners.

Behind the apparatus table is an evaporation niche closed by a sliding shutter; before this shutter hangs a wooden black-board, which, balanced by a counterweight, can be pushed up vertically. In the evaporation niche, lit by two gas-burners, is, at the height of table, a slate slab, on which are made the experiments by evaporation. A vapour pipe, warmed by a gas-jet with a "steatite" burner, leads the gases to the open air.

In order to exclude the light for optical experiments, and to protect the instruments, the windows of the physical-class and apparatus-room are fitted with blinds. In the middle of the apparatus-room stands a table for apparatus, and at the window a desk for the master. Against the walls are cupboards with glass doors.

The desks in the drawing-class-room are 12ft. 4in. long, 2ft. wide, and 2ft. 6in. high, for four pupils each, and stand 4ft. 7in. from back to back. Instead of fixed seats the pupils are provided with four-legged stools 18in. high, with seats 15in. by 12in., without backs. Eight inches above the top of the desk is an iron rail ½in. diameter, supported by five iron uprights, for the support of the copies, and in front is a groove to receive them.

The great hall is perfectly acoustic. From the most remote seat a speaker is distinctly heard. For musical performances it is also considered good. In the great hall and singing-class-room the seats measure from back to back 2ft. 4in., and have a height of 1ft. 8in. · Gas is laid on to all the rooms. The director's room has four burners, the teachers' room, six; each of the corridors, six; each of the staircases, three; the physical-class-room, six; the "Prima," four; and the drawing-class-room, ten.

In planning the scheme for heating, it was decided to warm the rooms constantly in use by pupils and teachers by hot water, and those only occasionally used—viz., the great hall, ante-room, singing-class-room, the director's and teachers' room, the vestibule, and the rooms for collections, by hot air. The caretaker's apartments are heated by stoves of Dutch tiles. Similar stoves were placed in the twelve rooms on attic floor.

A.—*Warm-water Heating.*—In each wing to the right and left of the main building has been placed a special heating apparatus, and the two boilers in connection therewith are placed centrally under the rooms to be heated. From each of the boilers a 5in. cast-iron pipe conducts the water, 70 to 75 degrees warm, into the expansion cistern, 2ft. diameter, and 2ft. high, placed in the roof, and connected with a cold-water cistern. A water-gauge ensures the water being always at one level.

The main blow-pipes, 3½in. diameter, branching off from each of the two expansion vessels, conduct the warm water by means of wrought iron 1½in. and 2½in. branch pipes (according to the size of the adjoining stoves) into the hot-water stove. The water, after passing through the stoves, is brought into the cast-iron 2ir. and 3in. return pipes (placed under the cellar floor) which lead it, after having been cooled from 30 to 40 degrees, to the lowest part of the boiler.

The two boilers are of a strength of 3° above pressure, and 9ft. 6in. and 10ft. 3in. long respectively, and 3ft. 10in. diameter. The sides are 5-16ths of an inch thick. Each boiler has two firepipes 1ft. 3in. diam. The fire-box contains, for the larger boiler, 190·5 cubic feet, and for the smaller boiler, 180·1 cubic feet.

The heating stoves belonging to this system are in the class-rooms, 7ft. 2in. high, and 2in. average diameter, and consist of a wrought-iron cylinder 8in. thick closed at the top and bottom by iron plates of the same thickness. In each of these cylinders are lengths of 3in. wrought-iron pipe, varying in number from 8 to 14, according to the sizes of the rooms to be heated, and soldered to the bottom. The cold air, passing through these pipes, and becoming warmed in the cylinder, is distributed from the top. The connection of the water-heating stoves with the supply pipes is by soldered tubes with cocks. The lower cock at the discharge pipe, 1½in. diameter, regulates the heating.

The stoves stand on a wrought-iron base, cased with zinc to represent a pedestal. Above they are ornamented with a zinc capping.

The physical-class-room, the drawing-class-room, and the corridors, are heated by so-called grate stoves, which are placed in niches. They consist of 3in. wrought-iron upright pipes in

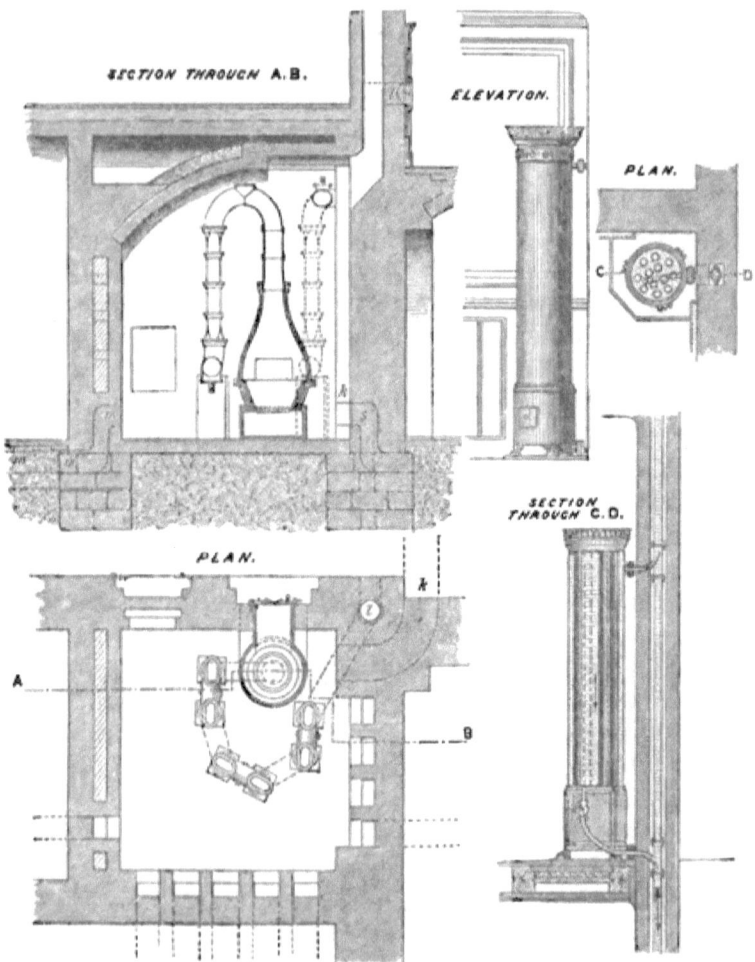

303.—WARMING APPARATUS.

two rows deep, 4½in. centre to centre, and run at top and bottom into a cast-iron box 9in. wide and 3¼in. high. The hot water circulates through these pipes. The in-flowing pipe communicates with the upper box, and the outflowing with the lower box, which is also supplied with a cock for regulating the temperature. Perforated zinc in wood frames are placed in front of the walls before these stoves.

In the class-rooms one cubic foot of heating has been calculated to each 60ft. of air to be heated.

The water-heating is considered to have proved generally satisfactory, except that the drawing-class-room and physical-class-room, both facing north, are difficult to heat in severe weather, although calculated at 1ft. of heating to 50 to 55 cubic feet of space to be heated.

The number of cubic feet which can be heated by this means is 213,168 at a cost of 1,423*l*., making a cost of 6*l*. 18*s*. per 1,030 cubic feet (German 1,000).

B.—*Air-heating.*—Four air-heating apparatuses are arranged in the basement, as shown on plan. The two chambers in front are 7ft. 8in. long and 7ft. 2in. wide; the two in the rear are 7ft. 2in. long and 6ft. 8in. wide, and all are 8ft. 2in. high in the centre. The fresh air inlets are 12in. above the level of pavement of courts. The air is conducted into the heating chambers by channels 12in. square. After being heated by the cast-iron pear-shaped stoves and heating-pipes, it rises through upright pipes, 5in. and 10in. in diameter respectively, in the main cross walls, to the rooms to be heated, where they discharge in vestibule, 3ft. 1in. above floor, and 10ft. 3in. above floor in director's room; by means of pipes of the same section, lying 2ft. above floor in director's room, the cooled air is reconducted into the heating-chamber.

To each heating-pipe there are two pipes for reconveying the cooled air to the heating-chambers. From each of the four chambers a heating-pipe leads to the great hall. The positions of the return-pipes are shown on basement-plan by dotted lines.

The air-heating stove is situated in the front part of the left wing, and conducts the warm air, by means of four pipes, to the rooms to be heated, viz., to the great hall, the ante-room, the director's room, and the vestibule. When not required to be heated, the supply can be shut off by means of a valve at the apertures in each room.

In all the rooms used for school purposes the windows have arrangements for ventilation. The upper lights have each six

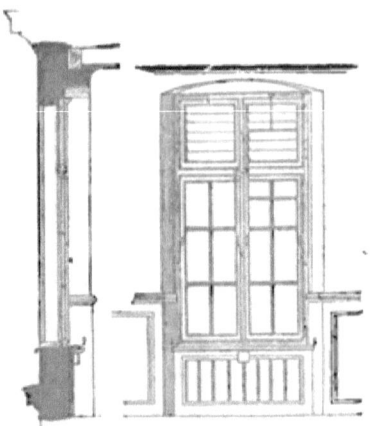

304.—CLASS-ROOM WINDOW.

glass louvres, 5in. wide, the opening and shutting of which is effected by means of a vertical sliding casement in an iron frame fixed inside the window. Above the class-room doors, also, are wood louvres, opened and closed by pulleys in the corridor (woodcut No. 305). The current of air necessary for the ventilation is, in the opinion of our German friends, caused by the difference of temperature between the shadow and sun side and the direction of the wind; and it is regulated by partly shutting the louvres.

Seven sunburners have been arranged for the great hall and the two adjoining rooms, viz., in the hall one with seventy

305.—CLASS-ROOM DOOR.

gas-jets, and four others, each with fifty-one jets, while in each of the adjoining rooms, the sunburners have thirty-four jets. The example given (woodcut No. 306) is one of the smaller kind in the great hall, of fifty-one jets, the burners being described as "Scotch." Connected with the gas-pipe, is an upright pipe terminating 2 feet below the ceiling in two concentric rings at the same level, with thirty-six burners, while at a height of 2 feet 6 inches above these rings is placed a third ring with fifteen burners.

The light from these fifty-one jets is reflected into the great hall by eight glass reflectors fixed in iron frames, and two rows of glass prisms. Above these reflectors is a six-inch iron plate pipe reaching up to the roof.

Partly in order to give an exit to the foul air, and partly with the view of isolating the sunburners, which become very hot, the inner pipe is surrounded by two pipes, 12 inches and 16 inches diameter respectively. The outer pipe is protected from rain by a cover at the top. To repair the reflectors and prisms, the lower part of the pipes can be removed by means of handles. The tap regulating the supply of gas is connected with a valve in the upper pipe in such a manner that the valve is closed when the gas is turned off. In order to protect the beams from fire, the outer cylinder is surrounded by a wall 5 inches thick supported on an iron frame.

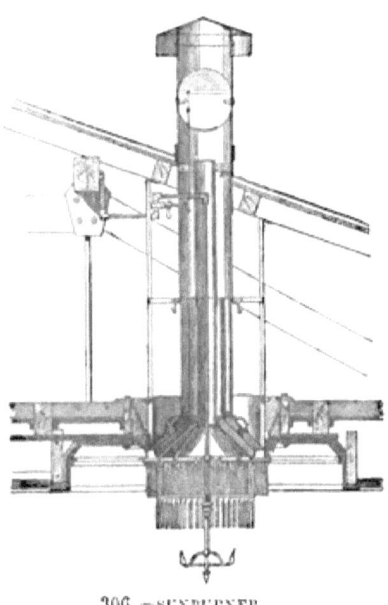

306. — SUNBURNER.

The cost of the building, including fittings (which have cost 7,000 thalers = 1,050*l*.), is as follows:—

		£	s.	d.
1. Earthwork		102	0	0
2. Masonry—A. labour		2,726	14	0
B. material		5,383	7	0
3. Wrought stonework		980	6	0
4. Palisading		111	12	0
5. Carpenter's work		2,702	6	0
6. Slater's work		57	9	0
7. Tinman's work		605	14	0
8. Works in terra-cotta, including £60 for "Borussia" and £153 for six statues		363	15	0
9. Stucco work		186	12	0

		£	s.	d.
10.	Joiner's work	2,293	4	0
11.	Iron bars and braces	517	10	0
12.	Locksmith's work	803	17	0
13.	Glazier	271	16	0
14.	Painter	865	7	0
15.	Hot-water heating apparatus	1,422	18	0
16.	Drainage	479	14	0
17.	Air-heating stoves	88	4	0
18.	Gas-work	230	2	0
19.	Seven sun-burners	154	1	0
20.	Potter's work	32	8	0
21.	Cast zinc-work	54	0	0
22.	Asphalte work	212	5	0
23.	Paperhanger's work and blinds	99	0	0
24.	Gilder's work	27	15	0
25.	Pavior's work	218	8	0
26.	Architect's commission	522	9	0
27.	Miscellaneous	171	9	0
	Total	£21,684	0	0

The portion of site covered by buildings is as follows, viz:—that covered by buildings 69ft. 2in. high is a square 14,077ft. 4in.; and that covered by buildings 14ft. 4in. high is square 1,024ft. 7in. Assuming the cost of the latter at 10s. 6d. per square foot of ground built upon, making 522l. 9s., it will be found that the actual grammar school, occupying a space of 14,077ft. 4in. square, covered by buildings, has a cost of 31s. 10d. per square foot. Comparing the cost of the building with the number of pupils, as the grammar school is capable of accommodating 960 pupils, it works out to 22l. 13s. per head.

In the above schedule of expenditure, amounting to 21,684l. are not included the following expenses:—

	£	s.	d.
A. Apparatus arrangements for summer gymnastic grounds	277	10	0
B. Adjusting the levels of playground and enclosing same with wall	774	18	0
C. Winter gymnastic-hall—estimated cost	2,100	0	0
D. Cost of site	9,000	0	0

In order to estimate the annual or working cost, it may be mentioned that the amount expended in the year 1866 for heating material, was 72l.; for lighting 18l.; and for water supply 27l.

APPENDIX B.

THE RULES OF THE EDUCATION DEPARTMENT.

THE "*Rules to be observed in Planning and Fitting up Schools,*" published by the authority of the Committee of Council on Education, have been frequently referred to in the foregoing pages,— especially at Page 15, Chapter II. Originally drawn up many years ago, they formed the first serious attempt in this country to reduce to something like order and scientific system the subject of school planning, and, on the instructions they contained, most of the national schools erected during the last twenty years (to say the least) have been modelled. No architect can fail to have them constantly in his mind when the plans of elementary schools are under consideration. They yet stand as the only authoritative directions for building, and the only official guide to the School Boards of the whole country. The new importance which educational matters have acquired, render their careful revision, at an early day, highly desirable. At the same time, their extremely valuable and suggestive nature, as well as their authoritative character, render it necessary to furnish a copy in connection with the present work.

PRELIMINARY REMARKS.

Before a school-room is planned,—and the observation applies equally to alterations in the internal fittings of an existing schoolroom,—the number of children who are likely to occupy it; the number of classes into which they ought to be grouped; whether the school should be "mixed," or the boys and girls taught in different rooms; are points that require to be carefully considered and determined, in order that the arrangements of the school may be designed accordingly.

Every class, when in operation, requires a separate teacher, be it only a monitor acting for the hour. Without some such provision it is impossible to keep all the children in a school actively employed at the same time.

The apprenticeship of pupil teachers, therefore, is merely an improved method of meeting what is, under any circumstances, a necessity of the case; and, where such assistants are maintained at the public expense, it becomes of increased importance to furnish them with all the mechanical appliances that have been found by experience to be the best calculated to give effect to their services.

The main end to be attained is the concentration of the attention of the teacher upon his own separate class, and of the class upon its teacher, to the exclusion of distracting sounds and objects, and without obstruction to the head master's power of superintending the whole of the classes and their teachers. This concentration would be affected most completely if each teacher held his class in a separate room; but such an arrangement would be inconsistent with a proper superintendence, and would be open to other objections. The common school-room should, therefore, be planned and fitted to realise, as nearly as may be, the combined advantages of isolation and of superintendence, without destroying its use for such purposes as may require a large apartment. The best shape is an oblong. Groups of benches and desks should be arranged along one of the walls. Each group should be divided from the adjacent group or groups by an alley, in which a light curtain can be drawn forward or back. Each class, when seated in a group of desks, can thus be isolated on its sides from the rest of the school, its teacher standing in front of it, where the vacant floor allows him to place his easel for the suspension of diagrams and the use of the black-board, or to draw out the children occasionally from their desks, and to instruct them standing, for the sake of relief by a change in position. The seats at the desks *and* the vacant floor in front of each group are *both needed*, and should therefore *be allowed for* in calculating the space requisite for *each class*.

The Committee of Council do not recommend that the benches and desks should be immoveably fixed to the floor in any schools. They ought to be so constructed as to admit of being readily removed when necessary, but not so as to be easily pushed out of place by accident, or to be shaken by the movements of the children when seated at them.

By drawing back the curtain between two groups of desks, the principal teacher can combine two classes into one for the purpose of a gallery lesson; or a gallery (doubling the depth of benches, and omitting desks) may be substituted for one of the groups. For simultaneous instruction, such a gallery is better than the combination of two groups by the withdrawal of the intermediate curtain; because the combined length of the two groups (if more than 15 feet) is greater than will allow the teacher to command at a glance all the children sitting in the same line. It is advisable, therefore, always to provide a gallery; but this is best placed in a class-room.

The master of a school should never be allowed to organize it so as to provide for carrying out the entire business of instruction without his own direct intervention in giving the lessons. He ought, as a rule, to have one or more of the classes (to be varied from time to time) in a group or in the gallery, under his own immediate charge. He must, indeed, at times leave himself at liberty to observe the manner in which his assistants or apprentices teach, and to watch the collective working of his school. But his duties will be very ill performed if (what is called) general superintendence forms the sum, or principal part, of them.

The reasons of the following rules will be readily inferred from these preliminary explanations.

RULES.

(*a.*) In planning a school-room, it must be borne in mind that the capacity of the room, and the number of children it can accommodate, depends not merely on its area, but on its area, its shape, and the positions of the doors and fireplaces.

(*b.*) The best width for a school-room intended to accommodate any number of children between 48 and 144 is from 16 to 20 feet. This gives sufficient space for each group of benches and desks to be ranged three rows deep along one wall, for the teachers to stand at a proper distance from their classes, and for the classes to be drawn out, when necessary, in front of the desks around the master or pupil teachers. (*No additional accommodation being gained by greater width in the room, the cost of such an increase in the dimensions is thrown away.*)

(*c.*) A school not receiving infants should generally be divided into at least four classes. (*The varying capacities of children between seven and thirteen years old will be found to require at least thus much subdivision.*)

(*d.*) Benches and desks, graduated according to the ages of the children, should be provided for all the scholars in actual attendance, and therefore a school-room should contain at least four groups.

(*e.*) An allowance of 18 inches on each desk and bench will suffice for the junior classes, but not less than 22 inches for the senior classes; otherwise they may be cramped in writing.

The length therefore of each group should be some multiple of 18 or 22 inches respectively.

Thus, at 18 inches per child,

A group	6ft.	0in.	long will accommodate	4	
,,	7	6	,,	5	
,,	9	0	,,	6	
,,	10	6	,,	7	
,,	12	0	,,	8	Children in a row.

At 22 inches per child,

A group	7ft.	4in.	long will accommodate	4	
,,	9	2	,,	5	
,,	11	0	,,	6	

In the annexed plans 18 inches have been taken as the allowance per child. The withdrawal of a child from each row of this dimension will practically answer the purposes of the other dimensions.

(*f.*) The desks should be either quite flat or *very slightly* inclined. The objections to the inclined desk are, that pencils, pens, &c., are constantly slipping from it, and that it cannot be conveniently used as a table. The objection to the flat desk is, that it has a tendency to make the children stoop. A raised ledge in front of a desk interferes with the arm in writing.

(*g.*) As a general rule no benches and desks should be more than 12 feet long; and no group should contain more than three rows of benches and desks (*because in proportion as the depth is increased, the teacher must raise his voice to a higher pitch; and this becomes exhausting to himself, while at the same time it adds inconveniently to the general noise*).

(*h.*) Each group of desks should be separated from the contiguous group, either by an alley 18 inches wide for the passage of the children, or by a space of three inches sufficient for drawing and withdrawing the curtains.

(*i.*) The curtains when drawn should not project more than 4 inches in front of the foremost desk. An alley should never be placed in the centre of a group or gallery, and the groups should never be broken by the intervention of doors and fireplaces.

(*j.*) Where the number of children to be accommodated is too great for them to be arranged in five, or at most six, groups, an additional school-room should be built, and placed under the charge of an additional teacher, who may, however, be subordinate to the head master.

---

1. The walls of every school-room and class-room, *if ceiled at the level of the wall-plate*, must be at least 12 feet high from the level of the floor to the ceiling; and, if the area contain more than 360 superficial square feet, 13 feet, and, if more than 600, then 14 feet.

2. The walls of every school-room and class-room, *if ceiled to*

the rafters and collar beam, must be at least 11 feet high from the floor to the wall-plate, and at least 14 feet to the ceiling across the collar beam.

3. The whole of the external walls of the school and residence, *if of brick*, must be at least one brick and a half in thickness; and, *if of stone*, at least 20 inches in thickness.

4. The doors and fireplaces in school-rooms for children above seven years of age must be so placed as to allow of the whole of one side of the school-room being left free for the groups of benches and desks.

5. There must be no opening wider than an ordinary doorway between an infant's and any other school-room, as it is necessary to stop the sound of the infant teaching.

6. An infant school should always be on the ground floor, and, if exceeding 80 children in number, should have two galleries of unequal size, and a small group of benches and desks for the occasional use of the elder infants.

No infant gallery should hold more than 80 or 90 infants.

7. The width of a boys' or girls' school-room must not exceed 20 feet.

The width of an infant school-room need not be so restricted.

8. The class-rooms should never be passage-rooms from one part of the building to another, nor from the school-rooms to the playground or yard.

9. The class-rooms should be on the same level as the school-room.

10. The class-rooms should be fitted up with a gallery placed at right angles with the window.

11. Framed wood partitions are not allowed between school-rooms and class-rooms. They must be separated by lath and plaster partition or a wall.

12. Infants should never be taught in the same room with older children, as the noise and the training of the infants disturb and injuriously affect the discipline and instruction of the older children.

13. The windows should be of glass set in wood or iron casements. Lead lights and diamond panes are not allowed.

14. The sills of the windows should be placed not less than 4 feet above the floor

15. A large portion of each window should be made to open.

16. The doors and passages from the school-room to the privies must be separate for the two sexes. So must also be the privies themselves. If they cannot be constructed entirely apart from each other, there should be between them a dust-bin, or other sufficient obstacle to sound as well as sight.

[*Waterclosets can now be provided at a very reduced cost, and they may be introduced with advantage wherever there is a sufficient supply of water to cleanse them thoroughly. Great attention must be paid to the drainage of them. Earthern pipes, measuring 4 or 6 inches in diameter, cemented at the joints, glazed and trapped, are the best for this purpose.*

Earth-closets are also frequently used with success.]

17. The privies must be subdivided, having a door and light to each subdivision.

18. The children must not have to pass in front of the residence on their way to their offices.

19. The *Residence for the Master or Mistress* should contain a parlour, a kitchen, a scullery, and three bedrooms; and the *smallest* dimensions which their Lordships can approve are—

		Ft.	Ft.		
(a) For the parlour		12 by	12	(e)	8ft. } in height to wall-plate. 8ft. }
(b) ,, ,, kitchen		12 ,,	10	of superficial area ;	8ft. if ceiled at wall-plate ; or 7ft. to wall-plate, and 9ft. to ceiling.
(c) ,, one of the bedrooms		12 ,,	10		
(d) ,, two other bedrooms		9 ,,	8	(f)	

20. The residence must be planned so that the staircase should be immediately accessible from an entrance-lobby, and from the parlour, kitchen, and each bedroom, without making a passage of any room.

21. Each bedroom must be on the upper story, and must have a fireplace.

22. The parlour must not open directly into the kitchen or scullery.

23. There must be no internal communication between the residence and the school.

24. There must be a separate and distinct yard, with offices for the residence.

25. The porch must be external to the school-room.

26. Iron or wooden buildings cannot be approved.

27. An infants' school must have a playground attached to it.

28. In the case of a mixed school there must be separate playgrounds for the boys and girls.

29. The playground should be properly levelled, drained, and enclosed.

APPENDIX C.

REGULATIONS OF THE SCHOOL BOARD FOR LONDON,

FOR THE MANAGEMENT OF ITS SCHOOLS.

I.—GENERAL REGULATIONS.

1. Infant Schools shall be mixed.

2. Senior Schools shall be separate.

3. Large schools shall be provided wherever it is practicable to do so.

4. As a general rule, Female Teachers only, shall be employed in Infant and Girls' Schools.

5. The period during which the children are under actual instruction in School shall be five hours daily for five days in the week. (This period may include the marking of Registers.)

6. During the time of religious teaching or religious observance, any children withdrawn from such teaching or observance shall receive separate instruction in secular subjects.

7. Every occurrence of corporal punishment shall be formally recorded in a book kept for the purpose. Pupil teachers are absolutely prohibited from inflicting such punishment. The head teacher shall be held directly responsible for every punishment of the kind.

8. Music and drill shall be taught in every School during part of the time devoted to actual instruction.

9. In all Day Schools provision shall be made for giving

effect to the following Resolution of the Board passed on the 8th March, 1871 :—

"That in the Schools provided by the Board the Bible shall be read, and there shall be given such explanations and such instruction therefrom in the principles of Morality and Religion, as are suited to the capacities of children : provided always—

1. That in such explanations and instruction the provisions of the Act in Sections VII. and XIV. be strictly observed, both in letter and spirit, and that no attempt be made in any such Schools to attach children to any particular Denomination.

2. That in regard of any particular School, the Board shall consider and determine upon any application by Managers, Parents, or Ratepayers of the district, who may show special cause for exception of the School from the operation of this Resolution, in whole or in part."

10. In all Schools provision may be made for giving effect to the following Resolution of the Board passed on July 26th, 1871 :—

"1. That in accordance with the general practice of existing elementary schools, provision may be made for offering prayer and using hymns in schools provided by the Board at the "time or times" when, according to Section VII., Sub-section II., of the Elementary Education Act, "Religious observances" may be "practised."

"2. That the arrangements for such "Religious observances" be left to the discretion of the Teacher and Managers of each School, with the right of appeal to the Board by Teacher, Managers, Parents, or Ratepayers of the District :

Provided always—

That in the offering of any prayers, and in the use of any hymns, the provisions of the Act in Sections VII. and XIV. be strictly observed, both in letter and spirit, and that no attempt be made to attach children to any particular denomination."

11. All the children in any one Infant, Junior, or Senior School, shall pay the same weekly fees.

12. The minimum weekly fee in Infant, Junior, and Senior Schools shall be one penny, and the maximum fee ninepence.

13. The half-timers attending any School shall pay half the weekly fees chargeable in that School, provided that such half-fees be not less than one penny.

14. The fees payable in Evening Schools shall be left to the discretion of the Managers, subject to the approval of the Board.

15. If exceptional circumstances should appear to render the establishment of a Free School, in any locality, expedient, the facts shall be brought before the Board, and its decision taken upon the special case.

II.—REGULATIONS FOR INFANT SCHOOLS.

16. In Infant Schools instruction shall be given in the following subjects:—

- (*a*) The Bible, and the principles of Religion and Morality, in accordance with the terms of the Resolution of the Board passed on the 8th March, 1871.
- (*b*) Reading, writing, and arithmetic.
- (*c*) Object-lessons of a simple character, with some such exercise of the hands and eyes as is given in the "Kinder-Garten" system.
- (*d*) Music and drill.

III.—REGULATIONS FOR JUNIOR AND SENIOR SCHOOLS.

17. In Junior and Senior Schools, certain kinds of instruction shall form an essential part of the teaching of every School; but

others may or may not be added to them, at the discretion of the Managers of individual Schools, or by the special direction of the Board. The instruction in Discretionary Subjects shall not interfere with the efficiency of the teaching of the Essential Subjects.

18. The following Subjects shall be Essential.

(a) The Bible, and the principles of Religion and Morality, in accordance with the terms of the Resolution of the Board, passed on the 8th March, 1871.

(b) Reading, writing, and arithmetic; English grammar and composition, and the principles of book-keeping in Senior Schools; with mensuration in Senior Boys' Schools.

(c) Systematised object-lessons, embracing in the six school years a course of elementary instruction in physical science, and serving as an introduction to the science examinations which are conducted by the Science and Art Department.

(d) The History of England.

(e) Elementary Geography.

(f) Elementary Social Economy.

(g) Elementary Drawing.

(h) Music and Drill.

(i) In Girls' Schools, plain needle-work and cutting-out.

---

19. The following subjects shall be Discretionary :—

(a) Domestic Economy.

(b) Algebra.

(c) Geometry.

20. Subject to the approbation of the Board, any Extra Subjects recognised by the New Code (1871) shall be considered to be Discretionary Subjects.

IV.—REGULATIONS FOR EVENING SCHOOLS.

21. The course of Instruction in Evening Schools shall be of the same general character as that already recommended for the Junior and Senior Day Schools.

22. Evening Schools shall be separate.

23. The Resolution of the Board on Religious Instruction, adopted 8th March, 1871, shall, in regard to Evening Schools, be interpreted as permissive, at the discretion of the Managers; provided, always, that any Religious Instruction, given in such Schools, shall be subject to the said Resolution.

24. Except in so far as the cost of Instruction has been defined by Regulations 21 and 23, the Managers of Evening Schools shall be left free to adapt the Instruction given in the Schools to local requirements.

25. The formation of Science and Art Classes, in connection with the Evening Schools, shall be encouraged and facilitated.

*** On August the 2nd, 1871, the Board passed the following Resolution:—"That, with respect to the methods of instruction, the books and apparatus to be used, the supply and salaries of Teachers, and the arrangements for Half-time Scholars, all the arrangements relating thereto be provisional."

INDEX.

A.

ABACUS, on stand, for infant schools, 397.
Air-heating in the König Wilhelm Gymnasium, Berlin, 413.
Aix-la-Chapelle, Polytechnic School at, 135–139; Chemistry School at, 139–141.
Aldenham-street School, 328–332.
Alphabet, or Infant School, 28.
America, schools in, 27–46; the Alphabet, 28–32; the Middle, the Normal, 32; Grammar, High, Latin, 33.
American double desks with single seats, 368; single desks and seats in primary schools, 372; double desks, with seats attached, 373.
— educationists, enthusiasm of the, 27.
— model primary school, plan of, 29.
— primary schools founded chiefly on those of Ireland, the higher schools based on the German model, 45.
Amsterdam, plan of school at, 25.
Andrew, St., Hildersheim, gymnasium of, 100–104.
Angel, the, standard desk, 369.
Angler's Gardens School, Islington, 317–320.
Apparatus, warming and ventilating, by H. C. Price, 285–287; warming, in König Wilhelm Gymnasium, Berlin, 412.
Architecture, style of, most suitable for schools, 321.

Arnold's, Mr. Matthew, Popular Education of France, 63; Report of the School Inquiry Commission, 1, 119; of the Secondary Schools in Switzerland, 24; remarks on the true culture in Germany, 145.
Aspect, the principal, of school-rooms, very important, 167.
Aula, the, or Examination Hall, 45, 77, 80, 108, 112; in the Imperial Gymnasium at Vienna, 156.
Austria, elementary schools of, 148; six Polytechnic schools in, 157.
Austrian education, theory of, 149, 150.

B.

BARNARD's excellent and well-illustrated work on school plans, 27.
Barnes' Home and Industrial School at Ardwick, near Manchester, 357, 358.
Bars, parallel, for infant schools, 193; horizontal, 194.
Bartley's, Mr., Schools for the people, 11, 17.
Beckar's, Dr. Theodore, programme of the Gymnasium at Darmstadt, 265.
BERLIN, a slight description of the schools in, 76; parish school in the Kurfürstenstrasse, 77; plans of, 78, 79; Polytechnic Institute at, 135; Friedrich-Wilhelm School at, 88; Cölnisches School at, 88; König Wilhelm Gymnasium at, 89-94; Victoria School for girls at, 143; Luisen Schule at, 143.

Bezirk, or district, each one must provide a school-house, 120.
Biven, Mr., his designs for the new building in Old Castle-street, Whitechapel, 294.
Blackboards, or large-framed slates in schools, 391, 392.
Black Hole at Calcutta, the, an extreme example of the want of ventilation, 265.
Blinds for schools, 387.
Board, a, for notices in schools, 388.
BOARD SCHOOLS OF LONDON, 291–350; Old Castle-street School, 293–296; Harwood Road School, 296–300; Johnson-street School, Stepney, 300–304; New North-street School, 305–310; Winstanley Road School, 310–314; Eagle-court School, Clerkenwell, 315–317; Angler's Gardens School, Islington, 317–320; style of architecture suited to, 321–324; Wornington Road School, 324–328; Aldenham-street School, 328–332; Orange-street School, 332–335; West-street School, Hackney, 336–338; Camden-street School, Camden Town, 339–342; Mansfield Place School, 342–346; Haverstock Hill School, 346–350.
Boston, the High and Normal Girls' School at, 37, 40–44.
Brick used as the building material for the new elementary schools in London, 396, 323.
Burdett Road, Tower Hamlets, certified Industrial School, 358.
Bürger, or Burgher Schools in Germany. 75.
Buss, Miss, Lady Principal of the North London Collegiate Schools for Girls, 239.

C.

Cabinet of objects in schools, 394; for weights and measures, 396.
Camden-street School, Camden Town, 339–342.

Capen Primary School, Boston, Massachusetts, 30; plans of, 31.
Carcer, or prison for refractory boys, 98, 117.
Caretaker, the, in schools, 222, 223; the presence of, necessary for the heating of schools, 273.
Chair, master's, 381.
Champneys, Mr., his designs for the Harwood Road School, Fulham, 296.
Cheltenham Training College, the, 13; pupil-teacher system first adopted in, 13.
Chemnitz, Königliches Gymnasium at, 129; the Höhere Webschule at, 133.
Church Education Society, the, 49.
Cistercian Abbey, the, of S. Mary Fforta, 119.
Class-rooms, 162–165; in infant schools, 182, 183; in graded schools, 197–201.
Cleanliness, absence of, in a German school, 103, 104.
Climbing frame, the, 249.
Climbing stand, the, for infant schools, 194.
Cloak-rooms in schools, 213, 214.
Clock, in schools, 389.
Coal-scuttles for schools, 386.
Code, the Revised, of the Education Department, 161, 162.
Cohn, Dr., on Myopia, 178; on school-desks and seats, 362.
Collége Chaptal, Paris, 62–68, 236; plans of, 66, 67.
Cölnisches Gymnasium, the, Berlin, 88.
Cologne Stadtische Realschulen, 111–114.
Committee of Council on Education,
Communal schools in the Eschen ach-Gasse, Vienna, 151.
Compactness of internal arrangement one of the first essentials of school-planning, 165.
Compendium métrique, used in French schools, 396; uniting abacus, syllabary and blackboard, 397–399.
Compulsory attendance at schools, 20,

INDEX.

45, 48, 49; first introduced in Saxony by Elector John George in 1573, 12); everywhere in Holland and Switzerland, 26; in Prussia, 71.
Cooking stove in girls' schools, 387.
Cork National Schools, 52.
Cost of schools, about 11*l*. per head, 350.
Cots provided in Continental infant schools, 401-402.
Cottbus, gymnasium at, 146.
Council on Education, Committee of, rules for planning and fitting up schools, 361.
Crèche, the, or day-nursery in France, 59.
Cupboards, 384; for boys' school, 385; mistress's, 385; for girls' school, 386.

D.

DADO, the, in schools, 231.
Davies, Miss Emily, the Girton College, Cambridge, under her auspices, 239.
Denominational, or Voluntary Schools, 4 n.
DESKS AND SEATS, immense improvement in shape and mechanism of, 16; importance in grouping, 168; the American plan, 170; in Holland, Germany, and Switzerland, 170; in graded schools, 172-174; the dual desk, 174; for infant schools, 189, 190; convertible or reversible, 362; height and breadth, 363; Cologne School, 364, 365; in a Gemeinde Schulhaus, Berlin, 367; double, with seats attached, 367; American double, with single seats, 368; Moss's patent school board, 369; the "angel" standard, 369; alternative design for desk standard, 369; the Sheffield, 370; from the König Wilhelm Gymnasium, and in the Victoria School, Berlin, 371; in Sweden, 371, 372; American, single, as used in primary schools, 372; American, double, with seats attached, 373; graduating desks at class, 374; Home and Colonial School Society's, 375; class-room, showing dual arrangement of, 376; code of drill, 377-379; head-master's, 381; teachers', 82; head-mistress's, 383; pupil-teachers', 384; in the König Wilhelm Gymnasium, Berlin, 409, 410.
Diagrams in schools, 394.
Drawing easels, 379; in the Victoria School, Berlin, 380.
Dresden, Bürgerschule at, 127, 128.
—, Bezirkschule at, 121-124.
—, Gemeindeschule at, 125, 126.
Drill, Code of, 377-379.
Droop, Herr, professor of gymnastics at Emden, 248, 249.
Dual arrangement of desks, 376; desks with flaps, 171-174.
Dublin, model national schools in, 52.
Dusters, towels, &c., for schools, 388.
Dutch double desks with seats attached, 367.

E.

EAGLE COURT SCHOOL, Clerkenwell, 315-317.
Easels for displaying maps and large diagrams in schools, 390.
Education Act, the Elementary, of 1870, 2.
Education Department, Revised Code of the, 161, 162.
Education, German, deficiencies in, 142; the exclusion of infant schools, 142; cannot fail to teach a useful lesson, 144.
EDUCATION, PHYSICAL, 244; an important feature among the ancient Greeks, 244; commencement of, in Germany in 1776, 244; made a part of the national system in 1842, 245; none of the playgrounds of our Board Schools properly furnished for, 247; the climbing frame, 249; gymnastic apparatus for open-air use, 250-253; the spring-board, 253; apparatus for deep jumping, 253; swing trees, 254; parallel bars, 254-257,
Elementary Education, Public, in England and Wales, an Act to provide for, 292.

Elementary instruction, want of, in France, 55.
Elementary schools, the pupil-teacher system the only scientific one, 20; of Germany, 144; theory of English, 159–167; difference between the German and English modes of conducting, 161.
Entrances to schools, 205, 206.
Evening schools, regulations for, 429.
Examinations, annual, in graded schools, 196.

F.

FALK, Dr., on school desks and seats, 362.
Fearon's, Mr., Report of the Schools Inquiry Commission, 1.
Feltham, Surrey, Certified Industrial School, 358.
Fforta, S. Mary, Cistercian Abbey of, 119.
Fireguard for schools, 387.
Flooring in schools, 229; wood-block, 230; plank, laid zig-zag, 230, 231.
Forster, the Right Hon. W. E., the Elementary Education Act, passed by his personal exertions, 2.
Foundation grammar schools, 232, 235.
FRANCE, well supplied with establishments for higher and classical education before 1789, 54; the wars of Napoleon not favourable to the cause of education, 54; a law passed in 1833 compelling the establishment of schools in every commune, 54; commission on technical instruction in, 55; want of elementary instruction, 55; mixed schools, 57; Salle d'Asile, 59.
Franké, Augustus Hermann, the old foundations at Halle commenced by, 117.
Frederick the Great gave his country the boon of education, 71.
Frey, Dr., on school desks and seats, 362, 367.
Friederich Wilhelm Gymnasium, the, at Berlin, 88.
Fræbel, originator of the Kindergarten system, 399.

FURNITURE, SCHOOL, and apparatus, 359–402; particular combinations of form exercise an important influence on the minds of the young and ignorant, 360; desks, 361–384; Dr. Wiese gives a synopsis of various writings on the best methods of uniting the seat with the desk, 362; double class-room, showing dual arrangement of desks, 377; drawing easel in Victoria School, Berlin, 380; head master's desk, 381; chair, 381; head mistress's desk, 383; mistress's work-table, 383; pupil-teacher's desk, 384; cupboard for boys' school, 385; mistress's cupboard, 385; cupboard for girls' school, 386; fireguard, 386, 387; coal-scuttle, 386; inkwells, 387; blinds, 387; board for notices, 388; dusters, rollers, towels, &c., 388; pegs for caps, bonnets, cloaks, 388, 389; mats, clocks, 389; thermometer, 389; easels, 389, 390; portable table, 391; blackboards, 391, 392; maps, globes, 393; cabinet of objects, 394; models, illustrative of mechanical powers, 395; cabinet of weights and measures, 396; Abacus on stand, 397; French compendium, 398; Kindergarten system, 399, 400.

G.

GALLERY, a large, the chief feature in a Wesleyan school, 13; for infant-schools, 185; of maximum size, 186, 187; plan of French, 188.
Gallery lesson, a, 359.
Gauger, M., on warming with open fires, 274.
GERMANY, principles of school system in, 69; useful hints for school-houses, 71; uniformity of teaching, 71; no private schools commenced without a licence from the authorities, 73; elementary schools of, include the Gemeindeschule, the Volkschule, the Bezirk, and the Vorschule, 74; in the secondary or higher schools we find the Burger,

INDEX.

the Realschulen, the Höhere Bürgerschule, and the Gewerbeschule, 75; secondary, trade, and technical schools in, present the finest models in the world, 145; attention to warming and ventilating in, 283.
Gerstenberg, Herr, 78, 81, 94, 107; official account of the König Wilhelm Gymnasium, Berlin, 402-416.
Gewerbeschulen, trade and practical schools in Germany, 133.
Girls' schools, 237; public boarding rare, 238; Miss Wolstenholme on the education of girls, 239; North London Collegiate, 239; the college formerly at Hitchin, now at Cambridge, under the name of the Girton College, 239; Milton Mount College, Gravesend, 240.
Girton College, Cambridge, formerly at Hitchin, 239.
Glasgow Normal Seminary, 13.
Glass, cooling power of, 267.
Government Statistics, Extract from, 163.
Graded schools, 163-165; desks for, 172-174; elementary, 195; have six standards of examination, 195.
Grammar schools, in America, 33; foundation, in England, 232, 235.
Grates, or open fires, 272, 273, 274; the Boyd, 276-278; the Galton, 276; the Manchester, 277; the Pierce, 278; the Longden, 279.
Gravel, the worst surface for playgrounds, 218.
Gross Münster School, the, at Zurich, 23.
Gymnasium, or the German grammar-school, 87; the most interesting is Friedrich-Wilhelm, Berlin, 88; König Wilhelm, Berlin, description of, 402-416.
Gymnasiums: at Liegnitz, 95-97; at Marburg, 97-100; of St. Andrew, Hildesheim, 100-104; the Sophien, Berlin, 107; Königliches at Chemnitz, 129; at Cottbus, 146.
Gymnastic apparatus for open air, 250-257; hall at Hof, 258-262.

Gymnastics first introduced in Germany, 244.

H.

Hackney, public recreation ground in school division of, 335.
Hague, the, plan of school at, 25.
Halberstadt, Realschule at, 109-111.
Halle, old foundations at, 117.
Hammock for infants, used in the French Asiles Communaux, 402.
Handel's Academy, Vienna, 151.
Harwood Road School, 296-300.
Hasenheide, near Berlin, a hall for gymnastics established by Jahn, 244.
Haverstock Hill School, 346-350.
Helfert, Baron, report of, 148.
High schools in America, 33.
Hof, gymnastic hall at, 258-262.
Holland, schools in, 24, 25.
Hollingsworth School, Philadelphia, U.S., 32, 33.
Home and Colonial School Society's desks (mounted), 375.
Hood's, Mr., work on ventilation and warming, 267.
Hot-water system of warming schools, 284.
Huxley's, Professor, committee, report in 1871, to the School Board for London, 19, 22.
Hygiene must be ever considered throughout the general principles and minor details, 6.
Hypocaust, the old Roman, 289.

I.

Industrial Schools, 351-358; the industrial school stands on the border land between vice and virtue, 351; five special officers devoted to the work of seeking out suitable cases for, 352; the classes of children who may be detained in, 353; a school for the *neglected*, 354; combine some of the features of workhouse, school, and

boarding house, 355; size the first necessary element, 356; Barnes' Home at Ardwick, near Manchester, 357, 358; at Feltham, Surrey, 358; at Burdett Road, Tower Hamlets, 358.
INFANT SCHOOLS, 180-194; the minimum size, 181; a separate room for babies, 182; suggested school-room for 300, 184; gallery of maximum size, 186, 187; plan of French gallery, 188; desks for, 189, 190; playgrounds, 191, 192; swings, bars, &c., 193, 194; climbing stand, 194; regulations for, 427.
Inkwells for schools, 387.
Inspection, public, of some kind necessary for English schools, 73.
Intime Club, the, transactions of, by M. E. Train, 64.
Ireland, education in, 47, 49-52; a favourite plan in, to give each pupil a separate seat, 12.
Irish National Board of Education, 52.

J.

JAHN, a high authority on gymnastics, 244.
John George, Elector of Saxony, first introduced compulsory primary instruction, 120.
Johnson Street School, Stepney, 300-304.
Junior and Senior schools, regulations for, 427.

K.

KINDERGARTEN model for schools, 28; the system not recognised as schools in Germany, 72; Mr. Mitchell's description of system of, 400; the "gift," or Kindergarten toys, 400; pictures of animals, &c., 401.
Kleiber, Dr., on school desks and seats, 362.
König Wilhelm Gymnasium, or Grammar School, Berlin, 89, 235; school desks from, 371; description of, 402-416.

L.

LAMPS in playgrounds of schools, 389.
Lancasterian school, plan of, 11.
Latin schools in America, 33.
Latrines in schools, 215, 216.
Lavatories in schools, 209-213.
Leibreich, Dr., ophthalmic surgeon of St. Thomas's Hospital, 176.
Liegnitz, gymnasium at, 95-97.
Lighting schools, 168-179.
Limoges, Salle d'Asile at, 60, 61.
St. Louis Public Schools, 36.
Lurgan Model School, Co. Armagh, 51; plan of, 51.
Luther took a considerable interest in the education of youth, 117; letter to the Elector of Saxony, 118.

M.

MANSFIELD Place School, Kentish Town, 343-346.
Maps, charts, globes for schools, 393.
Marburg, gymnasium at, 97; plan, 99.
Mats in schools, 389.
Mechanical powers, models illustrative of, 395.
Melon Street School, Pennsylvania, U.S., 35; plan of, 37.
Mezzanine floors for lavatories and cloak-rooms, 214; plan of, 302.
Middle schools in America, 32; middle or secondary, 232; Mr. Matthew Arnold on, 233, 234.
Milton Mount College, Gravesend, for the daughters of Congregational ministers, 240-243.
Mitchell, the Rev. M., description of the Kindergarten system, 401.
Mixed system, the, an expedient, not a principle, 17; advocated by Mr. Stow, 18; in high favour in America, 19; appears practically impossible in France, 58.
Moss's patent school-board desks, 369.
Munich, Polytechnic Institute at, 135.

INDEX.

N.

NATIONAL Education, by James H. Rigg, D.D., 46 n.
National Schools, the Educational Department mainly directed to the, 17; in Dublin and Cork, 52.
Naunyn-Strasse, Berlin, plan of parish school in, 247.
Neatness, instance of, in American Schools, 380.
New North Street School, Shoreditch, 305–310.
Newcastle's, Duke of, commission in 1861, 10.
Newton Primary School, Philadelphia, 29; plan of, 30.
Normal Schools in America, 32.
North London Collegiate School for girls, Camden Town, 239.

O.

OBJECT-LESSON stand in schools, 391.
Old Castle Street School, 291–296.
Orange Street School, Southwark, 332–335.
Owens' College at Manchester, a typical illustration of what English Secondary schools might become, 236.
Owens, Mr. John, founder of the Owens' College, Manchester, 236.

P.

PÆDAGOGY, the science of, 161.
Paukstrasse, Berlin, plan of parish school in, 246.
Parallel bars, 254–257.
Pardoux-les-Cars, a French mixed school, 57; plans of, 58.
Partitions, sliding, for dividing the classes, 30, 33; moveable in schools, 227; hanging or sliding, 229.
Pavement of playgrounds, 218; gravel the worst, 218.
Pedell, the, or caretaker, 116.
Pegs for caps, bonnets, and cloaks in schools, 388, 389.

Phipson's Mr., plan for maintaining ventilation, 332.
Plane, the double inclined, for infant schools, 192, 193.
PLAYGROUNDS, covered, 60, 61; plan of school at Berlin, showing, 82; the sun a necessary of life to, 167; for infant schools, 191–194; in schools, 216–219; pavement of, 218; in England, not fitted with suitable apparatus for gymnastics, 247.
Polytechnic institutes in Germany, 134; at Munich and Dresden, 135; at Berlin, 135; at Aix-la-Chapelle, 135; school at Zurich, 24.
Popular education of France by Matthew Arnold, M.A., 63.
Position of a school building with regard to light, 83, 84.
Préau couvert, the, or covered playground, 60, 61.
Price, Mr. H. C., his apparatus for warming and ventilating, 285–287.
Primary Education in England by the Rev. James H. Rigg, D.D., 20.
Primary schools, American model, 29; Newton, Philadelphia, 29, 30; Capen, Boston, Massachusetts, 30, 31.
PRUSSIA, has taken the lead in education as far back as the date of her adoption of compulsory primary education in 1763, 71; the system of public instruction military in spirit, 71.
Prussian, the, superior system of education, 45.
Pupil-teacher system, the, superseded that of the monitor, 11, 17; diametrically opposed to the German, 20–22; as laid down by the Code, 161, 162.
Pupil-teachers English method of teaching based on the employment of, 11, 15.

R.

RASCHDOFF, Herr, town architect of Cologne, 113.

Realschulen, the German or commercial schools for the middle classes, 104; the Sophien at Berlin, 105-108; at Halberstadt, 109-111; Stadische at Cologne, 111-114; Höhere Bürger, Wiesbaden, 115-117; Imperial, Vienna, 153-157.

Regulations of London School Board for the general management of its schools, 427; for infant schools, 427; for junior and senior schools, 427, 428; for evening schools, 429.

Religious teaching in public schools in Ireland, France, Germany, England, Scotland, and Australia, 48.

Report on the St. Louis Public Schools in 1871, 36.

Residence for master or mistress, 423.

Rigg, the Rev. James H., on Primary Education in England, 20 n.; on National Education, 46.

Robins, Mr., Milton Mount College erected from his designs, 240.

Rules of the Education Department to be observed in planning and fitting up schools, 361, 417-424.

S.

SALLE D'ASILE, the, of St. Etienne, Limoges, 59.

Sandon, Lord, at his instance the first batch of twenty schools were built in the most destitute districts of London, 293.

Saxon school system, the, 118.

Saxony, few provinces of Germany have so ancient a school history, 117.

Schinkel, his influence seen even in the works of less distinguished architects, 88, 90.

SCHOOLS, elevation of (infants and mixed), as suggested by the Education Department, 15, 16; elementary, 161, graded, 163-165, 172-174; infant, 180-194; elementary graded, 195-202; temporary, 203, 204; middle, 232; public boarding for girls, 238;

Board schools of London, 291-350; Inquiry Commission, 119; junior, 19-22; senior, 19, 22; primary or elementary, 23; private adventure, 9 n.; a large proportion of the buildings in which they are conducted quite unfit for the purpose, 9; secondary, great want of in England, 2.

SCHOOLS, Polytechnic, at Zurich, 24; American model primary, 29; Newton primary, Philadelphia, 29; Capen primary, Boston, 30; Hollingsworth, Philadelphia, U.S., 32; the Tasker, Philadelphia, 34; Melon-street, Pennsylvania, 35; Wood-street, Philadelphia, 33; George M. Wharton, Philadelphia, plan of, 39; the High and Normal girls, Boston, 37, 40; Lurgan Model School, county Armagh, 51; Pardoux-les-Cars, 57; Salle d'Asile, Limoges, 60, 61; Collége Chaptal, Paris, 62, 63; Gemeindeschule, Berlin, 77-80; Friedrich-Wilhelm, Berlin, 88; Cölnisches, Berlin, 88; König Wilhelm, Berlin, 89-94, 402-416; at Cottbus, 95; Bezirkschule, Dresden, 121-124; Gemeindeschule, Dresden, 125, 126; Bürgerschule, 127, 128; Kreuzschule, Dresden, 131-134; Polytechnic, Munich, 135; Polytechnic, Berlin, 135; Polytechnic, Aix-la-Chapelle, 135-139; Chemistry, Aix-la-Chapelle, 139-141; Victoria School for girls, Berlin, 143; Luisenschule, Berlin, 143; Polytechnic, Vienna, 157; North London Collegiate for girls, 239; Girton College, Cambridge, 239; Milton Mount, Gravesend, 240-243.

School Board for London, regulations for the management of its schools, 425.

School-buildings, those of Germany, especially of Saxony and Prussia, the best, 74.

Schoolmaster's house, village, 219-222.

Schulpforta, one of the most celebrated gymnasiums in Saxony, 119.

Scotland, education in, 47, 49; schools in, 53.

INDEX.

Seats in schools, 362 ; *see* desks.
Secondary schools in Switzerland, Report of the, by Mr. Matthew Arnold, 24.
Sheffield desk, the, 370.
Smith, Mr. T. R., his designs for the Johnson High School, Stepney, 300.
Sophien-Realschule, the, Berlin, 105–109.
Staircases to schools, 206–209 ; the double, 207 ; plans, 207, 208 ; flight, 209.
Standards of examination, 161.
Stanhope, Mr. Spencer, a sculptured panel of Knowledge strangling Ignorance, in Winstanley Road School, by, 314.
State control, desirability of, in schools, 234.
State interference, even private schools not exempt from in Germany, 8.
Stoves, 279 ; the Gurney, 279–281 ; the Gill, or studio, 282.
Stow, Mr. David, 52, 53 ; plan of a school on his system, 13 ; "The Training System," 18.
Sunburners in the König Wilhelm Gymnasium, Berlin, 414, 415.
Sunday Schools, first establishment of, in 1783, at Gloucester, 1.
Superficial feet for each child in Berlin, as in London, 9 or 10 feet, 83.
Swedish single-desk, with lifting-seat, 372.
Swing-trees, 254.
Swings, or see-saws, for infant schools, 193.
Switzerland, popular education in, 22 ; secondary schools in, 24.
Sympathy of numbers, the, 12, 13, 15, 170.

T.

TAR PAVEMENT the most satisfactory for playgrounds, 218.
Tasker School, the, Philadelphia, U.S., 33 ; plan of first floor, 34.
Teachers' rooms, 219.

Technical instruction, Commission of, in France, 55.
Temperature of school-rooms, 85 ; suitable means for producing and maintaining a, 266.
Temporary schools, 203, 204.
Theory of Austrian education in Baron Helfert's report, 149.
Theory of school-plans, 85.
Thermometer in schools, 389.
Train's, Mr. M. E., Transactions of the Intime Club, 64.
Training System, the, by Mr. David Stow, 18.
Turnhalle, or gymnastic room, 81 ; no public school in Germany complete without, 245, 248, 257 ; at the Bezirkschule, Dresden, 124.
Turnplatz, the, in all Gemeinde and Volkschulen, 248.

V.

VENTILATION, 263 ; *see* Warming and Ventilation.
Ventilation of schools, defective in Germany, 86.
Victoria School for girls, Berlin, desks for seating children in pairs, 371.
VIENNA, Handel's Academy at, 151 ; account of the French, by Imperial Commission, 152, 153 ; Imperial Gymnasium, 153–157 ; Stadische schools at, 157 ; Polytechnic school at, 157.
Virchow, Dr., on school desks and seats, 362.
Visual organs, the change in the function of the, developed during school-life, are threefold, 176–178.
Vögel, Dr., Rector of the Königliche Gymnasium at Chemnitz, 129.
Volkschulen, the, have the apparatus of the Turnplatz, or playground, 248.

W.

WALLS, the external, of sufficient thickness, 204.

Warm-water heating in the König Wilhelm Gymnasium, Berlin, 411.

WARMING AND VENTILATING, 263-290; difficulty of, 37; system of, in German schools, 86; must be treated as inseparable, 263; warming, the motive power of ventilation, 264; importance to schools, 265; cooling power of glass, Mr. Hood's work on, 267; the amount of heating power required in buildings, 268; methods to be avoided, 268; the most difficult subject in connection with school architecture, 270; the one great principle of *demand* and *supply* to be considered, 271; relative expense, 272; open fires, 272, 273, 274-279; stoves, 279-282; German method, 283; warming apparatus, 285-288.

Weaving, a technical school for, at Chemnitz, 133.

Weights and measures, cabinet of, 396.

Wesleyan Methodist Denomination Schools, the, deserve notice, 13.

Wesleyan Schools, a large gallery the chief feature in, 13; the class-rooms small, 14; plan of, 14; nearly all used for Sunday Schools, 14.

Wesleyan Training College, Westminster, 20.

West-street School, Hackney, 336-338.

Weyer, Herr, Stadt-baurath of Cologne, school desks introduced by, 363.

Wharton, the George M., school, 35; plan of, 39.

Wiesbaden, Höhere Bürgerschule at, 115, 116.

Wiese, Dr., minister of education in Prussia, 83, 95, 103, 115; on cloak-rooms in schools, 213; on comfortable sitting for school work, 366; his synopsis of various writings bearing on the best methods of uniting the seat with the desk, 362-364, 366.

Wimmer, Dr., the Saxon Gymnasium Schools called by him the hearths of classical learning in Germany, 120.

Windows in schools, 223-227; method of opening, 225; glazing always with clear glass, 226.

Winstanley Road School, 310—314.

Wolstenholme, Miss, essay on the education of girls, 239.

Women, education of, in Germany, 83, 142; condition and sphere of employment in Germany one of the greatest reproaches of the country, 142, 143.

Women teachers, employment of, 18.

Wood-street School, Philadelphia, U.S., 38.

Work-table, mistress's, with folding top, for sewing-classes, 383.

Wornington Road School, 324-383.

Wurtemberg, the ministry of, published a decree on the special sanitary regulations in schools, 141.

Z.

ZURICH, taken as the type of Swiss school-buildings, 23; Grossmünster School, 23; Polytechnic School at, 24.

Zwey, Herr, on school desks and seats, 362.

THE END.